Electrical Essentials for Powerline Workers

Second Edition

Wayne Van Soelen

THOMSON
DELMAR LEARNING

Australia Canada Mexico Singapore Spain United Kingdom United States

Electrical Essentials for Powerline Workers 2E

Wayne Van Soelen

Vice President, Technology and Trades SBU:
Alar Elken

Editorial Director:
Sandy Clark

Senior Acquisitions Editor:
Steve Helba

Development:
Dawn Daugherty

Marketing Director:
Dave Garza

Marketing Coordinator:
Casey Bruno

Production Director:
Mary Ellen Black

Production Manager:
Larry Main

Production Editor:
Thomas Stover

Art/Design:
Thomas Stover

Senior Editorial Assistant:
Dawn Daugherty

Cover Design
Jason Polkovitz

Printed in Canada
1 2 3 4 5 XX 06 05 04

For more information contact
Thomson Delmar Learning
Executive Woods
5 Maxwell Drive, PO Box 8007,
Clifton Park, NY 12065-8007
Or find us on the World Wide Web at
www.delmarlearning.com

Library of Congress Cataloging-in-Publication Data

Van Soelen, Wayne.
Electrical essentials for powerline workers / Wayne Van Soelen.-- 2nd ed.
p. cm.
Includes bibliographical references and index.
ISBN 1-4018-8358-3 (alk. paper)
1. Electric lines. 2. Electric power distribution. I. Title.
TK3226.V26 2004
621.319'2--dc22

2004021450

Notice to the Reader

Preface

Electrical Essentials for Powerline Workers is a comprehensive reference dealing with the electrical portion of the work that a powerline worker carries out. The book can be used as a reference book or as a training book. As a training manual, it needs to be studied in chronological order to understand the later chapters. Some of the information in this book is considered "must-know" information needed to work effectively and safely.

Prerequisites

This book is for a person working or training to work in an electrical utility. Reference is made to equipment and material with the assumption that a reader knows what they look like. The early chapters are intended for a person entering electrical-utility work. Later chapters assume some experience as a powerline worker.

Objective of Material

Electrical utilities have an abundance of manuals and reference material for the people in the line trade. However, most of the information supplied by the utility or employer involves work procedures, rules, and regulations. Electrical reference material can be found in widely scattered reference and training manuals. This book fills the need for a convenient single-volume electrical reference source on the operation of an electrical-utility system.

Existing sources of information tend to be very basic electricity and magnetism theory or complex theory at an engineering level. This book deals with equipment and situations to which powerline workers are exposed in their daily work.

There is a rapid growth in the technology affecting the operation of electrical utilities. Many customers have concerns about the quality of their power supply. The powerline worker has exposure, to some degree, to all aspects of an electrical-utility system. This book is intended to help a powerline worker meet the expectation that he/she will have a knowledge of the system being operated and maintained.

Acknowledgments

A special acknowledgment is due to the following people for helpful suggestions made during the development process of this second edition.

Kevin Wheeler
Linn State Technical College
Linn, MO

Mike Popko
Gogebic Community College
Ironwood, MI

Mark Holcomb
Vernon Regional Junior College
Wichita Falls, TX

Contents

CHAPTER 1

Electrical Power System Overview

1.1 Introduction

Three Systems within the Power System

1.1.1 This chapter gives an overview of an electrical power system. There are three main systems within an electrical utility power system:

- The *generation system* converts other forms of energy into electrical energy.
- The *transmission system* transmits energy over long distances. It includes the rights-of-way, transmission lines, switching stations, and substations.
- The *distribution system* distributes the energy to industry, commercial customers, farms, and residences. It includes subtransmission lines, distribution substations, distribution feeders, transformers, and services.

1.2 Electrical Energy

Energy

1.2.1 To do any kind of work requires energy. Energy has the ability to produce change or exert a force on something. There are many forms of energy, some of which are chemical, solar, potential, thermal, and electrical. Energy cannot be created or destroyed, but it can be converted from one form to another.

Many forms of energy can be converted to electrical energy. The generation of electricity is a process of converting other energy forms to electrical energy. Electrical energy is utilized when it is converted back to other forms of energy.

Utilization of Electrical Energy

1.2.2 Electrical energy is known as an energy source that is easily converted into power and light. The utilization of electrical energy comes from its four main effects:

- *Thermal effect:* The heat produced by electrical current is desirable for toasters, heaters, and ovens during the utilization stage, but it is wasted energy in the generation, transmission, and distribution stages.
- *Luminous effect:* Light is emitted when a filament is heated or an arc is generated. The design of incandescent, fluorescent, and sodium vapor lights takes advantage of this effect.
- *Chemical effect:* Electrical current can break down certain chemical molecules into their component atoms. For example, water (H_2O) can be broken down into hydrogen and oxygen through a process called electrolysis. Electrolysis is used in industry for electroplating and the manufacture of aluminum.
- *Magnetic effect:* The magnetic field around a wire can be increased by winding the wire into a coil around a core of magnetic material. This effect is used by a utility for generators, transformers, and reactors. At the utilization stage, the magnetic effect is used for motors, solenoid switches, circuit breakers, telephones, and stereo speakers.

Uncontrolled Energy as a Hazard

1.2.3 All work involves exposure to many forms of energy. For example, a power-line worker is exposed to electrical energy and the potential energy of gravity working at heights almost daily. For worker safety, these "wanted" energies must be controlled. When an energy source goes out of control, it can result in an accident.

A hazard can be defined as the potential for an "unwanted" energy flow to occur. Identifying energy sources that could cause serious harm helps to focus hazard identification on the job site.

Accidents happen when energy goes out of control. Injuries happen when a person is in the channel of the unwanted energy flow. Figure 1–1 shows that if unwanted energy is released and contacts a target, then there will be an injury or damage. Every accident can be analyzed in this manner.

Once an energy source (hazard) is identified, then some form of barrier should be placed on it to prevent it from being released. Figure 1–2 shows the three places where a barrier can be placed to prevent the next unwanted event from taking place.

Control barriers are placed to prevent the unwanted energy from being released in the first place. Lockout/tag-out procedures, cover-up, pole support, and ventilating a vault are examples of control barriers.

Once energy is released, safety barriers are needed to prevent the hazard from contacting the target or, in other words, to limit the damage or injury caused by

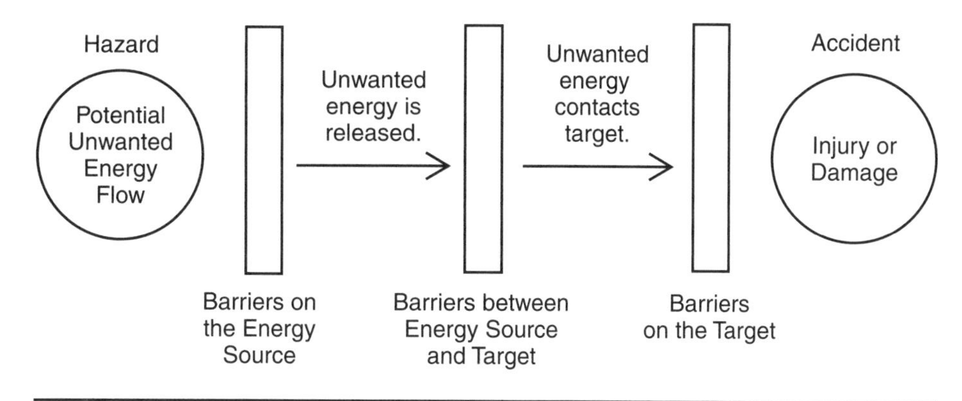

Figure 1–1 Energy as a hazard.

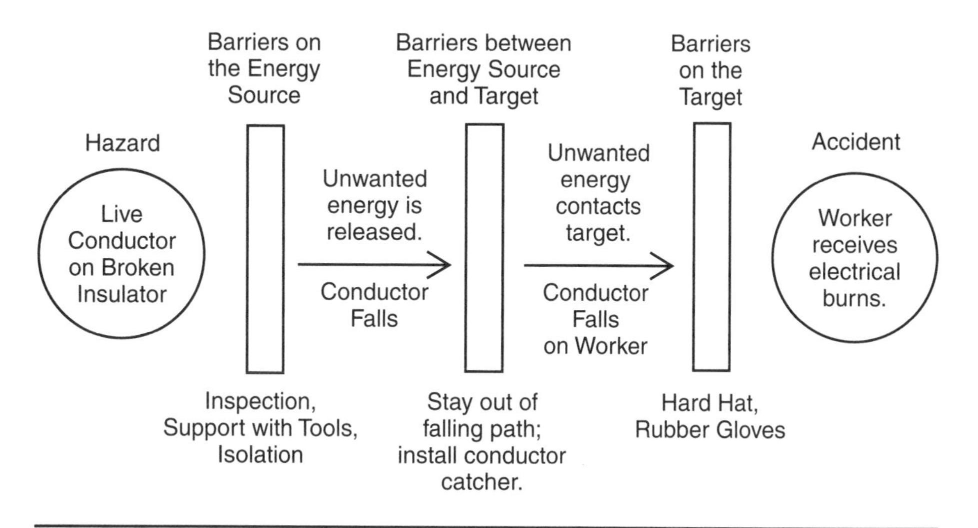

Figure 1–2 Placing barriers example.

out-of-control energy. Fall-arrest equipment, insulated platforms, vehicle grounds, ground gradient control mats, reel grounds, air quality monitoring in enclosed spaces, and trench shoring are examples of safety barriers that can prevent injury or damage after an unwanted energy flow is released.

Placing barriers on the target is a last resort control. Injuries and damage can be reduced by using safety barriers such as personal protective equipment: a hard hat, fire-resistant clothing, and electricity-resistant footwear.

1.3 Generation of Electrical Energy

Generation Basics

1.3.1 When a wire is moved within a magnetic field, an electrical charge is induced into the wire. Almost all commercially generated electricity involves the movement of wire coils in a magnetic field. In practice, this normally means that

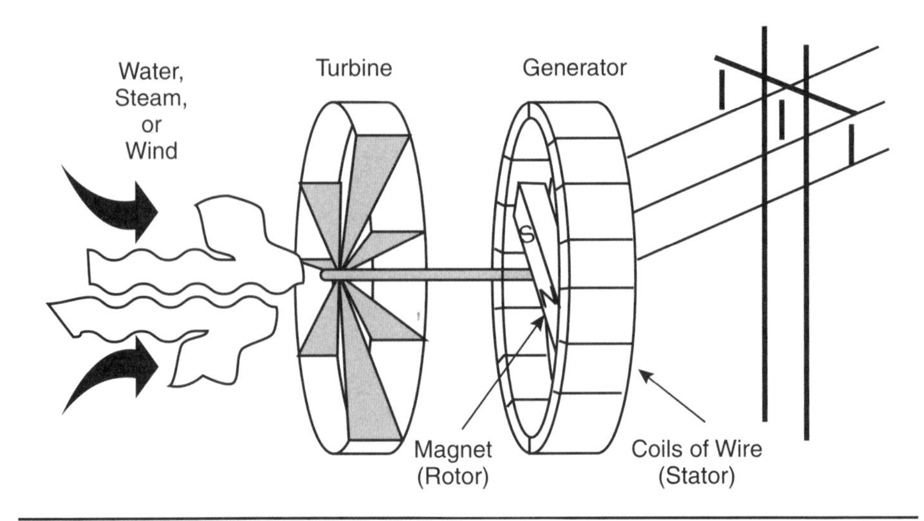

Figure 1–3 Simple generator.

many electromagnets are installed on a wheel or armature, which is turned inside a *stator* mounted with many wire coils.

The armature is connected to a turbine (Figure 1–3). The turbine is a wheel with blades mounted on it. The water, steam, or wind pushes against the blades and causes the turbine to turn.

Turning the Turbine

1.3.2 There are a surprising number of prime movers or sources of energy that can be used to spin a turbine. The earliest energy forms used for this purpose were falling water and wind. Almost all of the suitable falling water or *hydraulic* sites in the world have been harnessed, are spinning turbines economically, and are relatively pollution free. Harnessing the tides and winds to spin large turbines for commercial generation is a more recent development.

Most of the electrical energy produced in the world comes from the use of steam as a force to spin the turbines. The steam is converted from the heat energy of burning coal, oil, natural gas, wood chips, and garbage; or steam can come from the heat energy generated by a nuclear reactor or from geothermal (underground heat) sources.

Hydraulic Generation

1.3.3 Hydraulic stations are built where advantage can be taken of water at a higher level dropping to a lower level (Figure 1–4). This is normally accomplished by building a dam on a river with a suitable water flow and where a substantial difference in water level is created. The headwater formed by the dam is the potential energy that will be converted to electrical energy.

The headwater is funneled through a pipe called a *penstock*. The water rushes down the penstock and hits the turbine blades with a force that comes from the water's speed and weight. The turbine spins, which in turn spins the generator. The water continues out through the tailrace and back into the river.

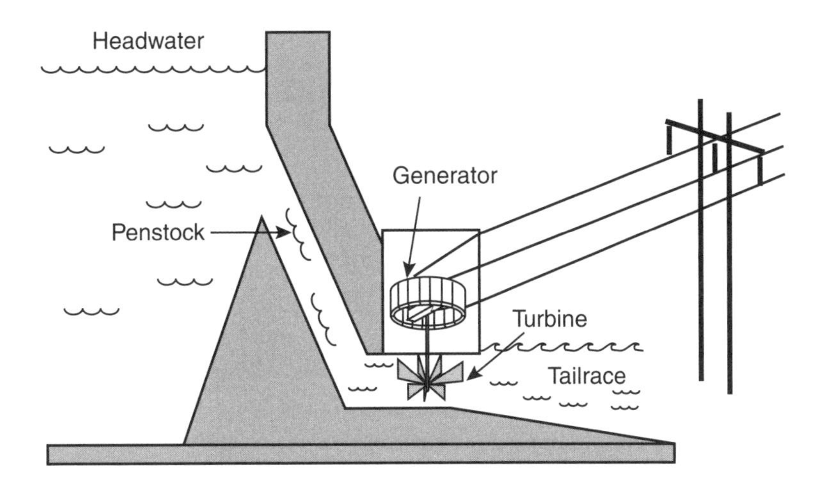

Figure 1–4 Hydraulic generator.

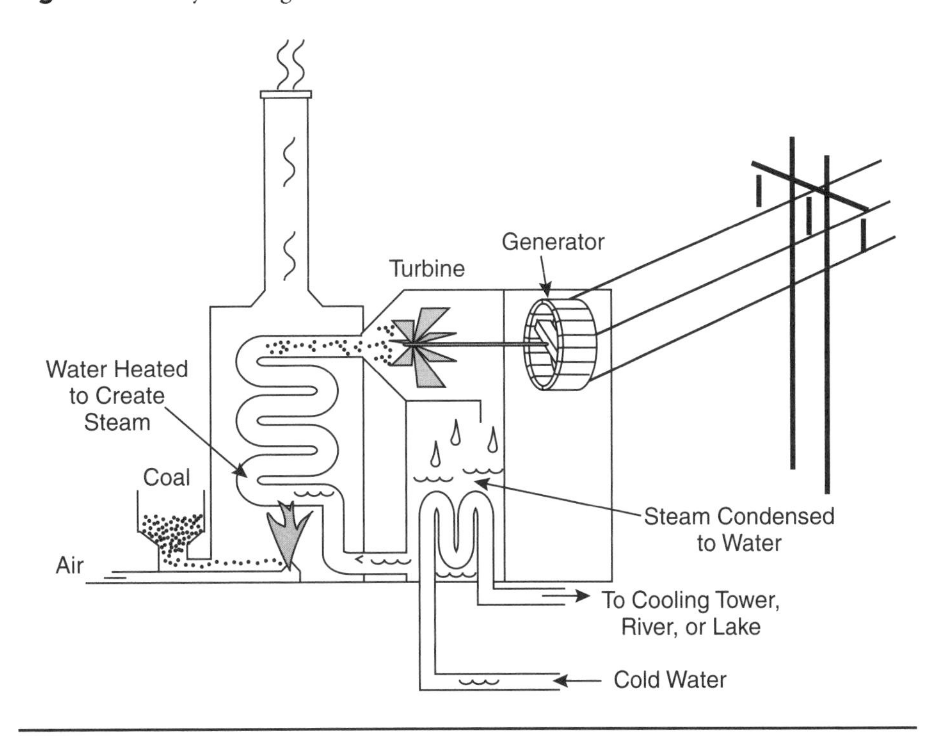

Figure 1–5 Steam turbine.

Generation from Steam

1.3.4 Generating heat from the burning of fossil fuels such as coal, oil, and natural gas, or from a nuclear reactor, is the most common commercial method of creating steam. The steam expands and pushes against the turbine blades, causing the turbine to turn.

Like a hot-water heating system in a house, the water in a steam plant is in a closed loop that is heated and cooled over and over again (Figure 1–5). A heat exchanger in the boiler heats the water in the closed loop, and another heat exchanger uses water from an ocean, lake, or river to cool the steam and condense it back into water. The water is then pumped back to the boiler to be reheated.

Nuclear Generation

1.3.5 A nuclear generating station is similar to a conventional steam plant except it uses a nuclear reactor to create heat for making steam. The heat comes from uranium atoms splitting in a controlled reaction.

Uranium is a dense, unstable element. Neutrons, which are particles within the nucleus of an atom, are easily knocked free from a uranium atom nucleus. A uranium atom splits if it is struck by a free neutron given up by another atom. When an atom splits, more neutrons are released that in turn hit other atoms, splitting them, therefore causing a chain reaction. A nuclear *fission* (splitting) of atoms generates a huge amount of heat.

Gas Turbines

1.3.6 The hot exhaust gases from the burning of oil or natural gas in a high-pressure combustion chamber can spin a turbine when the exhaust gases expand through the turbine blades, much like the way a jet engine operates. High-pressure air is added to the combustion chamber to add more force to the escaping gases.

The most efficient way to use the gas turbine is in a *combined-cycle* system (Figure 1–6). After the hot exhaust gases spin a gas turbine, the still hot gases heat water to make steam and spin a steam turbine. Usually, several gas turbines feed hot exhaust to one steam turbine.

Cogeneration

1.3.7 Cogeneration plants are generating stations used to generate electric power and heat. The electric power can be sold to the grid, and the heat, which would otherwise be waste heat, is sold to a central heating plant or manufacturer. Because it is impractical to transport heat over any distance, cogeneration stations are built close to their heat users.

Cogeneration stations (Figure 1–7) are fired by fuels such as natural gas, wood, agricultural waste, or peat moss. Steam pressure generated by burning the fuel turns the turbines and generates power. Normally, about one-third of the energy in the original fuel can be converted to steam pressure to generate electricity. The excess heat supplied to the customer is usually in the form of relatively low-temperature steam exhausted from the turbines.

Alternative Electrical Energy Sources

1.3.8 A small percentage of the total electrical supply is generated from alternative sources. Some of these alternative sources provide the most economical way to supply electrical energy to remote locations where it is not practical to build a line to the customer.

- There are commercial generation stations that use the *ocean tide* to spin turbines. These are hydraulic stations that take advantage of unusual high tides in some locations in the world. For example, the high tides in the Bay of Fundy in New Brunswick are used to generate power.

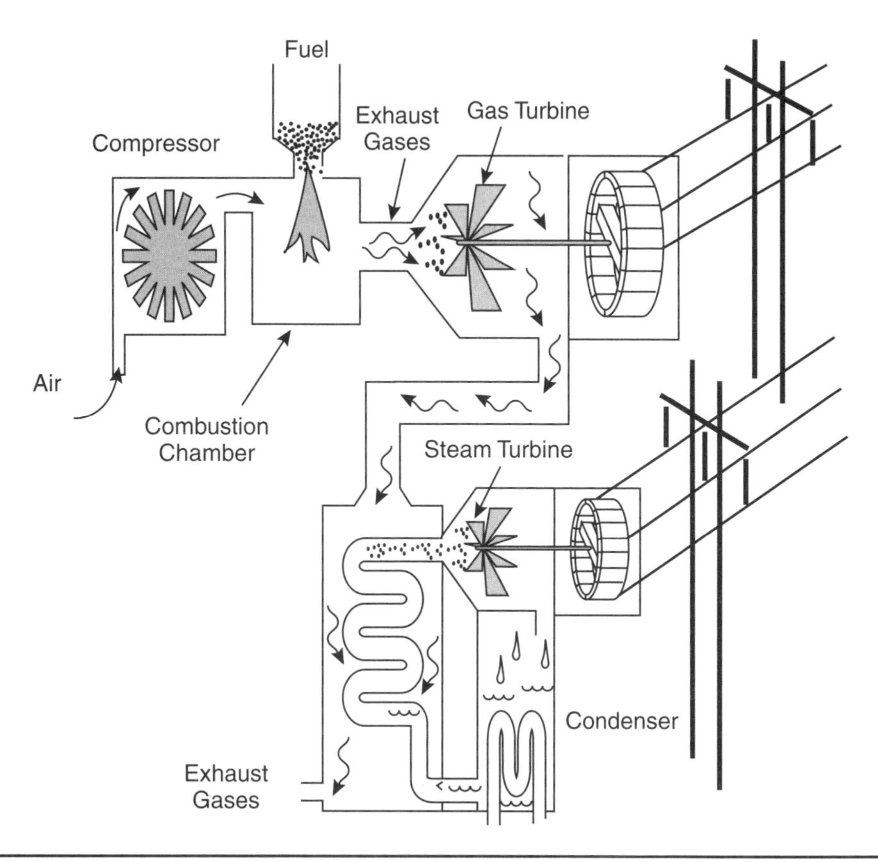

Figure 1–6 Combined-cycle generation.

Figure 1–7 Cogeneration plant.

- There are locations, especially in California, where hundreds of *windmills* (wind turbines) form large-scale "wind farms." Wind energy varies with the cube of the wind speed so that when the wind speed doubles, there is eight times more energy.

- There are *geothermal* generating stations that generate steam from the heat below the surface of the earth.
- There are *engines* (diesel, gasoline, gas) used to run generators in remote communities and as standby units.
- *Solar energy* is responsible for many of the alternative energy sources. Solar energy is more than just sunlight. It is the heat from the sun that indirectly gives us the winds to drive a wind turbine and the rain to give us the water to turn the hydraulic turbines. A *photovoltaic cell* (solar cell) can generate electricity directly from sunlight without the need for turbines. When sunlight strikes a photovoltaic cell made from material such as silicon, electrons are dislodged and caused to move. Free electrons collect on one surface of the cell, causing an imbalance between the front and back surfaces. An electron flow (electrical current) is produced when the front and back of the cell are put into a circuit.
- The burning of *biomass* can produce steam to spin steam turbines. Products such as sawdust and bark from the lumber industry, wood from fast-growing plantations, ethanol from corn, or methane from the decomposition of vegetation and garbage are burned in relatively small generating stations in many areas.
- *Fuel cells* generate electricity through an electrochemical process. The system converts the chemical energy of hydrogen or hydrocarbons and oxygen into electrical energy. In a fuel cell, hydrogen and oxygen are combined to form water and electricity (the opposite of the old experiment where hydrogen and oxygen are produced when electricity is passed through water). The hydrogen needed for a fuel cell can be found in natural gas, coal-derived gas, ethanol, gasoline, and other fuels.

1.4 Transmission of Electrical Energy

Transmission of Electricity

1.4.1 Electrical energy can be economically transported over long distances. Electricity is transmitted from the generating station to the customer load centers on high-voltage transmission lines. A transmission line can be compared to a water pipe: the higher the pressure and the larger the pipe, the more water will flow through the pipe. Similarly, the higher the voltage and the bigger the wire, the more electrical energy will be able to flow through the transmission line.

Typical Transmission Line Construction

1.4.2 The vast majority of transmission lines are overhead because underground lines are prohibitively expensive for long-distance transmission. Overhead conductors are suspended on structures such as lattice steel towers (Figure 1–8), wood poles, concrete poles, or steel poles. The purpose of a structure is to keep high-voltage conductors insulated from ground in all kinds of weather and out of reach of accidental contact. Tall structures allow long spans and, therefore, fewer structures.

The insulator length or size is dependent on the voltage: the higher the voltage, the longer the string of insulators. Conductors are usually stranded aluminum with

Figure 1–8 Transmission line corridor.

a steel core. Aluminum is a good conductor of electricity, and the steel core gives the conductor tensile strength. A strong, lightweight conductor can be strung with less sag over long spans.

Transmission Line Voltage

1.4.3 Commercial stations generate power at a voltage ranging from 13,800 volts to 24,000 volts. A step-up transformer station next to the generation station boosts the voltage (pressure) so that it can be transmitted efficiently. Generation voltages are boosted up to common transmission line voltages such as 115,000 volts; 230,000 volts; 345,000 volts; 500,000 volts; and 765,000 volts. The high voltages are normally expressed in kilovolts (kV) so that a 500,000-volt line becomes a 500-kilovolt line. As a rule of thumb, if the voltage is doubled, the energy that can be transmitted is quadrupled without an increase in line losses.

Extra high voltage (EHV) lines, such as 500-kilovolt circuits, use bundled conductors, which are two, three, or four conductors tied together with spacer dampers. The bundling of conductors is to counter certain problems caused by extra high voltage; however, the increased conductor capacity plus the high voltage allows a single 500-kilovolt circuit to carry the equivalent of eight 230-kilovolt circuits.

Transmission System Substations

1.4.4 The terminals of transmission lines are at substations and switchyards. Substations (Figure 1–9) are voltage-changing stations. Transformers can step up the voltage to allow for the efficient high-voltage transmission of power or can step down the voltage to allow for a more manageable voltage to distribute the power down roadways and streets.

Figure 1–9 Transmission substation.

Figure 1–10 shows a typical layout of a small transmission substation. Note that the layout is designed so that the station can continue to feed out on each subtransmission line when any one component is out of service.

Switchyards

1.4.5 Switchyards are found at the terminals of transmission lines. A switchyard has disconnect switches, circuit breakers, relays, and communications systems to provide circuit protection. Switchyards allow the routing of power through various circuits to ensure that customers continue to receive service even when some parts of the power system fail.

A switchyard ties the many circuits coming into the yard to a common circuit called a *bus*. The term *bus* comes from the word *omnibus,* which means a collection of numerous objects or, in this case, a collection of numerous circuits. A bus must be able to carry very high current and, therefore, usually consists of large, rigid aluminum or copper pipe or very large conductors. Switchyards are usually within the same fenced area as the transformer and form part of the substation.

Communications between Stations

1.4.6 Operators in a control room monitor meters and alarms that indicate the condition of the substations and lines within their zone of control. An operator can open and close switchgear in remote generating stations and substations. This "supervisory control" of the system depends on communication systems between stations.

To transmit information and signals from station to station, utilities use telephone lines, utility-owned fiber-optic cable, power-line carrier systems, microwave systems, or satellites. Because continuous communication is critical, there is usually more than one system in place in case one of the systems fails.

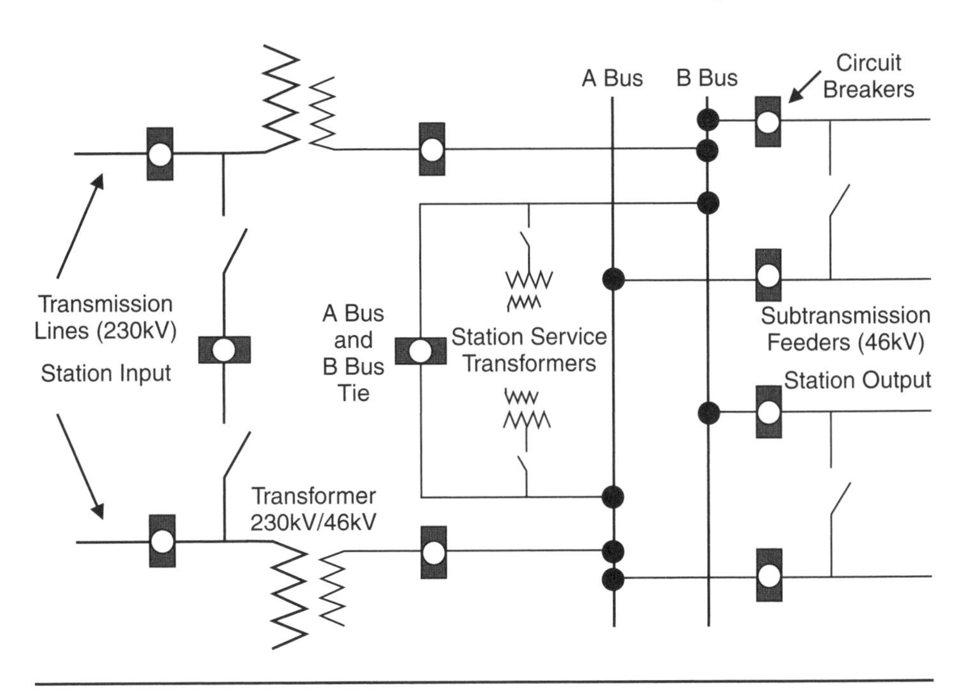

Figure 1–10 A typical transmission station layout.

Telephone lines are a common link between stations. The use of fiber-optic cable strung in the shield-wire position on transmission lines is becoming a popular mode of communication.

A power line carrier system uses the power line conductors to transmit information. The communication signals are sent on to and received from the power conductors by a device that looks like a potential transformer but is a coupling capacitive voltage transformer (CCVT). To keep the transmitted signals within the desired sections of the power line, wave traps (Figure 1–11) are installed. The wave trap, which looks like a large cylindrical coil, stops the signals from continuing farther down the line.

Microwave communication between substations requires towers (Figure 1–12) with microwave antennas in each station. Microwave-sending and -receiving antennas need to have a direct line of sight with no obstacles between them. Microwave towers are located on hills, where possible, and are about 60 to 100 kilometers (35 to 60 miles) apart to relay the signals between towers.

Electrical Power Pools

1.4.7 Generating stations are interconnected by transmission lines into giant regional pools or grids that cross utility boundaries. The flow of electric power in these grids goes where it is needed. The flow could go south during a heat wave to feed the peak air-conditioning loads or go north during a cold snap to feed the peak heating loads.

Figure 1–11 Wave trap.

Metering at line terminals or substations determines the amount of energy that crosses utility boundaries and what payment needs to be exchanged. Sometimes a utility's transmission line only transfers or *wheels* power from one neighboring utility to another. The utility receives payment for supplying this wheeling service.

Blackouts and Brownouts

1.4.8 There was a huge *blackout* in the central and northeastern United States and Canada on August 14, 2003. The failure in one element of the power pool started a chain reaction that led to a loss of most of the transmission grid. A fault on one transmission line caused another transmission line to overload and fail, which caused the generation that fed the line to be isolated from the grid, which overloaded neighboring lines and generation. Protection schemes are in place to isolate failures to the offending location. However, these protection schemes can be overwhelmed and fail.

Along with improved protection schemes, utilities have procedures that intentionally lower the voltage on or shed load from the system when the customer demand is greater than the system can supply.

Figure 1–12 Microwave tower in substation.

When customer demand on the power pool is higher than available generation or transmission lines can provide, shedding or dropping load is a last resort. Before any load is shed, the voltage is lowered on the grid, which reduces the total energy supplied to the customers. Customers may notice that their lights become a little dimmer and their motors run hotter. About twice a year some utilities conduct tests by reducing the voltage on the system. These *brownouts* are generally only noticed by customers who are already receiving below-normal voltage in ordinary times.

After intentionally reducing the voltage on the system, and if there is still not enough to meet customer demand, some large industries have their load dropped from the system first. These industries have a contract with the utility that allows the dropping of their load in exchange for a better rate.

In the unusually cold winter of 1993–1994, there was difficulty meeting the demand in Washington, D.C. Instead of implementing rotating blackouts, demand was reduced by closing the federal buildings on the coldest days.

When all else fails, electrical load is shed on a rotational basis to the general population for a preset time. The deliberate dropping of load on a rotational basis results in *rolling blackouts* to specific geographical areas for specified periods of time, usually thirty or sixty minutes.

1.5 Electrical Distribution

Distribution Basics

1.5.1 The *transmission system* brings electrical energy close to the load center and transforms the voltage down to a subtransmission voltage or directly to a distribution voltage.

The *distribution system* (Figure 1–13 and Table 1–1) consists of subtransmission lines feeding distribution substations, which transform the voltage down to distribution feeder voltage. Distribution feeders deliver the energy to a transformer at the customer's premises and transform the voltage to a utilization level. By far, the biggest volume of line work involves the distribution system.

Distribution System Designs

1.5.2 A distribution system can be laid out to give varying degrees of service continuity. A system with a high degree of service continuity is more expensive and is found where the customer density is high, namely, in a city.

Radial System

1.5.3 The layout of a radial system (Figure 1–14) is much like the design of a tree. The main trunk is one of the three-phase feeders going out from a substation. Three-phase or single-phase branches or lateral taps feed customers along the circuit. The conductor in the main trunk carries the most load, and the branches get smaller as they feed out from the trunk. The length of a feeder is usually limited by the voltage and connected load.

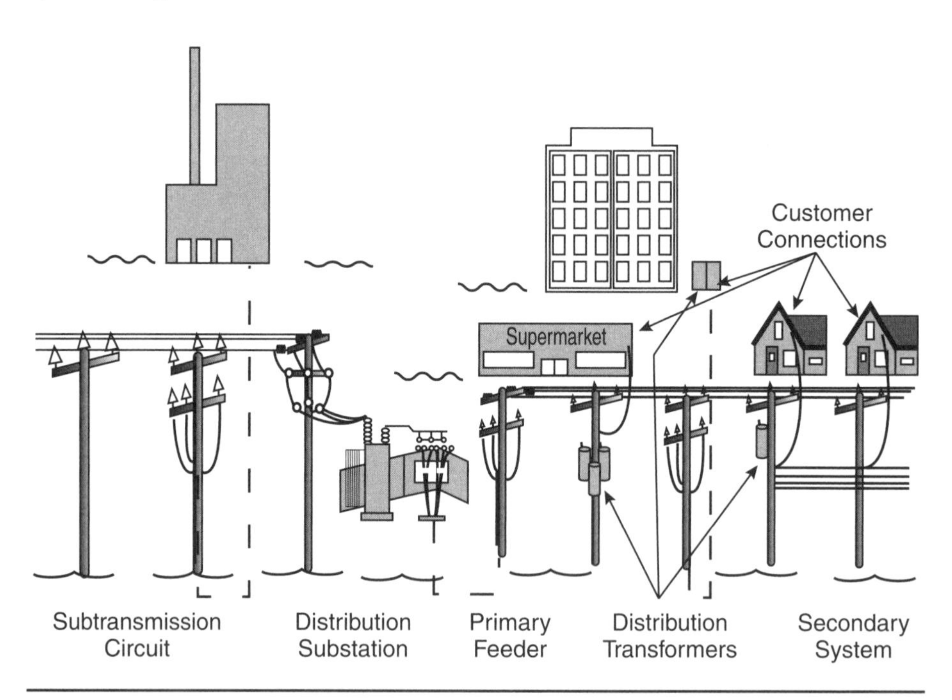

Figure 1–13 A distribution system.

TABLE 1–1 Six Main Segments of a Distribution System

Segment	Function
Subtransmission Circuits	Subtransmission circuits transmit power from the large transmission substations to the distribution substations. Examples of subtransmission voltages are 13.8kV, 27.6kV, 34.5kV, 46kV, and 69kV. At these voltages, the structures and insulation are small enough to allow building the lines along roadways. Some utilities may consider subtransmission lines to be part of the transmission system.
Distribution Substation	The transformer in the distribution substation steps down the subtransmission voltage to a distribution voltage. The substation consists of: • Switchgear on the subtransmission circuit • A transformer • Voltage regulation equipment • A distribution voltage bus • Multiple feeders connected to the bus • Distribution voltage switchgear for each feeder Many distribution substations are operated remotely from a central control room. The central control room has access to substation data such as feeder voltage and loading and has the ability to operate the substation switchgear. Supervisory Control and Data Acquisition (SCADA) is the communications technology used to operate a distribution substation remotely.
Primary Feeders	The primary (low-voltage) feeders leaving the station can be underground or overhead and are normally three-phase. A distribution feeder can be a radial feeder that branches off and ends at the end of a street or road. A distribution feeder can also be networked into a grid with other feeders allowing it to be fed from two directions. The loop between feeders can have a normally open switch keeping them separate or can be looped with automatic switchgear.
Distribution Transformer	The distribution transformer feeding the customer steps down the primary feeder voltage to a utilization voltage. Depending on the type of distribution system, the transformer can be overhead, on a concrete pad, or below grade in a vault.
Secondary Systems	A secondary system can range from a single service fed from one transformer to a secondary bus network fed from many transformers.
Customer Connections	The service into the customer can come directly from the transformer or from a secondary bus. The service can be overhead or underground. The utility responsibility for the service usually ends at the electric revenue meter.

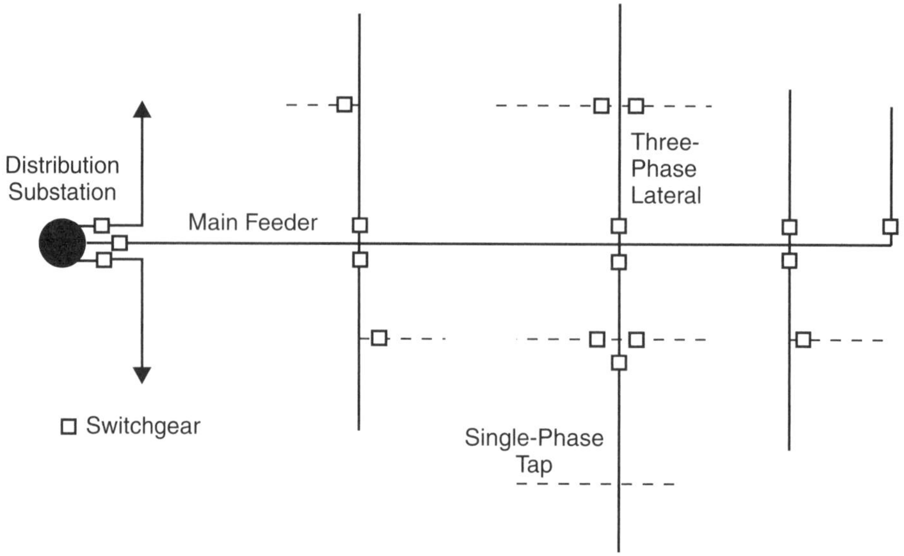

Figure 1–14 Radial feeders.

Loop Primary

1.5.4 A typical loop circuit (Figure 1–15) starts at the distribution substation, makes a loop through the area to be served, and ends up returning to the substation. It is similar to two radial circuits with their ends tied together. A loop primary keeps most customers on automatically when there is a fault on the line.

Circuit breakers are installed in the loop so that sections of the loop can be automatically isolated with the opening of any two breakers. Relays sense an overload situation and cause circuit breakers on each side of the fault to open.

Lateral taps from the loop are usually radial. An underground tap is usually in a loop, but an open switch within the loop keeps the two sides fed radially.

Primary Network

1.5.5 A primary network (Figure 1–16) is used for heavily loaded downtown areas in a city. It is similar to a loop primary except that the loop is fed from more than one substation and from more than one feeder from each substation.

Overhead and Underground Systems

1.5.6 A distribution system is either overhead, underground, or a combination of both. Urban centers tend to have underground systems, and rural areas tend to have overhead systems.

The advantages of an overhead system are:

- Lower costs for conductor and associated switchgear and transformers
- Easier and quicker detection and repair of a breakdown to the system
- Much lower cost to upgrade an existing overhead system because there is less need to dig up finished streets, curbs, and lawns

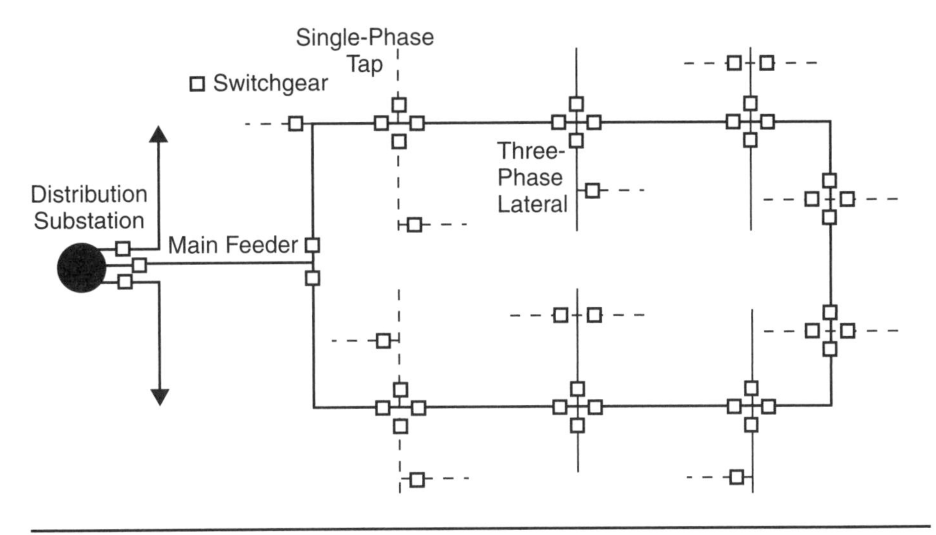

Figure 1–15 Loop primary.

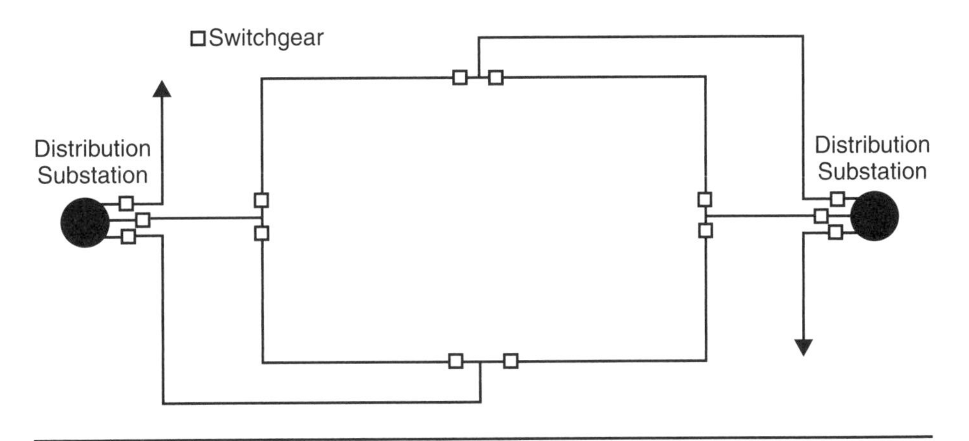

Figure 1–16 Primary network.

The advantages of an underground power system are:

- Almost no exposure to storms, trees, automobile accidents, insulator breakage, and insulator contamination
- More aesthetically acceptable to the public
- A necessity around airports, or where local laws require cable
- Long runs across water as submarine cable
- Less public exposure to the possibility of electric shock
- Generally, a longer system life expectancy

Two Types of Underground Systems

1.5.7 Generally, there are two types of underground systems: the duct and maintenance hole (manhole) system and the direct-bury system. The duct and maintenance hole system is used in cities where the presence of concrete and pavement would require very expensive digging for maintenance or upgrading. The cables are in concrete ducts, and equipment such as transformers and switching units is below the surface in maintenance holes.

The direct-bury system is used mostly in residential subdivisions where most of the cable is buried under grass. The cable must have an envelope of sand around it to prevent any pressure points on the cable because pressure points are often sources of cable failure. Ducts suitable for direct burial are also used to provide mechanical protection for the cable (e.g., at road crossings). The transformers and switchgear are often "pad-mount design" and sit on the surface on a concrete pad.

Managing a Distribution System

1.5.8 Most line work is in the distribution system. There are a variety of skilled people needed to keep a distribution system functioning, as shown in Table 1–2.

TABLE 1–2 Skilled Workers in a Distribution System

People	Work
Design Engineers and Technicians	Set the design standards for structures, equipment poles, etc., considering their strength, electrical clearances, radio interference, lightning protection, and insulation.
Planning Engineers and Technicians	Monitor the system voltage and load. Decide on funding priorities for betterments. Carry out studies for fuse coordination, voltage regulation, voltage flicker, and load growth.
Customer Service Staff	Depending on the individual utility organization, customer service staff is the front-line customer contact regarding: • High bills • Meter reading • Conservation information • New service and service upgrade requirements • Collections for nonpayment of bills
Line Crews	Construct, maintain, and troubleshoot the overhead and underground distribution system. There are numerous individual tasks that could be listed here.
Forestry or Tree Crews	Responsible for keeping the lines clear of vegetation by trimming trees that are in close proximity to live circuits while keeping customers relatively happy with the quality of their work.

Review Questions

1. The utilization of electrical energy comes from its four main effects. What are three of the four effects electrical energy can produce?
2. What is the definition of a hazard?
3. What physical act is required to generate power in almost all commercial generators?
4. What are three common energy sources used to drive a turbine?
5. If the voltage on a transmission line could be doubled, how much more energy would it be able to transmit?
6. Name three advantages of an overhead distribution system.
7. Name three advantages of an underground distribution system.

CHAPTER 2

Electrical Units

Topics to Be Covered	Section
Introduction	2.1
Electrical Potential	2.2
Electrical Current	2.3
Electrical Resistance	2.4
Electrical Power and Electrical Energy	2.5

2.1 Introduction

What is Electricity?

2.1.1 All materials are made up of atoms. Each atom has a nucleus, and each nucleus has electrons in orbit around it in the same way as the planets orbit the sun. The positive charge of the nucleus and the negative charge of the electrons keep the electrons in orbit and keep the electrical charge of the atom neutral.

The electrons of good conducting material, such as copper or aluminum, are dislodged fairly easily. With an external force, the electrons can be bumped from their own orbit into the orbit of the next atom. An atom that loses an electron will then have a net positive charge and will be susceptible to gaining another electron. The atom that gains an electron will then have a net negative charge and will be susceptible to losing an electron. A charged atom has potential and is called an ion.

The transferring of electrons from one atom to the next is electrical current. In other words, electrical current is the flow of electrons.

Summary of Electrical Units

2.1.2 Electricity is a current or flow of electrons. Electrical current will only flow in a circle (circuit) and must always return to its source. The *ampere* is the unit used to measure the rate of flow.

When there is an electrical current, it must have a pressure pushing it. Pressure, an electron moving force, which is also called an electromotive force (emf), is measured in *volts.* Pressure is needed to overcome any resistance, which impedes the current flow in a circuit. Resistance or impedance is measured in *ohms.*

A combination of electromotive force (volts) and current (amperes) is a measure of the rate of work being done. The unit of work is a *watt* (one volt × one ampere), which is more commonly measured in blocks of 1000 watts or kilowatts.

When the rate of work is at one kilowatt and it lasts one hour, then one *kilowatt-hour* of work is completed. The quantity of electricity used is measured in kilowatt-hours.

2.2 Electrical Potential

Voltage Basics

2.2.1 To introduce electrical concepts, reference is often made to the properties of a water system. For example, water flows in a garden hose when there is some force or pressure pushing the water from a high pressure area to a low-pressure area. Electricity also needs a pressure or a potential difference to have a current flow. Water pressure comes from a water pump, and electrical pressure comes from an electrical generator.

Electrical potential is measured in *volts.* The symbol for electrical potential is E (from electromotive force) or V (from volts).

Measuring Voltage

2.2.2 To measure the amount of potential or voltage that is available in a circuit, a voltmeter is used. In the lines trade, voltage checks are frequently made because an improper voltage is the first indication that there is something wrong. To measure voltage, the voltmeter leads need to be connected across (parallel) two different potentials. Most voltmeters are rated up to 750 volts. Measurement of higher voltages is done at substations using voltage transformers (VT) to bring a representative voltage into the control room.

A powerline worker has no real reason to measure higher voltages but does need to check whether a circuit is hot by testing for voltage. A potential tester is used to determine if the circuit is alive or dead. A potential test is an essential step before placing protective working grounds on a circuit.

Safety with Electrical Potential

2.2.3 Where there is a potential difference, there is also a possibility of receiving an electrical shock. One electrical potential is always looking for a path to a different potential.

If a person puts one hand on each post of a car battery, they would normally not feel anything because the potential is only 12 volts, which is not high enough to overcome the resistance of a person's skin. Most people can feel 40 volts from hand to hand. A common voltage, such as 120 volts, is a high enough potential to drive a fatal current through a person's body.

Powerline workers are exposed to much higher voltages. Safe contact can only be made with a high-voltage circuit when the resistance between the circuit and a

TABLE 2–1 Voltage Standard

Nominal Voltage	Extreme Low Voltage	Normal Low Voltage	Normal High Voltage	Extreme High Voltage
		Single Phase		
120/240	106/212	110/220	125/250	127/254
240	212	220	250	250
		Three-Phase Four-Wire		
120/208Y	110/190	112/194	125/216	126/220
277/480Y	240/418	258/446	288/500	293/508
347/600Y	306/530	318/550	360/625	367/635
		Three-Phase Three-Wire		
240	212	220	250	254
600	530	550	625	635

person is high enough to prevent a current flow. Such resistance is provided by live-line tools or rubber gloves.

Maintaining Good Voltage

2.2.4 Voltage in a circuit is susceptible to many influences that cause it to sag or surge. The quality of electrical service requires that a customer's voltage is kept within an acceptable range. Voltage that is too high or too low will damage a customer's motors, appliances, and electronic equipment.

The American National Standards Institute (ANSI) standard for proper voltage is a range of + 6 percent or − 13 percent. Table 2–1 shows a typical voltage standard for a North American utility. When a voltage is found to be lower or higher than extreme voltage, power should be shut off to avoid damaging customer equipment.

Voltage Drop

2.2.5 When a circuit has no load on it, there is no current passing through it, and there is no significant drop in the circuit. When a load is added to the circuit, some of the voltage is "used up" in pushing the current through the resistance.

A voltage drop is equal to I × R, where

$$I = current$$

$$R = resistance$$

Typical Utility Voltages

2.2.6 There is a large variety of standard voltages used in the electrical utility business. A person working on distribution needs to be alert when choosing transformers, surge arresters, switches, and insulators.

Power line workers can differentiate between familiar voltages within their utility by referring to utility operating diagrams. When the voltage of a circuit is in doubt, reference can always be made to a nameplate on an existing transformer.

TABLE 2–2 **Typical Utility Voltages**

Transmission Line Voltages	Subtransmission Line Voltages	Distribution Voltages	Utilization Voltages
765kV	13.8kV	34.5/20kV	240/120V
500kV	23 kV	27.6/16kV	208/120V
345kV	27.6kV	25/14.4kV	416/240V
230kV	34.5kV	13.8/8.0kV	480/240V
138kV	46 kV	12.5/7.2kV	480/277V
115kV	69 kV	8.3/4.8kV	600/347V
69kV		4.16/2.4kV	

Insulators and cutouts are not reliable indicators of the system voltage because of standardization of materials and pre-building for future voltage conversions.

When a circuit voltage is given, it is normally the "nominal" phase-to-phase voltage. Common North American voltages shown in Table 2–2 are phase-to-phase voltages. Where applicable, a phase-to-ground voltage follows the slash.

Note that distribution voltage equipment is also commonly called medium voltage equipment, and utilization voltage equipment under 1000 volts is commonly called low-voltage equipment.

2.3 Electrical Current

Current Basics

2.3.1 The flow of electrical current can be compared with the flow of water in a garden hose. A garden hose conducts water while a wire or conductor conducts the flow of electrical current. Just like a large pipe can conduct more water than a small pipe, a large-diameter electrical wire can conduct more electrical current. It is the flow of current that does the work. For example, when electrical current meets resistance, heat is produced.

Electrical current is actually the flow of electrons jumping from one atom to the next. While electricity is known to travel at the speed of light, which is 186,000 miles per second (300,000 kilometers per second), the electrons themselves do not actually travel at that speed. The actual speed of electron travel in a conductor is about 0.003 millimeters per second. It is the electrical charge or voltage that travels at the speed of light.

The symbol for electrical current is *I*, from the French word *intensité.* The unit of measure is the *ampere,* which is represented by the symbol *A.*

Measuring Current

2.3.2 Current is measured with an ammeter. An ammeter is usually described as a meter that is connected in series with a circuit and measures the current flowing through it. To break into a circuit so that the current could be measured would create an unnecessary hazard.

A convenient clip-on type ammeter should be used because it can measure the current without having to connect the meter in series with the conductor. The magnetic field around the wire induces a representative current into the clip-on ammeter coil. A clip-on ammeter is not voltage sensitive and can be used on all voltages as long as the meter is used with rubber gloves or on a live-line tool suitable for the voltage of the circuit.

Electrical Current Needs a Circuit

2.3.3 Electrical current will not flow unless it is in a circuit. The current that leaves the source must make a complete "circle" and return to the source. An electrical circuit, therefore, needs a return path to the source. Depending on the type of circuit, the return path can be ground, a neutral, or another "hot" wire. A break anywhere in the circuit, including in the return path, opens the circuit, and the current flow will stop.

Opening a ground wire or a neutral on a live circuit is dangerous because an electrical current could be interrupted. When a current is interrupted, a voltage will appear across the open point in the circuit.

Safety with Electrical Current

2.3.4 It takes a certain amount of voltage to break down the initial skin resistance of a human body before a current path is established. Once a current path is established, it is the amount of current and the path the current takes through the body that does the damage.

In Figure 2–1, it can be seen that one milliampere can cause damage to a human body. Considering that a typical household circuit is fused at 15 amperes or 15,000 milliamperes, the potential for a lethal electrical shock is available on all electrical circuits.

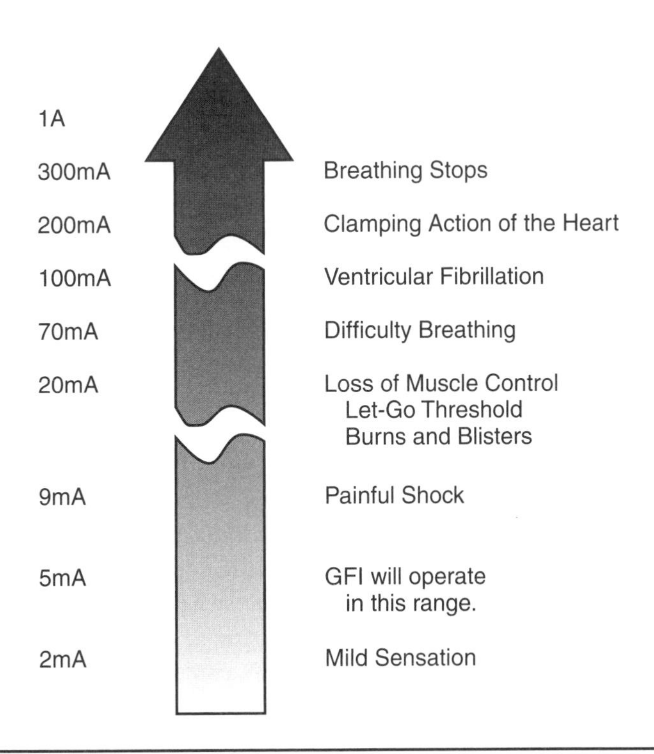

Figure 2–1 Typical effects of electrical current on the human body.

Note that a ground fault interrupter (GFI) on a utilization circuit will open a circuit before any dangerous current can flow. A device as sensitive as a GFI is not available on utility circuits.

Load Current

2.3.5 Load current in a circuit is the current needed to supply the load demands of the customer. When an ammeter is used to take a reading on a circuit, the reading measures the load current at that moment along with a small amount of current due to line loss.

Voltage is *supplied* by an electrical utility, and load current is *drawn* by the customer. A utility can control the voltage, but it cannot control the load current because load current is based on customer demand.

Peak Load Current

2.3.6 The load current in a system fluctuates with the seasons and the time of day. Peak load current is measured by recording ammeters found at substations and portable recording ammeters installed on lines. Peak demand for power tends to occur on the coldest and hottest days of the year at the time of the evening meal.

An electrical system must be built to meet these peak demands even though they may occur infrequently. More trouble calls occur during peak load conditions because any weakness—such as an overloaded transformer, an overloaded circuit, or a low voltage—will appear.

Utility conservation programs try to reduce peak loads by having reduced rate structures during off-peak hours, by installing demand meters that penalize customers with high peak loads, or by encouraging the efficient use of electricity to bring down the total base demand.

Cold Load Pickup

2.3.7 When trying to restore power, especially during peak load periods, a fuse or breaker will sometimes trip out even though the cause of the original outage has been fixed. It is difficult to pick up a cold load after a circuit has been out of service due to the loss of diversity and the initial inrush current into the circuit.

Diversity in a circuit refers to the normal situation in which everyone on the circuit is not using the furnace, air conditioner, appliance, or water pump at the same time. After a circuit has been out of service for awhile, much of this equipment could be set to come on at the same time when power is restored. The load current on the circuit could be as much as two times the diversified load current. An outage of 30 minutes may be enough to lose load diversity, and it could take as much as 45 minutes to restore the circuit to normal diversity.

There is a short period of high inrush current to load such as transformers, motors, and heaters. The initial inrush current is considerably higher than the current needed to maintain these loads.

Educating customers to switch off much of their electrical load when the power goes off would go a long way to solving the cold load pickup problem. After power is restored, the customers could switch their appliances back on one at a time. Meanwhile, without customer help in reducing load, a trouble crew has to rely on a time-consuming process of sectionalizing the circuit to pick up the load in smaller chunks.

Fault Current (Short-Circuit Current)

2.3.8 When a circuit is faulted or short-circuited, all the current which the electrical system is able to supply to the faulted location goes to that fault. Fault current can be explained by using an automotive electrical system as an example. The normal usage of the starter and lights would draw load current for which a battery is designed to deliver. However, if a wrench was dropped across the battery terminals, an explosive fault current goes from one terminal to the other. All the current available in the battery would feed the fault.

Similarly, in an electrical system, a fault current due to a short circuit can be extremely high. The resultant flash and heat generated by the fault current can be very dangerous to a worker in the vicinity. Eyes are the most vulnerable to a large flash, and safety glasses should be worn any time there is a potential for a flash.

The highest fault current is near the source of a circuit, such as close to a substation or secondary conductors close to a distribution transformer. Using the car electrical system as an example, a short circuit at the battery is much more explosive than a short circuit at a taillight. The electrical system cannot supply nearly as much power to a fault at a taillight because the small wire and the distance to the taillight add resistance, preventing a large current flow.

2.4 Electrical Resistance

Resistance Basics

2.4.1 Just as a garden hose provides resistance to the water flowing in it, an electrical wire also provides resistance to the flow of electrical current. Resistance in a conductor causes electrical energy to be converted into heat. This will be useful or wasteful, depending on whether the heat is a desired product or a line loss.

Some of the current intended to run a motor or light a building is converted to heat because there is resistance in every part of an electrical circuit. Resistance can also be added to a circuit intentionally, such as when a heating element is used in an electric range.

The symbol for electrical resistance is *R*. The unit of measure is *ohms,* which is represented by the Greek letter Ω (omega). A circuit has a resistance of one ohm if one volt causes a current of one ampere to flow.

Measuring Resistance with an Ohmmeter

2.4.2 Resistance is measured with an ohmmeter, which is normally found as one function of a multimeter. The ohmmeter is used often in the lines trade, but it can be useful to check continuity at a meter base or to check for a dead short or open circuit at a customer location. Do not use an ohmmeter on a live circuit. It will destroy the ohmmeter/multimeter as well as expose the operator to a serious flash.

When an ohmmeter is hooked to two different wires and the meter reads "zero," then there is no resistance between the two wires and they must be connected together or shorted somewhere. When an ohmmeter reads "infinity" (an infinity symbol on a meter is ∞), then the resistance is high and the two wires can be considered insulated from one another.

Measuring Earth or Insulation Resistance

2.4.3 *Earth resistance testers* or *insulation resistance testers* are like an ohmmeter except they put out a higher voltage than a regular ohmmeter. A higher output voltage is needed to test the relatively higher resistance of earth or the very high resistance of an insulator or cable insulation. These testers are commonly called *meggers,* which is easier to say than megaohmmeter.

Earth resistance testers can have various scales, such as 0–1 ohm, 0–300 ohms, or 1–1000 ohms. It is important to have a low-resistance grounding network at station sites, transformer locations, and customer premises. Ground rods are driven until the earth resistance tester verifies that the grounding network meets specifications. An ideal grounding network at a transformer should read 15 ohms or less before connecting it to the neutral.

Insulation resistance testers can have scales that read megaohms and infinity. For example, when a 500-volt insulation tester is used at a meter base to check between two wires or between a wire and ground, a person would know that there is no short or partial breakdown when the insulation tester reads infinity.

Ohm's Law

2.4.4 The relationship between volts, amperes, and ohms is expressed in an equation known as Ohm's Law (Figure 2–2). The current flowing in a circuit is equal to the circuit voltage divided by the resistance. This is the most used electrical theory formula for the lines trade. The equation is generally shown in a pie format as a visual aid in remembering how it is used, where

E represents voltage (*E*lectromotive force)
I represents current (*I*ntensité)
R represents *R*esistance

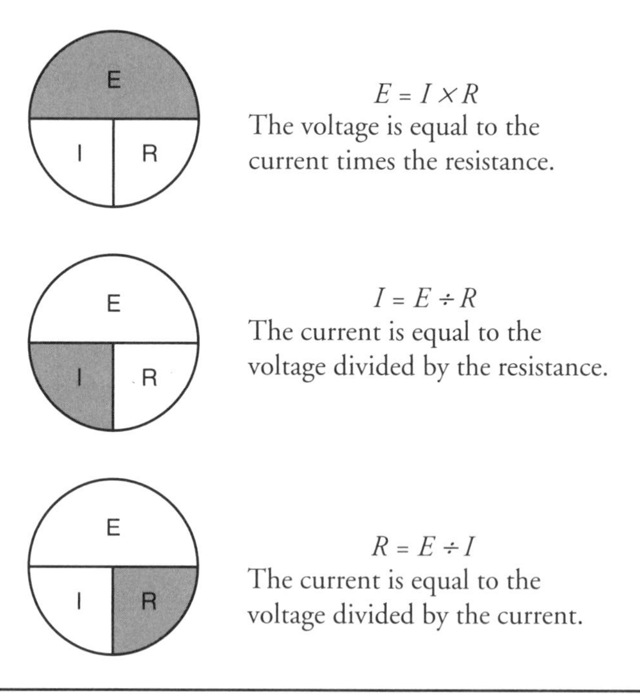

Figure 2–2 Ohm's Law.

Conductance

2.4.5 Conductance is the opposite of resistance. The term is sometimes used in the lines trade for objects that are not true conductors, such as semiconducting tape, or in reference to the conductivity of a wooden pole.

The symbol for conductance is *G*. The unit of measure is *mhos* (reverse of ohms). A mho is rarely used in any calculations involving the lines trade.

$$\text{Conductance } G = \frac{1}{R} \text{ resistance}$$

$$\text{Ohms resistance} = \frac{1}{\text{mhos conductance}}$$

Impedance

2.4.6 Impedance is another term for the opposition to current flow. In the lines trade, the terms *resistance* and *impedance* tend to be used interchangeably. Technically, the flow of current in an AC circuit is impeded by reactance as well as resistance. Reactance is explained in the chapter on AC power.

In a heater or incandescent bulb, there is only resistance to the flow of electrical current. When AC current flows into a coil, there is additional opposition to the current flow caused by inductive reactance. Some electrical loads, such as motors, electronic equipment, and fluorescent lights, are inductive loads. A portion of the load current flows in and then back out of the load without doing any work. The inductive reactance created by these loads causes additional impedance to current flow.

Similarly, capacitive loads such as the capacitive reactance created by long lengths of paralleling conductors also oppose current flow. Inductive reactance, capacitive reactance, and resistance provide the total impedance to current flow in a circuit. The symbol for impedance is *Z* and the unit of measure is ohms. Impedance can be substituted for resistance in Ohm's Law as $E = I \times Z$.

In an electrical utility circuit, resistance makes up the bulk of the total impedance in the circuit. The reactive current usually makes up less than 10 percent of the current in a distribution circuit. The opposition to current flow in an AC circuit, therefore, is resistance and reactance. The combination of resistance and reactance is called impedance, but in this book you will find the terms *impedance* and *resistance* used interchangeably.

Conductor Resistance

2.4.7 In an electrical system, a short run of a large conductor provides the least resistance to electrical current. A wire table that shows the current-carrying capacity of different sizes of conductor is not used often in the utility business because there are other factors to be considered. The expected load current to be carried, the fault current the conductor can be exposed to for short periods of time, the length of the feeder, acceptable line loss, and expected voltage drop all have to be balanced with affordability when choosing a conductor size.

High Resistance and Insulation

2.4.8 Electrical current flows because there is a pressure (voltage) pushing it through any resistance towards an area where there is less pressure. It is like the common saying, "electricity is always trying to find a path to ground." For example, electrical current flows through the resistance of a heating element because the pressure from the supply voltage pushes the current through the element to the low or zero pressure of the neutral end. A high resistance is needed to stop a current flow to ground. Current flows in the intended circuit as long as there is no low-resistance path to any other objects at a different pressure or voltage. High-resistant materials are used to insulate an electrical circuit from ground and other conductors.

Overhead conductors are generally bare and insulated from other objects by porcelain glass or polymer insulators. The longer the path through which the current has to flow and the higher the resistance of that path, the lower the current. Insulators have a number of curves along their surface. This increases the length of the path that current has to flow to ground. Increasing the size of an insulator increases its resistance and allows it to withstand a higher voltage.

Underground conductors are insulated by a rubber or polyethylene covering. The insulation needs to be well protected because even a little damage will cause the voltage (pressure) to stress the damaged location and eventually cause a short circuit in the cable.

High voltage can stress any insulation and cause it to fail. For example, air is normally an insulator, but it can become a conductor when it is electrically stressed and becomes ionized. Air is ionized when electrons in orbit around the atoms are displaced because of being stressed by voltage. An electrical arc and lightning are visible examples of air that has become conductive.

The Effect of Rain on Insulators

2.4.9 Contrary to popular belief, water is not a good conductor of electricity. When it rains and the water flows on clean insulators, the resistance of the insulators is not reduced substantially.

When it rains on contaminated insulators, the dirty and wet insulators are less resistive and can eventually short out the circuit. Insulators can become dirty when near industrial areas, saltwater, or roadways spread with salt in the wintertime. The failure of contaminated insulators is delayed due to the irregular shape and skirts on insulators, which keep parts of the insulator dry and make a longer leakage path for current to flow.

Insulators are cleaned using high-pressure water or corncob and nutshell blasting to mechanically remove grime from the insulators. This work is done with the circuit energized.

Rain or water around electricity is very dangerous if a live conductor has fallen to the ground. Water will absorb the salts in the earth and become very conductive. Dry ground is not a good conductor, but when the same ground gets wet, it becomes a very good conductor. Water will greatly reduce the resistance of any dirty surface.

Safety and Electrical Resistance

2.4.10 If an ohmmeter is used to measure a person's body from one dry hand to another dry hand, the resistance would be approximately 100,000 ohms. Most of this resistance comes from the skin. For some people, a 120-volt source is not high enough to overcome the resistance of dry, calloused hands. If a person is perspiring, the resistance is reduced to approximately 35,000 ohms. Once contact is made, the skin resistance can break down in a very short time. After skin resistance breaks down, there will be much less resistance through the internal organs of the body. The internal resistance of a body ranges between 100 and 400 ohms.

There can be many variables involving a person's resistance when electrical contact is made. The clothing and gloves being worn can increase the contact resistance and make a difference when electrical contact is made. When studies and calculations are made involving the resistance of a human body during an electrical contact, an average of 1000 ohms is normally used.

Because people may have made contact with a 120-volt source and were not hurt, they mistakenly think that 120 volts will not hurt them; however, there are fatalities every year when contact is made with that voltage. The amount of resistance being imposed, often hand to hand, is not enough to prevent a small amount of current flow, considering that 100 milliamperes can be fatal.

Second Point of Contact

2.4.11 A human body does not have enough resistance to prevent a fatal current flow through it when contact is made with a primary voltage. A more important consideration is what other objects a person is touching when contact is made with a live circuit.

Current must have a circuit before it can flow. Before current will flow through a human body, it must enter the body at one contact point and leave the body at another contact point that is at a different potential. An electrical accident usually occurs because one part of the body is in contact with a live circuit and another part of the body is in contact with earth, a neutral conductor, a deenergized load, or another live wire. If the body is not touching a second point of contact, there can be no current flow through it.

For example, some people will claim that they can touch a 120-volt wire and not feel anything. The source may be 120 volts, but a person is not exposed to all the available voltage unless another part of the person is well grounded. A person may not feel anything if standing on a wooden floor or wearing good boots. The same person standing barefoot on wet ground would probably receive a fatal shock.

Keeping away from a second point of contact is a fundamental rule for working on or near live circuits. A body cannot become part of a circuit unless there is a potential difference across it. If a power line worker makes contact with a live conductor while not in contact with any second point of contact, there cannot be a current flow through the body. This is the basis for barehand work.

2.5 Electrical Power and Electrical Energy

Heat from Current Passing through a Resistance

2.5.1 When current passes through a resistance in a circuit, heat is released. The amount of heat produced varies with the amount of current squared (I^2). If the current doubles, the heat produced quadruples. A low resistance draws more current, and more heat is produced.

Joule's Law of electric heating states that the amount of heat produced during each second by electrical current in a conductor is proportional to the resistance of the conductor and to the square of the current.

$$J = I^2 \times R \times t$$

where J = the amount of heat in joules
I = the amount of current flowing
R = the amount of resistance in the circuit
t = the time the current is flowing

The same formula can be converted to the basic unit for electrical energy which is a watt-second.

One watt-second = One Joule

When current travels through the resistance of a human body, heat is also released. Catastrophic electrical burns are immediately apparent by visibly blackened skin where contact is made and where current leaves the body.

The consequence of a less-serious electrical burn may appear as a small entry burn and a small exit burn on the body. However, the current flow that caused these burns flowed through the body along the lower-resistance blood vessels and nerves. The heat from the current flow will cause damage that can be much more serious than the external injuries might suggest. In other words, most of the damage is not visible and even so-called minor electrical burns should have medical attention.

Electrical Power and Energy Basics

2.5.2 Electrical energy is the product measured at a customer's meter. Electrical energy is electrical power × time. In other words, electrical energy and electrical power are not the same thing.

Electrical power is the product of volts and amperes and is measured in *watts* (W). The common equations for calculating power are:

$$W = E\,I$$
$$W = I^2R$$
$$W = \frac{E^2}{R}$$

where W = watts
E = voltage
I = current

Electrical energy is the product of watts and time. One watt times one second is a watt-second. One kilowatt times one hour is a kilowatt-hour.

Power Formulas Derived from Ohm's Law

2.5.3 There are many electrical equations based on Ohm's Law. The equation wheel in Figure 2–3 shows the interrelationship of these equations.

P = power in watts
I = intensity of current in amperes

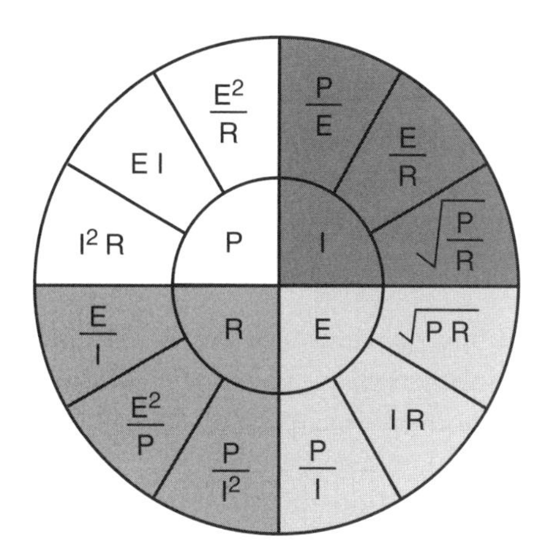

Figure 2–3 Ohm's Law equation wheel.

TABLE 2–3 Putting Power into Perspective

One Kilowatt (kW) = 1000 Watts	Hardware store portable generators have ratings such as 750W, 2.5 kW, or 3.5 kW. A typical toaster oven is 1.5 kW.
One Horsepower (HP) = 746 Watts	With 100HP = 74.6kW, a 75kVA transformer is needed to run a 100HP motor. However, in reality a motor is only about 80% efficient, and a 100kVA transformer would be needed.
One Megawatt (MW) = 1000 Kilowatts	Some of the older hydraulic stations on small rivers have generating units rated at about 1MW. The utilization station at a small factory is often rated at 1, 2, or 3MW. 1MW = 1 million joules per second.
One Gigawatt (GW) = 1000 Megawatts	The largest generating units are about 1GW and are found in newer thermal or nuclear plants. Hydraulic plants such as the R.H. Saunders/Robert Moses generating station across the St. Lawrence river have 32 units, 16 in Canada and 16 in the United States. Each unit is rated at 57MW which means the total plant output is 57MW × 32 units = 1.8GW. The summer peak load of New York City is about 11GW.

R = resistance in ohms
E = electromotive force in volts

Large Units of Electrical Power

2.5.4 A watt is the basic unit of power, but it is very small. In the electrical utility business, kilowatts (kW) and megawatts (MW) are the most common units (see Table 2–3).

In the lines trade, the terms kilovolt-ampere (kVA) and megavolt-ampere (MVA) are used when discussing power. To transmit AC power, more volts × amperes (VA) are needed to deliver the actual wattage used by a customer. In an AC circuit, there is a counterforce causing a reactance that impedes the flow of electrical current. In other words, slightly more than one kilovolt-ampere is needed to deliver one kilowatt of power. The chapter on AC power will explain this in greater detail.

The Kilowatt-Hour

2.5.5 As a unit, the watt is the amount of power being used at a given instant. It is also necessary to know how long the power is used to determine the amount a customer is charged for energy. Customers are billed based on the kilowatt-hour (kWh), where the kilowatt is the rate at which energy is used, and the hour is the length of time the power is used.

$$kWh = kW \times hours$$

Kilowatt-hour meters are installed to measure the kilowatt-hours used by a customer. There is a large variety of revenue metering used to measure other variables, but the primary charge for power used is the kilowatt-hour.

Kilowatt-Hours in Perspective

2.5.6 One kilowatt-hour is a small unit (see Table 2–4). When utilities and even countries are compared, the amount of energy generated is shown in billion kilowatt-hours.

TABLE 2–4 Putting Kilowatt-Hours into Perspective

One Kilowatt-Hour	One kilowatt-hour will run a 1000-watt microwave oven or a 1000-watt hair dryer for one hour. One liter of gasoline has the energy equivalent of approximately 10kWh. (One U.S. gallon of gasoline has the energy equivalent of approximately 37kWh.)
1000 Kilowatt-Hours	An average household in North America is considered to use 1000kWh hours per month, and this figure is often used when comparing electric bills between utilities. One cord of dry hardwood has the energy of approximately 6000kWh. One barrel of crude oil has the energy equivalent of approximately 1700kWh.
One Billion Kilowatt-Hours	Annual statistics for the sale or generation of electricity are expressed in billions of kWh. The total generating capacity of the world is approximately 12,000 billion kWh. Some statistics from countries in the 1990s are: United States 2300 billion kWh Canada 500 billion kWh France 410 billion kWh

Review Questions

1. What physical matter is flowing in an electrical current?
2. How is high voltage, over 750 volts, measured safely?
3. Why does a person normally not feel anything when hands are placed on each post of a car battery?
4. How can powerline workers differentiate between voltages within their utility?
5. Does electrical current stop once it gets to the load it is feeding?
6. Why is opening a ground wire or a neutral on a live circuit dangerous?
7. If a circuit is carrying 100 amperes, what portion of that current could induce ventricular fibrillation and probably death?
8. How would a line crew normally restore power when the switchgear will not pick up the load because of cold load pickup?
9. Where along a line would an accidental short circuit be most explosive?
10. When an ohmmeter is hooked to two different wires and the meter reads "zero," what does that indicate?
11. Why does keeping away from a second point of contact reduce the risk of an electrical accident?
12. What is the cause of damage when current flows through a body?

CHAPTER 3

Alternating Current (AC)

Topics to Be Covered	Section
Introduction	3.1
Characteristics of AC	3.2
Reactance in AC Circuits	3.3
AC Power	3.4
AC or DC Transmission	3.5

3.1 Introduction

Why Alternating Current?

3.1.1 Alternating current (AC) became the standard form of electrical power over direct current (DC) in the pioneer days of electrical power. There were major arguments about which form of electrical power should be delivered to customers. Thomas Edison promoted the case for DC power.

Alternating current has one major advantage: the easy transformation from one voltage to another. Easy transformation allows the voltage to be stepped up for efficient transmission of electrical energy over long distances.

Another advantage to AC power is that every time the voltage and current reverse direction, the magnitude of the voltage and current is zero. This assists in extinguishing arcs when opening switchgear. AC does, however, introduce some complicating phenomena to electrical circuits that are not found in DC circuits.

3.2 Characteristics of AC

AC Basics

3.2.1 Current flows in a conductor as long as there is a potential difference present. To have a potential difference, one end of the circuit is at an opposite pole (polarity) to the other end. These polarities are labeled as positive and negative. The direction of the current flow in a circuit is determined by the polarity of the source terminals.

With DC, the polarity does not change and the current flows in one direction only. With AC, the polarity at the source alternates between positive and negative and the current direction changes with every change of the source polarity (see Figure 3–1).

For example, the current in a single-phase circuit flows toward the load while the current in the neutral flows away from the load. In the next moment, the roles are reversed; the current in the neutral flows toward the load, and the current in the phase wire flows away from the load. The voltage on the phase wire is positive with respect to the neutral when it flows in one direction, and the voltage is negative when it flows in the other direction. The voltage on the neutral is unchanged and is close to zero with respect to a remote ground.

Frequency

3.2.2 In North America, alternating current supplied by electrical utilities travels 60 times in each direction in one second. In some other parts of the world, 50 cycles per second is common. The term "cycles per second" has been replaced by the international standard term for frequency, which is *hertz,* and is represented by the symbol *Hz.*

Unlike voltage or current, the frequency in a circuit stays constant right from the generator to the customer. When the frequency starts to drop, it is an indicator that the generator supplying the electrical system is overloaded and slowing. A small reduction in frequency will trigger the electrical system to trip out of service. A typical range for frequency is 59.97 to 60.03 hertz. Some systems are set up to start load shedding when the frequency reaches 59.3 hertz.

Generation of AC Power

3.2.3 When a loop is rotated within a magnetic field, an electric current is induced in the loop. With AC generation, one half of a loop travels in one direction through

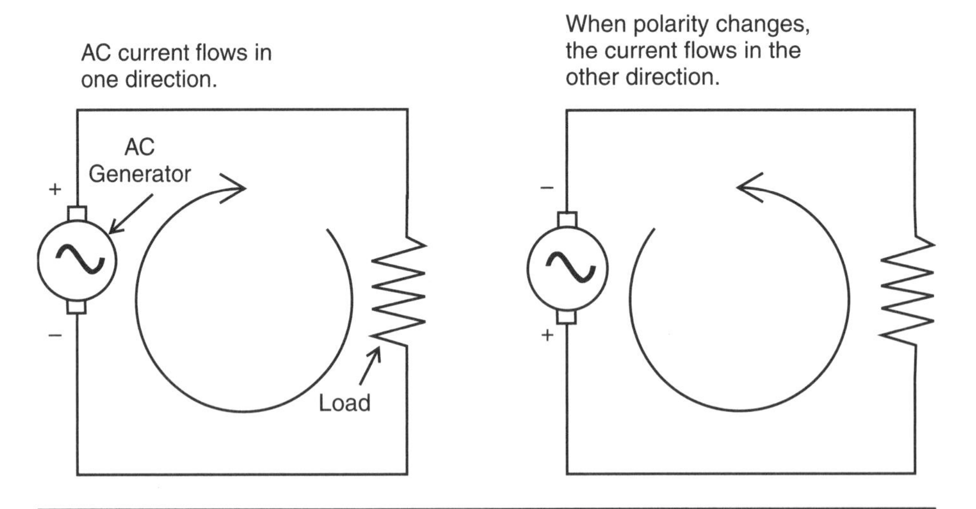

Figure 3–1 Direction of current flow.

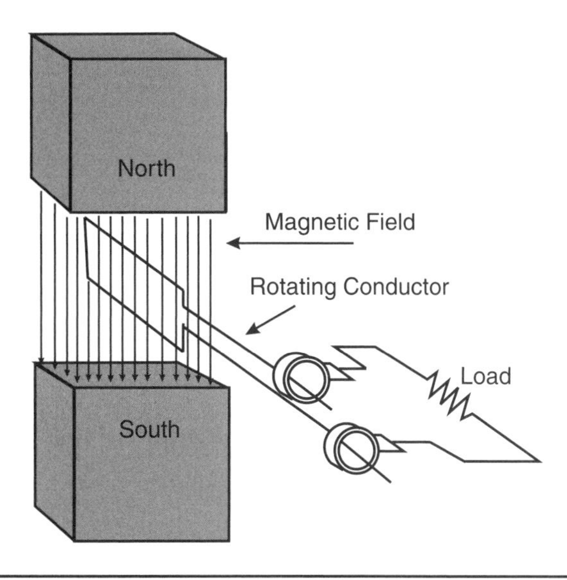

Figure 3–2 Simple AC generator.

the magnetic field while the other half travels in the opposite direction. The current flow induced in the two halves of the loop, therefore, also travels in opposite directions within the magnetic field but in the same direction in the loop.

In large commercial generators, there are many electric magnets mounted on the rotating part (rotor), and there are many loops (coils) mounted on the fixed part (stator). The coil shown in Figure 3–2 is traveling through the magnetic field at right angles, which is where the maximum current is generated.

Rise and Fall of Voltage and Current

3.2.4 When a loop or coil travels through a magnetic field, the induced current is at maximum when the coil cuts across the magnetic field at right angles. When a coil travels in parallel with the magnetic field, there is no induced current.

The current and voltage rise from zero to maximum value and drop back to zero while they travel in one direction and then repeat the zero-to-maximum rise on the return (Figure 3–3). In other words, AC and voltage change in both polarity and magnitude.

When a conductor finishes one rotation at 360 degrees, the conductor will be traveling parallel with the magnetic lines again. No current will be generated, as shown in the drawing illustrating the conductor at 0 degrees.

AC Represented by a Wave

3.2.5 The rate of the rise and fall of the voltage and current and the direction of flow can be represented on paper by drawing a sine wave (Figure 3–4). A sine wave is a wave shape that follows a mathematical form.

The wave above the zero current line represents the value of the current traveling in the positive direction, and the wave below the zero current line represents the value of the current traveling in the negative direction.

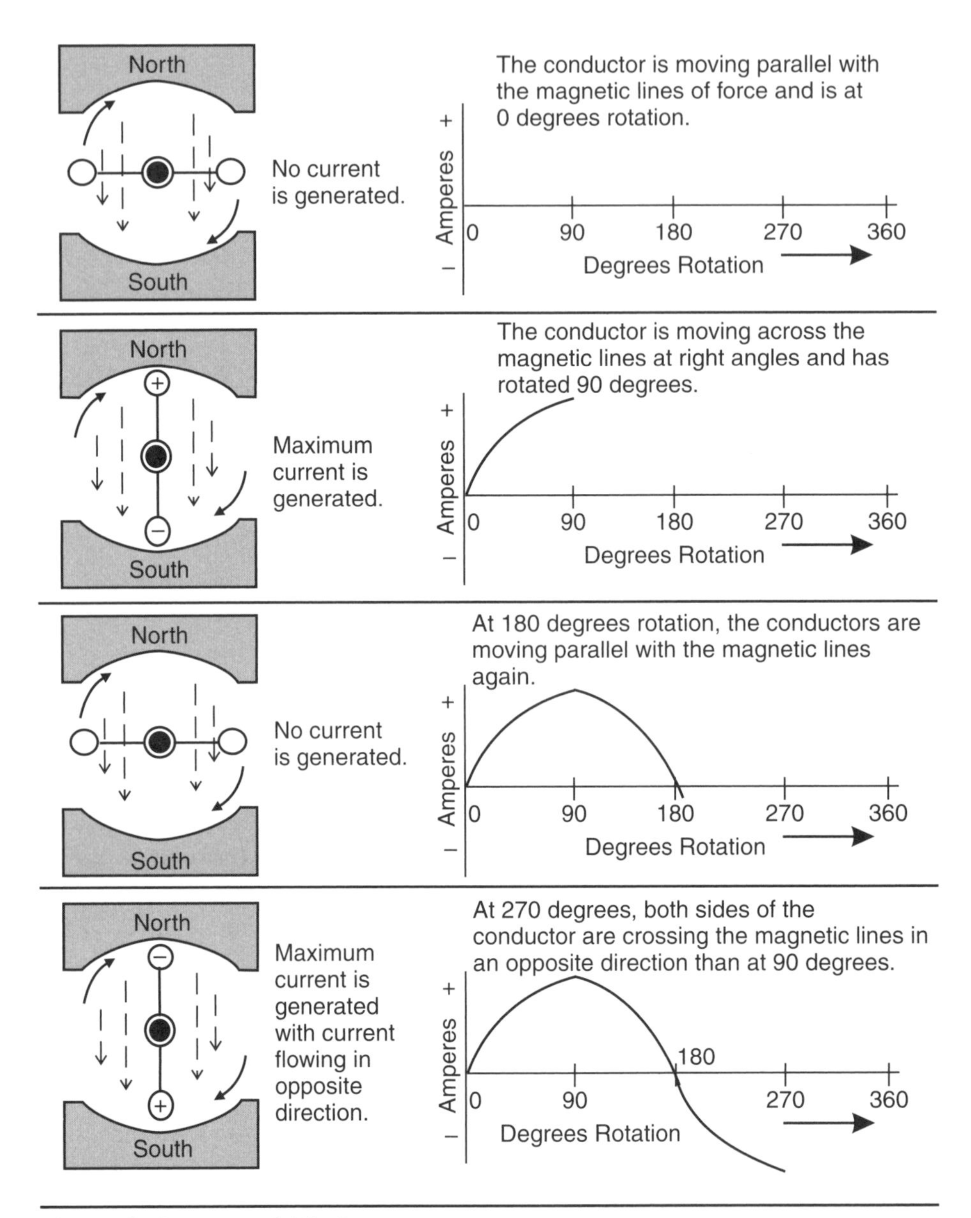

Figure 3–3 Rise and fall of current.

Sine Wave as a Graphical Representation

3.2.6 Because a sine wave is used to display how AC power behaves, there is sometimes the idea that AC travels in waves. The current actually behaves more like a pulse that builds up from zero to maximum magnitude and drops back to zero. It does this while traveling in one direction and then repeats this pattern in the reverse direction. The sine wave is only a graphical representation of the rate that the current and voltage values rise and fall and of the direction the current travels during one cycle.

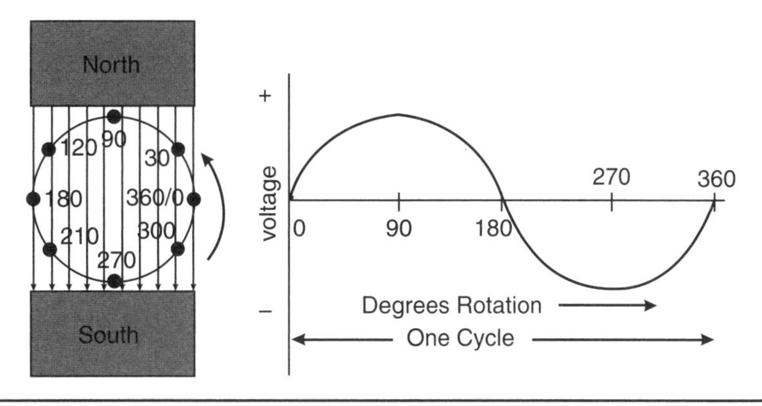

Figure 3–4 AC current represented by sine wave.

Values of AC and Voltage

3.2.7 With the values of AC and voltage continuously changing, what values do we actually read on an ordinary ammeter or voltmeter?

- The *instantaneous value* is the actual value of the voltage and current at each instant in the cycle. This value is not measured with an ordinary meter.
- The *peak value* is the highest instantaneous value that the voltage and current reach in both directions during the cycle. While this value is not measured by an ordinary meter, this value must be considered by design engineers when planning the insulation needed for different voltage systems.
- The *average value* is obtained by calculating the average of all the instantaneous values in half a cycle. The average value works out to be 0.636 of the peak value. Surprisingly, this also is not the value measured with an ordinary meter.
- The *effective value* of AC and voltage is the value of a DC and voltage that would have the same heating effect. The effective value is 0.707 of the peak value (Figure 3–5). This is the value measured with the ordinary ammeter and voltmeter. The effective value is often referred to as the root-mean-square (rms) value. In AC applications, the effective value of voltage and current is the value used for calculations and measurement.

Voltage and Current in Phase with Each Other

3.2.8 When both the voltage wave and the current wave reach their maximum and zero values at the same time, then the voltage and current are considered to be *in phase* with each other. In an electrical system, the voltage and current are generally not exactly in phase.

When current is flowing into certain loads, such as a motor load, the current wave will peak after the voltage wave peaks. This current is *out of phase* and *lags* the voltage.

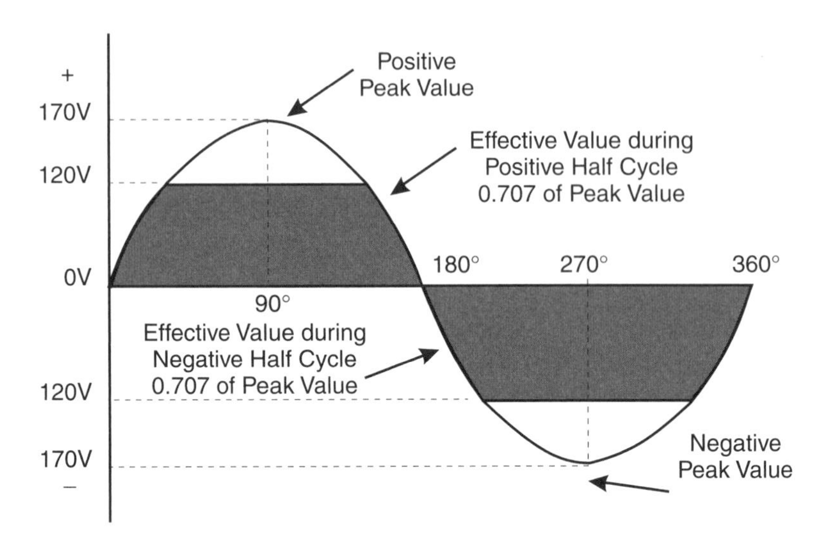

Figure 3–5 Effective value in AC.

When the current is flowing into a capacitor, the current wave will peak before the voltage wave peaks. This current is *out of phase* and *leads* the voltage.

Circuits in Phase with Each Other

3.2.9 An electrical power system of a utility is part of a large power pool or grid fed from many sources and locations. A generator must generate power so that the magnitude and the direction of voltage and current are synchronized with the other generators supplying power to the same grid. In other words, the voltage from all power sources must be in the same position on the sine wave before feeding power into the grid. All generators must be in phase with each other. If an out-of-phase generator is connected into the grid, it will act as a short circuit.

3.3 Reactance in AC Circuits

Loads Fed by AC Circuits

3.3.1 There are three kinds of loads fed by an AC circuit:

1. Heating and lighting are *resistive* loads. A resistive load can do work. The amount of work being done can be measured by a kilowatt-hour meter.
2. The energy used to magnetize a motor or a transformer is an *inductive* load. An inductive load does not generate heat or light and is not measured by the kilowatt-hour meter.
3. The energy used to supply a capacitive effect at capacitors, paralleling conductors, or in cables is a *capacitive* load. A capacitive load does not generate heat or light and is not measured by a kilowatt-hour meter.

Reactance in an AC Circuit

3.3.2 A resistive load opposes the flow of current, but so do inductive and capacitive loads. All three types of loads oppose the flow of current and add to the total opposition to current flow in a circuit. This disturbance imposed by inductive and capacitive loads in a circuit is called *reactance*. The symbol for reactance is *X*.

Reactance can be an *inductive reactance* (X_L) caused by loads such as motors, transformers, fluorescent lights, and computers, or it can be a *capacitive reactance* (X_C) caused by capacitors or paralleling conductors.

Impedance in an AC Circuit

3.3.3 Resistance is the only opposition to the current flow in a DC circuit. In an AC circuit, the opposition to current flow consists of resistance *and* reactance. This combination of resistance and reactance opposing the current flow is referred to as *impedance.* Impedance is measured in ohms and is represented by the symbol Z. Impedance can be used interchangeably with resistance in calculations using Ohm's Law

$$Z = IR \qquad Z = \frac{E}{I} \qquad I = \frac{E}{Z}$$

where E = effective voltage in volts
I = effective current in amperes
Z = total impedance in ohms

Resistance in an AC Circuit

3.3.4 Light and heat are resistive loads and do not cause any other disturbance to the circuit. Either AC or DC can supply a resistive load. Ohm's Law applies to a resistive AC circuit in the same way that it applies to a DC circuit. For practical applications in the lines trade, resistance can be used for most calculations involving AC circuits, and the result will generally have less than a 10 percent error.

Induction in an AC Circuit

3.3.5 Voltage can be generated into a conductor:

- By moving a conductor within a magnetic field.
- By having a conductor near a moving magnetic field.

Voltage (electromotive force) is induced into a conductor when a conductor is moved through a magnetic field. There is no voltage generated unless the conductor is moving.

Voltage can also be induced into a conductor when a magnetic field from a nearby live AC circuit moves through the conductor. Only an AC circuit would have a moving or fluctuating magnetic field. Induced voltage is always in a direction that opposes the direction of the current flow. Induced voltage is, therefore, the opposite polarity of the source circuit.

With AC, the magnetic field is moving (expanding and collapsing) and any nearby stationary conductor will have a voltage induced on it. This phenomenon is familiar to the powerline worker, because induction from live circuits is a concern when grounds are installed on an isolated circuit. Some of the magnetic field around a live conductor cuts through the live conductor itself and induces some voltage onto itself. This *self-inductance* creates a counter-electromotive force (voltage) within the conductor.

When the conductor is part of a coil, such as in a motor or transformer, the magnetic field around the conductor cuts through adjacent wires in the coil, which increases the self-inductance in the conductors forming the coil. This opposition to the current flow delays the rise and fall of the current but not the voltage. The

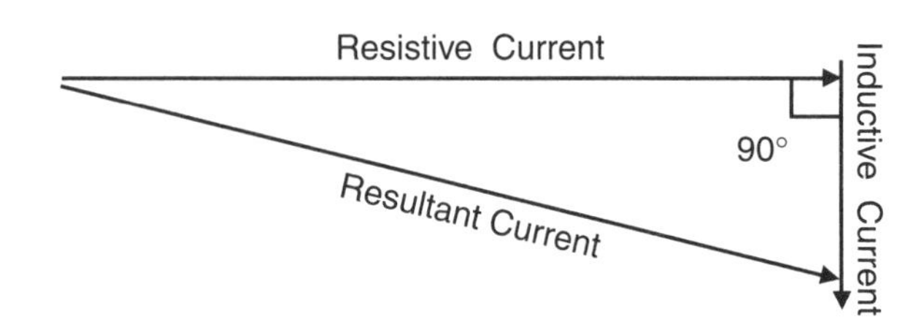

Figure 3–6 Vector drawing showing inductive current acting at 90°.

counter-electromotive force generated by the continuously changing voltage and current is called *inductance.*

Figure 3–6 shows the relationship between resistive current, inductive current, and the total current in a circuit. Inductive current acts on a circuit as though it flows at 90 degrees to the resistive current. When the resistive current is added to the inductive current, the resultant current vector is longer, which means that the resultant current is higher.

The symbol for inductance is *L*. The unit of measurement for inductance is a *henry,* but a henry is not a unit a powerline worker would use.

Inductive Reactance

3.3.6 The opposition to current flow within a conductor or coil is called *inductive reactance,* which is represented by the symbol X_L and measured in ohms. The formula is shown only to illustrate the factors that influence the magnitude of inductive reactance. It is not a formula a line worker would need.

$$X_L = 2\pi f L \text{ ohms}$$

where X_L = inductive reactance in ohms
π = 3.14
f = frequency of the circuit in hertz
L = inductance in henrys

Current Lagging Voltage

3.3.7 Distribution feeders tend to have a combination of resistive load and inductive load (electric motors). The inductive load sets up an inductive reaction in the circuit, and this opposition to the current flow causes it to lag behind the rise and fall of the voltage. In other words, as represented by a sine wave, the voltage will peak before the current peaks. The power output of the circuit is reduced because

$$Power = volts \times amperes$$

Figure 3–7 shows the current lagging the voltage by 30 degrees, which means the current is 30 degrees out of phase with the voltage.

Applications of Inductive Reactance

3.3.8 Two applications of inductive reactance used in substations are the current-limiting reactor and the shunt reactor. A current-limiting reactor is a coil of cable that sets up an inductive reaction in a circuit. The current-limiting reactor is

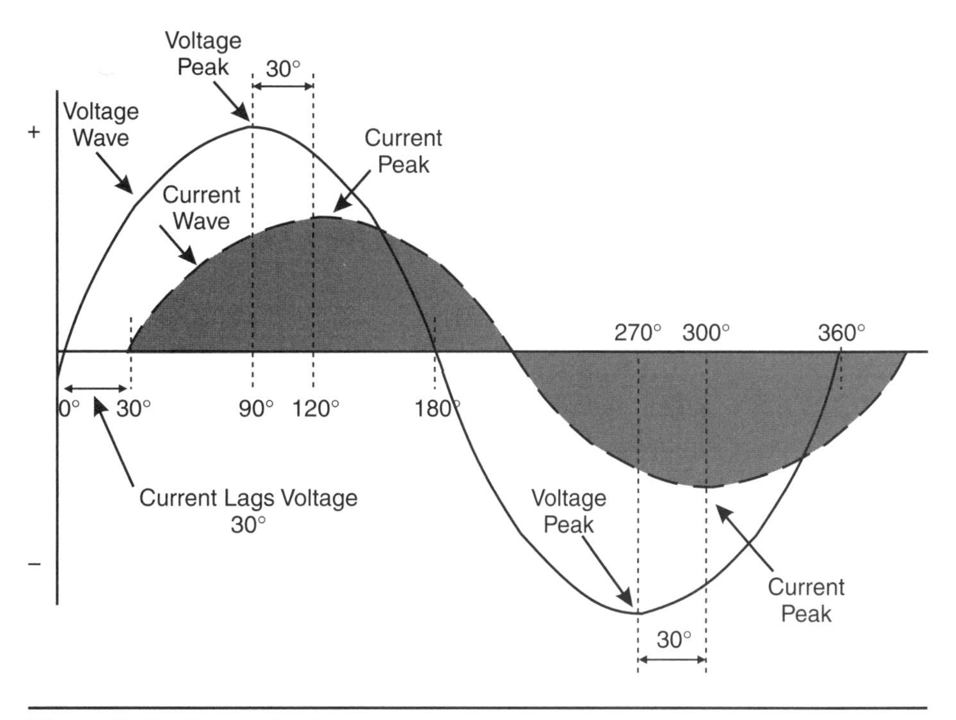

Figure 3–7 Current lagging voltage.

designed so that a large inductive reactance is created during a high fault current condition. The inductive reactance impedes the flow of damaging fault current in the substation.

To balance large capacitive loads caused by parallel conductors of a long transmission line, shunt reactors are installed in substations. The reactors produce an inductive reactance that cancels out an equal amount of capacitive reactance in the transmission line. A reactor looks like a power transformer and is connected into the circuit as a shunt (a parallel load). The only connections to the shunt reactor are to high-voltage bushings. There are no output connections such as the secondary of a transformer.

Capacitance

3.3.9 The major sources of capacitance in a system are paralleling conductors on long transmission lines, underground cable, and capacitors installed specifically to put more capacitance into the electrical system. The tendency of a circuit to store electricity when a potential difference exists between conductors is called *capacitance.* Capacitance occurs when an electrically charged conductor (plate) electrostatically induces an equal in magnitude but opposite polarity charge on a nearby conductor (plate). The two conductors are not in contact but separated by some kind of insulation.

When a capacitor is part of an AC circuit, the plates are charged and discharged alternately. The capacitor tends to store the acquired charges and cause a counter-electromotive force (voltage) to oppose the continuing voltage change.

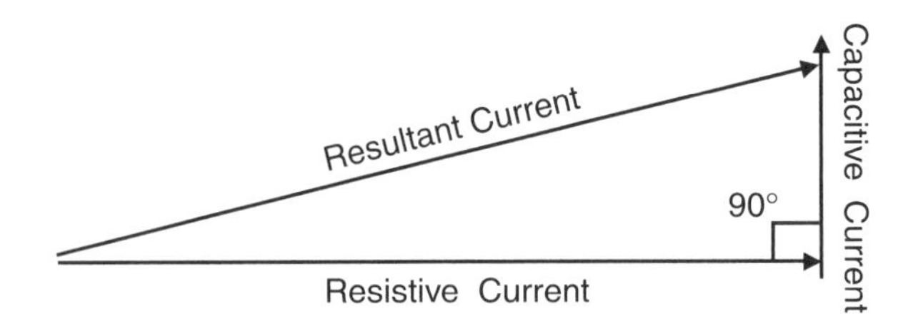

Figure 3–8 Vector drawing showing capacitive current acting at 90°.

Figure 3–8 shows the relationship between resistive current, capacitive current, and the total current in a circuit. Note that capacitive current is in the opposite direction to the inductive current shown in Figure 3–6.

Capacitive current acts on a circuit as though it flows at 90 degrees to the resistive current. When the resistive current is added to the inductive current, the resultant current is higher. When capacitive current is added to a circuit with inductive current, the two types of reactance cancel each other because they act in opposite directions. The symbol for capacitance is *C.* The unit of measurement for capacitance is a *farad,* but a farad is not a unit a line worker would need.

Capacitors

3.3.10 Capacitance normally occurs in a circuit. Parallel conductors are like a capacitor where the two conductors are separated by air and the conductors electrostatically induce charges on each other. Underground cable is like a capacitor where the inner conductor is separated from the outer sheath by insulation and a charge is built up between the conductor and the sheath.

A capacitor, as illustrated in Figure 3–9, is constructed to induce capacitance in a circuit. It consists of two strips of aluminum foil rolled up together with insulating oiled paper in between. As a capacitor, one electrically charged aluminum roll of foil (plate) will electrostatically induce an opposite polarity charge in the other foil (plate).

Capacitive Reactance

3.3.11 When a capacitor is charged up, a counter-electromotive force (emf) equal to the source voltage is built up on the opposite plate. Current will flow into the capacitor only while the voltage is rising. When the voltage approaches peak value, the counter-electromotive force on the opposite plate is also approaching peak value, which causes the current flow to decrease. There is no current flow when the voltage is at its peak (90 degrees). In a capacitor, the current reaches its peak before the voltage. A capacitor opposes a *change* in voltage, which is what the voltage is doing continually in an AC circuit.

This delaying or opposing of changes to the voltage is called *capacitive reactance.* Capacitive reactance has the symbol X_C and is measured in ohms. The formula is shown only to illustrate the factors that influence the magnitude of capacitive reactance.

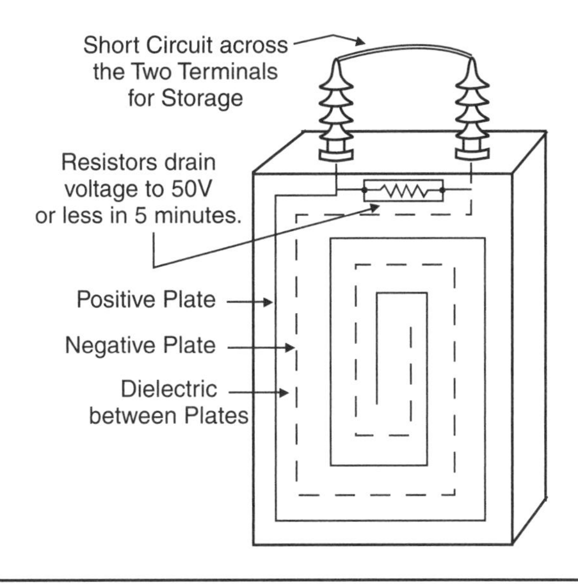

Figure 3–9 The construction of a capacitor.

$$X_C = \frac{1}{2\,\pi f\, C}\, ohms$$

where X_C = capacitive reactance in ohms
$\pi = 3.14$
f = frequency
C = farads

Current Leads the Voltage

3.3.12 Because change to the voltage across the capacitor plates is delayed, this capacitive reaction causes the current wave to lead the voltage wave. Figure 3–10 shows the current wave leading the voltage wave by 15 degrees.

Applications of Capacitive Reactance

3.3.13 Capacitors are installed on electrical systems to draw a leading current from the circuit to counteract the more predominant lagging current in utility circuits. When the lagging current, due to motor load, is partially balanced by the leading current, due to the installed capacitor, there is less total reactive current. With less reactive current, there is less total current; and, therefore, there would also be less voltage drop and line loss. The most common use of capacitors, therefore, is as a voltage booster.

Because most customer load has a heavy inductive element to it, an increased load on a transmission or distribution line means that there is an increase in the inductive load. As the voltage starts to decrease, an increase in capacitance is needed to keep the voltage up.

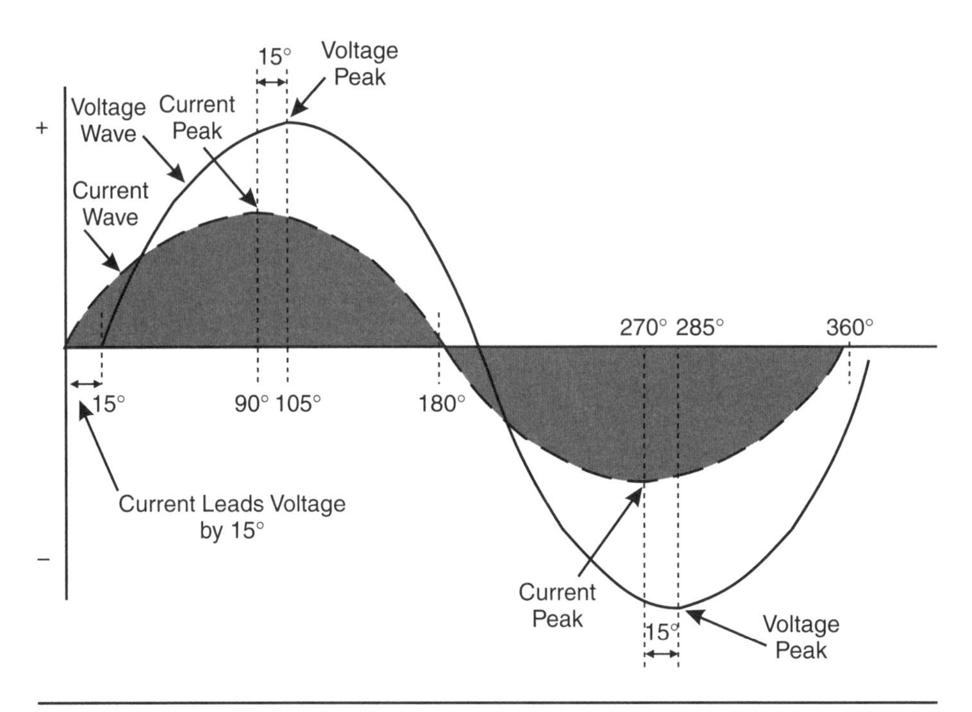

Figure 3–10 Current leading the voltage.

Effect of Frequency on Reactance

3.3.14 The formulas for inductive reactance and capacitive reactance show that the inductive reactance increases when the frequency goes up and capacitive reactance decreases.

$$X_L = 2\,\pi f L \textit{ ohms}$$

$$X_C = \frac{1}{2\,\pi f C} \textit{ ohm}$$

There is somewhat less inductive reactance in a 50-hertz circuit than in a 60-hertz circuit. A 60-hertz transformer has 10 to 15 percent less material than a 50-hertz transformer. A higher frequency sets up a greater magnetic flux linkage, and, therefore, a smaller iron core can be used. The electrical system in an aircraft is 400 hertz, which allows even less iron and less weight for electrical equipment using an iron core. Anyone in the lines trade who has worked with transformers in the few locations that still have 25 hertz can verify that a 25-hertz transformer is much heavier than the 60-hertz transformer of the same kilovolt-ampere rating.

A 50-hertz circuit would have more capacitive reactance than a 60-hertz circuit. The increased capacitance could be significant on long transmission lines.

Summary of Resistance, Inductance, and Capacitance

3.3.15 Resistance opposes the *flow* of current in an electric circuit. Inductance opposes a *change* in current. The nature of an AC circuit has its current always changing by rising and falling in magnitude and reversing in direction. Loads such as those with wire coils in them are inductive and increase the amount of inductive reactance in a circuit.

Capacitance opposes any change in *voltage.* The nature of an AC circuit is that its voltage is always changing by a rising and falling in magnitude and its polarity is continually reversing. Capacitive loads such as capacitors increase the amount of capacitive reactance in a circuit.

3.4 AC Power

Active Power

3.4.1 The total power produced in a DC circuit is calculated by multiplying total voltage by total amperes. Because there is no reactance in a DC circuit, all of the power the circuit supplies is useful power or *active power.*

Active power is also referred to as effective power, true power, or real power, because this is the power that gives light, gives heat, and turns motors. Real power can also be expressed as voltage × resistive current.

In AC circuits, power alternates at the same rate as the voltage and current are alternating. The power measured at a customer's meter is referred to as active power. Therefore, the total active power supplied by a circuit is equal to total effective current × total effective voltage, or

$$P = IE$$

Active power is measured at a customer meter and is measured as *watts* or in blocks of 1000 watts, which is *kilowatts.*

Apparent Power

3.4.2 The total power supplied by a circuit is called *apparent power,* because in an AC circuit all of the power does not perform actual work. Apparent power is a combination of *active* power and *reactive* power.

$$\text{Apparent power} = \sqrt{(\text{active power})^2 + (\text{reactive power})^2}$$

Apparent power is measured in volt-amperes or kilovolt-amperes (kVA). The term kVA is common terminology for the lines trade when referring to transformer sizes. The term for apparent power (kVA) is more suitable than the term for active power (watts) when referring to transformer size, because a transformer has to supply the apparent power needed by the load.

Reactive Power

3.4.3 *Reactive power* is the element in the apparent power formula that does not do any work. In a circuit, reactive power is transferred back and forth between the reactive load and the circuit. Reactive power is not used up. Reactive power is equal to voltage × reactive current or I^2X. It is measured in volt-amperes reactive or kilovolt-amperes reactive (*VAR* or *kVAR*).

In most distribution circuits, the net reactive power is inductive due to a portion of the motor loads, fluorescent lighting, and electronic loads. Reactive power is normally very small compared to the active power available in the circuit.

Inductive reactive power is necessary to create electromagnetic fields in equipment such as transformers and motors. Capacitive reactive power is necessary to create electrostatic fields in capacitors. Volt-amperes reactive will increase current

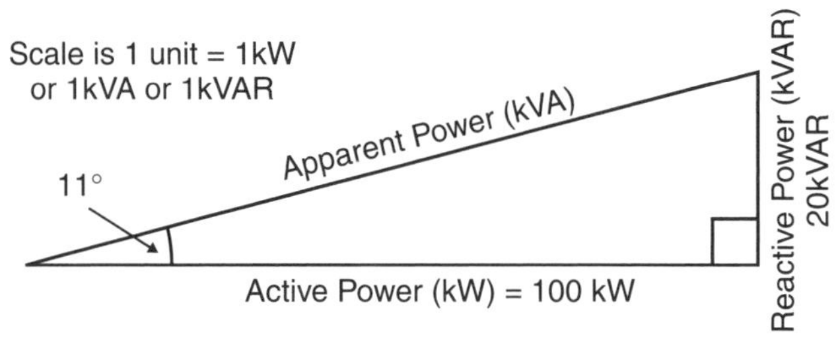

Scale is 1 unit = 1kW
or 1kVA or 1kVAR
Apparent Power (kVA)
11°
Active Power (kW) = 100 kW
Reactive Power (kVAR)
20kVAR

Figure 3–11 Power triangle.

flow in a circuit but do not represent energy consumption. Volt-amperes reactive store energy in one part of the cycle and return it in the next part of the cycle.

The Power Triangle

3.4.4 The mathematical relationship among the three kinds of AC power can be illustrated by a vector diagram (Figure 3–11). In a vector diagram, the length of each side represents the magnitude of each type of power.

The sides of the triangle in Figure 3–11 represent the magnitude of active power and reactive power. The hypotenuse represents the magnitude of apparent power. If two sides are known, the remaining side can be measured or calculated. In Figure 3–11, active power is represented by 100 units, and the reactive power is represented by 20 units. The resultant can be calculated, as shown in the following equation, or measured from the vector drawing.

$$Apparent\ power = \sqrt{100^2 + 20^2} = 102\ kVA$$

Power Factor

3.4.5 The ratio of active power to apparent power is called the *power factor* and is normally expressed in percent. If there is no reactive power in the circuit, the power factor is 100 percent. Keeping the power factor high in a circuit reduces line loss, voltage drop, and even generation. The power factor is measured at some customers' locations by revenue meters, and customers are penalized for a low power factor.

The angle between apparent power and active power represents the amount of reactive power in a circuit. A power factor can be measured from the power triangle or calculated in two ways:

$$Power\ factor = \frac{active\ power}{apparent\ power}$$

or

$$Power\ factor = \frac{R}{Z}$$

In the example in Figure 3–11, the angle is 11 degrees. The power factor is equal to the cosine of 11 degrees = 0.98, or 98 percent, or, when using a formula:

$$\frac{Active\ power}{Apparent\ power} = \frac{100}{102} = 0.98 = 98\%$$

Power Factor Correction

3.4.6 In a circuit with a low power factor, the apparent power needed to supply the load on the circuit becomes unacceptably high. With too much reactive current in a circuit, more power must be generated to supply the load. More current must flow in the conductor, which causes more line loss (I^2R) and a greater voltage drop.

Because the reactance in most distribution circuits is due to inductive loads, the easiest way to correct the power factor is to install a nearly equal amount of capacitive volt-amperes reactive to balance the inductive volt-amperes reactive. Installing capacitors on the circuit would reduce the apparent power needed to supply power to the customer and, therefore, reduce line loss and raise the voltage.

In industries with large inductive loads, the industry is billed for the apparent power used as well as the active power. These industries do their own power factor correction in the plant by installing capacitors.

Figure 3–12 illustrates that adding 5 kilovolt-amperes reactive capacitors to a circuit with 8 kilovolt-amperes reactive inductance will reduce the total inductive reactance in the circuit to 3 kilovolt-amperes reactive.

3.5 AC or DC Transmission

AC or DC

3.5.1 Most transmission lines transmit AC. Alternating current transmission became universal because AC was easily transformed to higher or lower voltages. For long distances, however, DC high-voltage transmission can be more economical because there is no reactive current being generated or transmitted.

The transformer made it possible for generation, transmission, and distribution to be performed at ideal voltages. High-voltage transmission lines made long-distance power transmission possible from remote hydraulic generation sites.

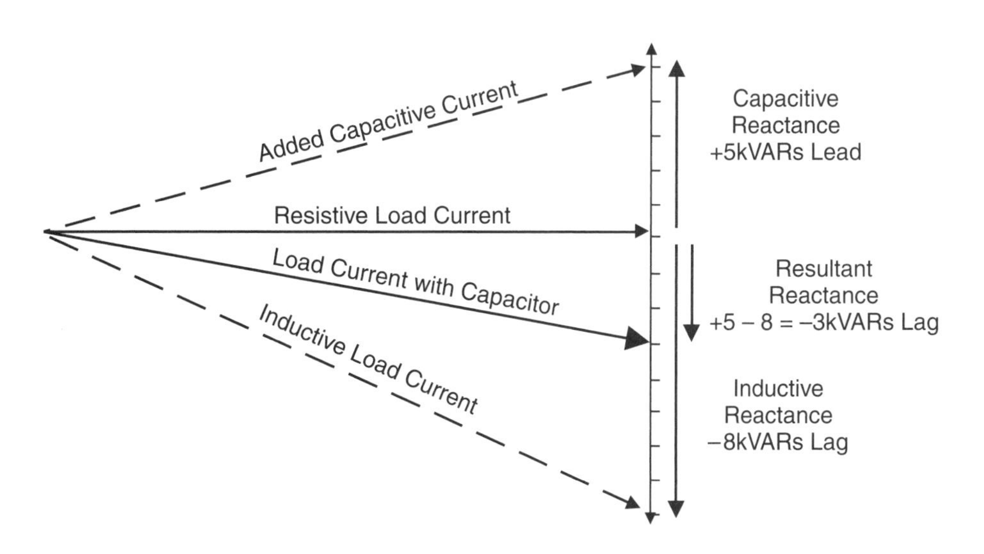

Figure 3–12 Capacitors reduce load current in circuit.

High-voltage direct current (HVDC) transmission has become popular for transmitting a large amount of power over long-distance overhead lines and long submarine cable crossings. Improvement in converter equipment has made DC transmission an increasingly popular option. Converters are called *rectifiers* when changing AC to DC and *inverters* when changing DC back to AC.

Effect of AC Frequency on Transmission of Electricity

3.5.2 The frequency of AC has an effect on the ability to transmit power. A standard 50- or 60-hertz circuit can transmit power for long distances if allowances are made for the inductive reactance and capacitive reactance by installing capacitors and reactors, respectively.

If the frequency was increased to an impractical 1000 hertz, the inductive reactance would be so high that it would block the transmission of power after a very short distance. At 1000 hertz, the skin effect on the conductors would also increase and cause the current to be concentrated on the outer surface of the conductor. The amount of current the conductor can carry would be severely limited.

Direct current transmission has no frequency and, therefore, has the opposite effect of high frequency AC. Induction and skin effect are eliminated, and, therefore, DC is a very efficient way to transmit power.

Main Applications of DC Transmission

3.5.3 The main applications of DC high voltage transmission are:

1. Long-distance transmission of large amounts of power.
2. Long lengths of underground or submarine cable.
3. Tying together two AC systems with different frequencies.
4. Tying together two different AC systems at locations where it is difficult to maintain control of voltage, frequency, and power flow.
5. A long length of distribution line to feed a small remote load, which would otherwise need a high-voltage transmission line and substation or its own generation.

Advantages of DC Transmission

3.5.4 The advantages of DC transmission are:

1. The power factor of the line is at unity, and, therefore, there is no need for reactors or capacitors to correct the power factor.
2. There is less corona loss and no skin effect in the conductor.
3. There is no charging current to energize a DC line.
4. More power can be transmitted per conductor. Ground return can be used instead of a conductor because the ground return is not subject to the impedance an AC circuit would have. If a two-conductor circuit needs work, work can be carried out on one conductor and the other conductor with ground return can continue as an independent circuit.
5. There is less short-circuit current on a faulted DC line, and the fault does not add to the fault current in the connected AC system.

6. A DC line can feed longer distances because there is no reactance in the circuit.

Disadvantages of DC Transmission

3.5.5 The disadvantages of DC transmission are:

1. Converters are expensive, and they need reactive power to operate. This requires the installation of capacitors to keep the AC system feeding the DC circuit at an acceptable power factor. Modern electronic converters solve many of the earlier problems involving conversion.
2. Converters generate harmonics, and, therefore, filters need to be installed to limit the effect on the AC system feeding the converters.
3. The steady state of DC current makes design of high-voltage DC switchgear difficult. This limits the ability to operate between DC circuits. Alternating current switchgear takes advantage of the fact that the current wave goes through its zero point and there is momentarily no energy to maintain an arc.

Description of DC Transmission Lines

3.5.6 A DC transmission circuit requires a conductor from the source to the load and a return path back to the source. The circuit, therefore, usually consists of two conductors. A two-conductor DC line is referred to as a two-pole circuit. Some underground or submarine DC circuits consist of one conductor and use the earth or seawater as a return path.

There is a converter at each end of the circuit. The converter at the sending end is a rectifier, which changes the AC to DC, and the one at the receiving end is an inverter, which changes the DC back to AC. A converter can be used as a rectifier or an inverter, which allows the circuit to feed in either direction.

Circuit breakers are installed on the AC side of the converters. It is easier to break the arc of AC because the magnitude of the voltage and current becomes zero, 120 times a second, on a 60-hertz circuit.

Resistance to Current Flow on DC

3.5.7 Direct current is fairly constant and is surrounded by a constant or stationary electric field. The energy of this field is in the form of potential energy. There is no electromagnetic energy or its associated induced current. There is no reactance on the DC line itself. The only impedance to current flow is the resistance of the conductor. In other words, the power factor of the DC circuit is 100 percent.

The converter on each end of a DC line draws reactive power from the AC system, and capacitors are needed for power factor correction.

DC Underground or Submarine Cable

3.5.8 The length of AC underground or submarine transmission lines would be limited because reactors would be needed every 25 to 50 miles (40 to 80 kilometers) to offset the capacitance generated by the cable. Similarly, when underground distribution is used in rural areas over long distances, reactors are installed at regular intervals to offset the capacitance of the cable. There is no similar concern about reactance in a DC cable. Direct current cable is the only choice for long-distance submarine power transmission.

Working on DC Circuits

3.5.9 On a two-pole transmission circuit, one pole can be grounded and worked on while the circuit stays in service. The other pole and the earth make up a circuit. The location and number of grounds must be such that the ground current does not flow back up the grounds through the circuit to get back to the source. Standard potential testers used on AC circuits will not work on a DC circuit because potential testers measure the electric field around a conductor.

Review Questions

1. What is the one major advantage of AC power?
2. Unlike voltage or current, _____ in a circuit stays constant right from the generator to the customer.
3. What changes occur to AC as it goes through a cycle?
4. In relation to the peak value of AC power, what value does an ordinary ammeter and voltmeter measure?
5. What kind of AC load uses the energy needed to magnetize a motor or a transformer winding?
6. What kind of AC load uses the energy needed to supply a capacitive effect at capacitors, paralleling conductors, or in cables?
7. What three types of loads oppose the flow of current and form the total impedance to a circuit?
8. Distribution feeders tend to have a combination of resistive load and inductive load, creating a lagging power factor. What equipment is installed to counter a lagging power factor?
9. What are the three types of power in an AC circuit?
10. What are three applications of DC transmission lines?

CHAPTER 4

Three-Phase Circuits

4.1 Introduction

Three-Phase Basics

4.1.1 Electric current needs a circuit (a complete circle) before it can flow. A single-phase wye circuit can be a phase and a neutral with the current flowing through the "hot" wire and returning to the source through the neutral.

Most utility circuits consist of three phases. A three-phase circuit is not three different single-phase circuits, but one circuit with all three phases interconnected. Each phase helps to complete the circuit by acting as the return path for the other two phases.

A live conductor in a circuit tends to be called a *phase* because when three-phase voltages and currents are generated, each of the three conductors gets its voltage and current at a certain phase of a cycle. An electrical phase is often represented in drawings and text by the Greek letter phi (**Φ**).

4.2 Characteristics of Three-Phase Circuits

Why Three Phase?

4.2.1 The effect of three-phase power as compared to single-phase power is similar to the effect of a six-cylinder engine as compared to a single-cylinder engine. A six-cylinder engine produces a smoother six small pulses per cycle while the single-cylinder engine produces one large pulse per cycle.

The values of the voltage and current in each of the three phases overlap with the other phases; therefore, three interconnected phases provide a smoother power than the relatively more pulsating power of a single phase. Three-phase current supplies a rotating magnetic field. Even though the power on each individual AC phase pulsates when it goes through the AC cycle, the sum of the power in all three phases at any point is constant.

Large generators and large motors are more efficient and are considerably smaller as three-phase units compared to equivalently powered single-phase units. In a three-phase motor, the magnetic field automatically rotates, bringing the rotor along with it.

With the same voltage and current per phase, a three-phase system needs only one additional wire (without a neutral, there is a 50 percent increase in conducting material) over a single-phase system but increases the circuit capacity by 73 percent. A three-phase circuit can carry twice as much load as a single-phase circuit while maintaining the same voltage.

Generation of Three-Phase Power

4.2.2 A simplified three-phase generator, as illustrated in Figure 4–1, shows three coils mounted on the armature at 120 degrees apart. Each coil generates an AC and voltage, but the power generated in each coil reaches its peak and direction at 120 degrees apart.

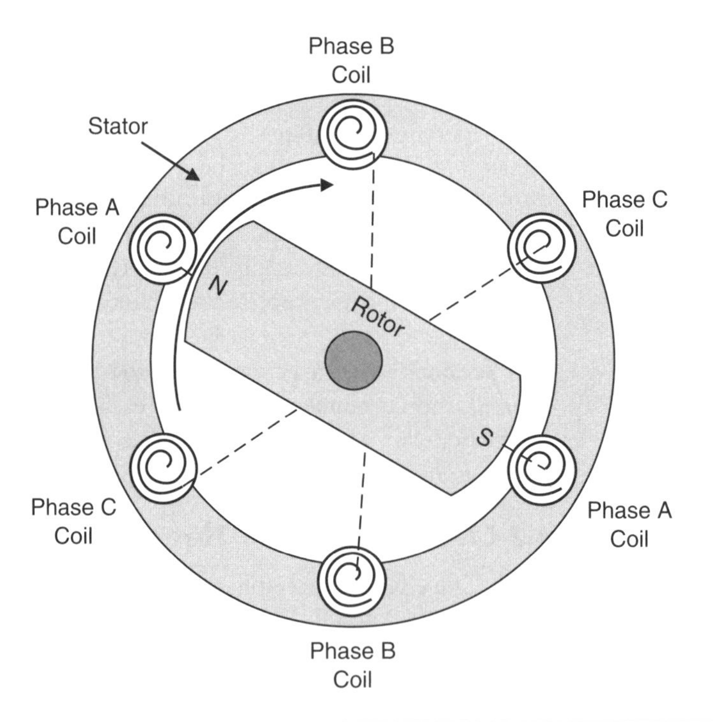

Figure 4–1 Simplified three-phase generator.

Commercial generators mount many coils on the stator and many magnets on the armature. The individual coils are wired so that they are connected together as three circuits 120 degrees apart. Each of the three circuits becomes a phase of a three-phase circuit.

Phases 120 Degrees Apart

4.2.3 When three phases are 120 degrees apart, as shown in Figure 4–1, the values also can be shown in graph form as in Figure 4–2. The first vertical line in Figure 4–2 shows the values being generated in Figure 4–1. Phase A, at 90 degrees, is generating at the maximum value, phase B, at 210 degrees is climbing towards the zero value, and phase C, at 330 degrees, is approaching the maximum return or negative value. The second vertical line shows what is happening to the power generated in each phase when phase A is 120 degrees (one-third of the way) into the cycle, phase B would be 240 degrees (two-thirds of the way) into the cycle, and phase C would be at 360 degrees (at the end, which is also the beginning) of the cycle.

A three-phase circuit is like having three separate AC single-phase circuits with identical voltage that reach their peak values at a different time. At 60 *hz,* the second phase reaches its positive peak at 1/180 (0.00556) seconds after the time the first phase reaches a positive peak, and the third phase reaches its positive peak 1/180 (0.00556) second later. The first phase again reaches a positive peak 1/180 (0.00556) seconds after the third phase, starting the next cycle. Even though each phase has the same voltage, they are out of phase with each other and there is a voltage difference between them.

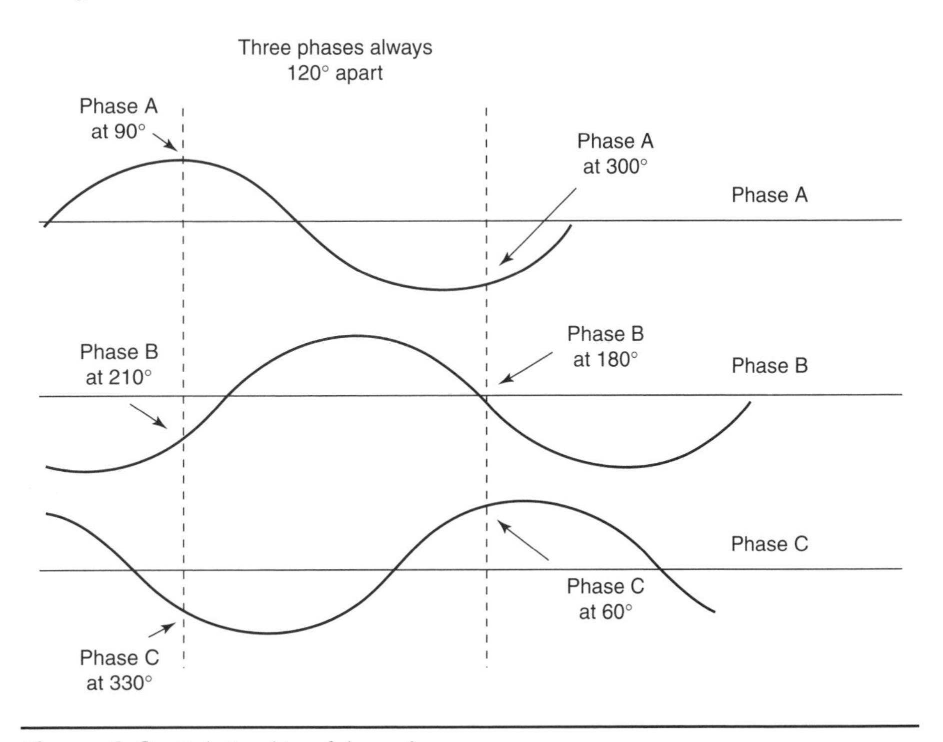

Figure 4–2 Relationship of three phases.

When the load on each phase is identical, the instantaneous power output of the three phases added together is constant. When one phase of a three-phase circuit reaches a peak voltage, another phase is close to zero volts and the third phase is on the return flow. Each phase is 120 degrees out of phase with the other phase. The voltage and current in each phase are 1/180 second, or one-third distance in a cycle behind another phase.

Phase Designations

4.2.4 In the lines trade, it is frequently necessary to trace individual phases to ensure that the correct phase is connected to the correct terminal. Individual phases are named and marked at various locations on the system to keep them apart. Utilities use various designations and markers, some of which are:

1. Red phase, white phase, and blue phase
2. Red phase, yellow phase, and blue phase
3. A phase, B phase, and C phase
4. #1 phase, #2 phase, and #3 phase
5. X phase, Y phase, and Z phase

Phase Rotation

4.2.5 When a three-phase motor is part of a customer's load, it is necessary to have the three phases in a sequence, or the motor will run backward. If the hookup to the customer had a B phase, A phase, C phase sequence, the motors would run backward. Switching any two phases at the transformer will put the phase rotation back in sequence. For example, with a B, A, and C sequence, switching A phase and C phase would give a proper B, C, A (which is A, B, C) sequence.

A test can be made with a rotation meter before the customer load is connected. If the rotation meter shows the supply phases are in proper sequence and the motor runs backward anyway, then the customer will need to reverse two leads on the motor.

Three-Phase Connections

4.2.6 A three-phase circuit starts at a generator. After generation, these three phases are connected to the input and output sides of transformers throughout the transmission and distribution system. A three-phase circuit is interconnected at the transformers in two ways (Figure 4–3).

1. The three phases can be connected in parallel. One end of each of the three coils is connected together to a common point, and the other ends of the three coils are the three-phase connections. This is referred to as *wye-*connected (named after the letter Y). A vector drawing showing the three phases interconnected is in the shape of a Y.
2. The three phases can be connected in series, which is referred to as *delta-*connected (named after the Greek letter delta **Δ**). A vector drawing showing the three phases interconnected is in the shape of a delta.

Wye and Delta Systems

4.2.7 All circuits worked on by the lines trade are fed from a transformer bank somewhere in the system. When the circuit source is from a transformer bank with

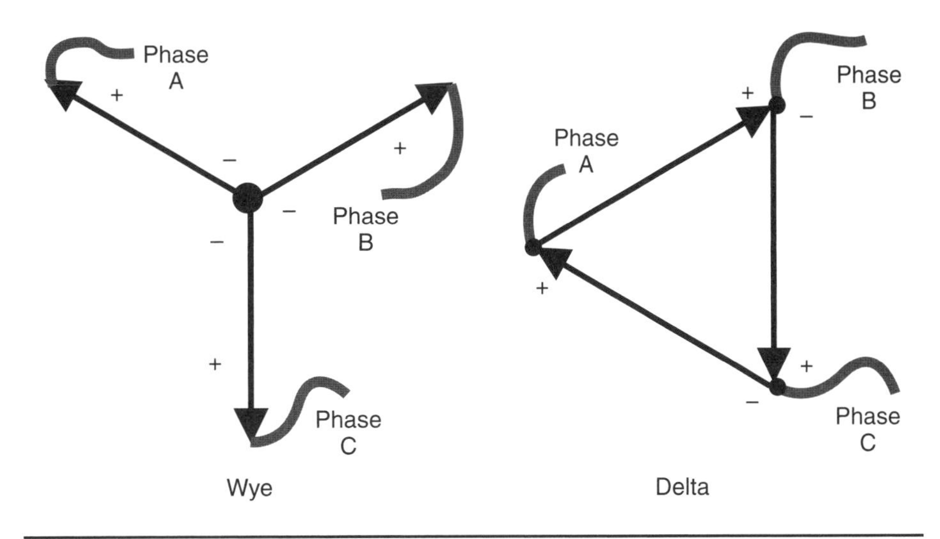

Figure 4–3 Wye and delta configurations.

a delta-connected output, the circuit is a delta system. Similarly, a wye circuit comes from a transformer bank with a wye-connected output.

A three-phase delta circuit consists of three wires. Each wire is a phase 120 degrees out of phase with the other wires. Transformers are connected phase to phase. Lateral taps branching off with two phases are used as single-phase circuits, which feed single-phase transformers and customers.

A three-phase wye circuit consists of four wires. Three wires are phase wires 120 degrees out of phase with each other, and the fourth wire is the connection to the common point, which is the neutral. Lateral taps branching off, consisting of a phase and a neutral, are single-phase circuits, which feed single-phase transformers and customers.

Connections to the three coils of three-phase transformers or motors are always made based on the standard wye or delta shape, as shown in Figure 4–4.

4.3 Delta-Connected Systems

Delta- or Series-Connected Three-Phase System

4.3.1 A delta system is a circuit fed from a delta-connected secondary of a three-phase transformer bank. The transformer bank provides a three-phase delta output with the three phases interconnected as one circuit.

In order to operate, each coil of a transformer must have a potential difference across it. On a delta connection, this can be achieved by connecting each coil phase to phase. The phase-to-phase connections are not at random, but each coil is connected so that the end of one coil is connected to the end of the other until all three are connected all the way around. In other words, the three phases are connected in series.

The current does not circulate around the delta. The current in each leg is traveling in a different direction at 120 degrees out of phase with each other.

Figure 4–5 shows how delta connections are made when the coils are in a delta shape and when the coils are side by side.

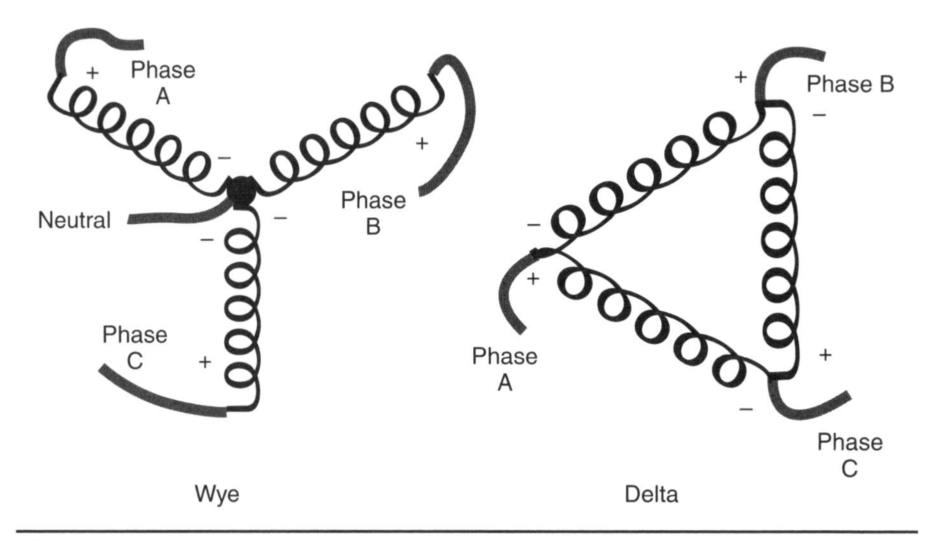

Figure 4–4 Wye and delta connections in wye and delta shape.

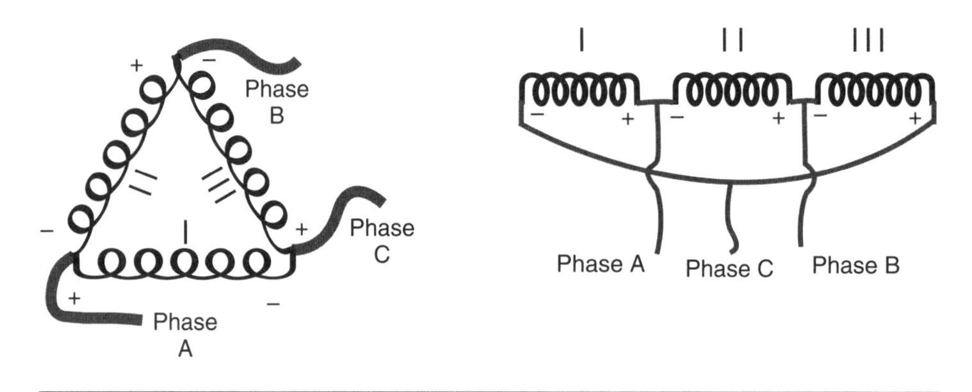

Figure 4–5 Delta configurations.

Voltage in a Delta System

4.3.2 The voltage across each coil is the same as the voltage measured from phase to phase. There is a voltage across each coil or between any two phases because each phase is in a different part of the AC cycle. The three phases are, in fact, always 120 degrees apart.

Current in a Delta System

4.3.3 For this discussion, the phase current is the current in each coil and the line current is the current in each conductor leaving the transformer bank. The current leaving each junction point is the resultant of the current in two phases 120 degrees apart. The current through each coil is equal to the full-phase current, but the line current leaving each junction point is 1.73 × the phase current in each of the two coils feeding the phase.

Return Flow in a Delta Circuit

4.3.4 To complete a circuit, electrical current must return to the source. In a delta circuit, the return flow to the source is in the opposing phase conductors. There is no return flow through the ground. There is a potential difference between the three phases of a delta circuit, but there is no "theoretical" potential between a phase and ground because going from phase to ground does not complete a delta circuit.

Using the same theory, a metal car body is the return path to the battery in an automobile. Theoretically, if a person avoids touching the car body, that person can touch the spark plug and not get a shock. If some of the return current takes a path to ground down one tire and up another tire, however, a person would still be exposed to a shock.

If one of the phases of a delta circuit became grounded, the voltage between the other two phases and ground becomes equal to phase-to-phase voltage. Unless there is a circuit breaker equipped with a ground fault relay, a grounded conductor on a delta circuit will continue to be at a phase-to-phase voltage with the other two phases.

Ground Fault Protection in a Delta Circuit

4.3.5 Many, but not all, delta circuits have protection against ground faults. The nature of a delta circuit would not notice a phase-to-ground fault; therefore, a grounding transformer and relays need to be installed.

A grounding transformer is installed in a substation between one phase of a delta circuit and ground. There is normally no voltage difference between a delta phase and ground, so there would be no current flow through the transformer. During a phase-to-ground fault, the current flowing back to the source will flow through the transformer and send a signal to the ground fault relay. The relay will detect this current and trip out the circuit.

4.4 Wye-Connected Systems

Wye- or Parallel-Connected Three-Phase System

4.4.1 A wye system is a circuit fed from a wye-connected secondary of a three-phase transformer bank. The transformer bank provides a three-phase wye output with the three phases interconnected as one circuit.

When connected in a wye configuration, each coil of a three-phase transformer is connected between a phase and the neutral. One end of each coil is connected together at a common point, which is the neutral. A voltage will be available at the other ends of the three coils when measured between any two phases. There is also a voltage between each phase and the common point. The three coils are connected in parallel.

Figure 4–6 shows how wye connections are made when the coils are in a wye shape and when the coils are side by side.

Voltage and Current in a Wye System

4.4.2 There are two different voltages available in a wye system: phase-to-phase voltage and phase-to-ground voltage. The phase-to-phase voltage is $\sqrt{3} \times$ the phase-to-neutral voltage. The phase-to-phase voltage is the resultant (vector or algebraic) sum of two phase-to-neutral voltages that are 120 degrees out of phase with each other.

The phase current going through each coil in a transformer is the same as the line current that flows into each phase of the wye circuit. The current flow in a wye circuit is $1 \div \sqrt{3}$, or 58 percent less than in an equivalent delta circuit.

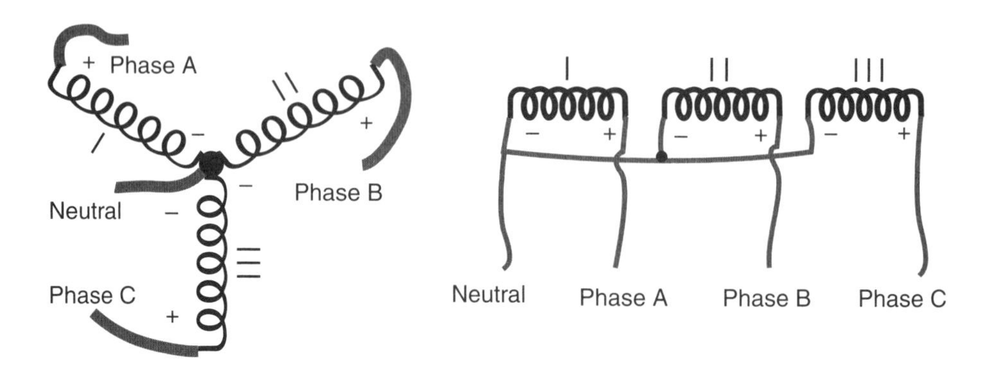

Figure 4–6 Wye configurations.

The Neutral in a Wye System

4.4.3 The neutral is the common point of each transformer winding in a wye system. If the loads on the three phases were equal to each other, there would be no current on the neutral. The neutral carries the sum (algebraic or a vector sum) of the three phase currents back to the source. With the three phases at 120 degrees, the three currents cancel each other out. The neutral wire can be smaller than the phase wires because it is sized to carry the unbalanced load between phases.

The worst-case scenario for an unbalanced wye circuit would be a phase-to-ground fault. Most neutrals are multigrounded, so fault current that enters the ground would return to the source through the ground and the neutral. For the duration of a phase-to-ground fault, the neutral current gets very high and the voltage between the neutral and ground also rises. On a single-phase line, the neutral—and to a lesser extent the ground—is the only path back to the source to complete the circuit.

Some three-wire circuits are actually wye circuits, which do not carry a neutral. The need for a neutral is minor when the load is balanced among the three phases. Transmission and subtransmission lines have relatively balanced loads and are often wye circuits without a neutral. In some utilities where there is good earth, the neutral is omitted on a distribution feeder because the earth is able to complete the circuit for unbalanced load and fault.

Earth Return System Neutral

4.4.4 There are some locations, mostly in rural, thinly populated areas, where there is no neutral conductor strung on the wye-distribution system. A single-phase circuit would consist of only one conductor strung on a pole. This system uses the earth as the neutral. A similar approach is used in the electrical system of a car. The metal body of the car is used as the ground wire and is attached to the negative pole of the battery.

The advantage of the earth return system is that with fewer conductors, fewer pole-top fittings, and ease of construction, these lines require less capital to build. One disadvantage is that the down-ground wire on a distribution transformer pole is the transformer primary neutral. Anyone getting across a break in the downground is exposed to full primary voltage. It is important to ensure that the downground has good physical protection installed where it is accessible to the public.

Converting from Delta to Wye

4.4.5 Many delta circuits have been converted to wye circuits. There are economical and safety differences between the two systems.

When the output connections of a distribution substation transformer are converted from delta to wye, the phase-to-phase voltage will be increased 1.732 ($\sqrt{3}$) times. For example, a delta-connected transformer bank that has a phase-to-phase voltage of 2400 volts across each coil of the transformer secondary can be converted to a wye connection by having one end of each 2400-volt coil connected to a common point. Now, the phase-to-phase voltage of the newly connected wye circuit will be 4160 volts, and the phase-to-neutral voltage will be 2400 volts. *Note:* Usually a conversion involves going to an even higher voltage fed from a new substation transformer.

For the same load, the current in each wye-connected phase is reduced by 1 ÷ 1.732, or 58 percent. A lower current results in less line loss and less voltage drop.

During conversion, an existing distribution transformer that was connected as phase to phase on the delta circuit can now have the same primary voltage by connecting it phase to neutral on a wye circuit. *Note:* Usually a conversion involves going to an even higher voltage and installing dual-voltage transformers to reduce the time of customer outage during the changeover.

Single-phase wye-connected transformers require only one cutout and surge arrester instead of the two needed on a delta transformer. Similarly, a single-phase lateral on a wye system needs only one line switch.

The multigrounded neutral on a wye system is an excellent ground and is available for transformers and secondary services regardless of possible poor local grounding conditions. On a delta system, the ground at a transformer is dependent on the driven ground rod at the transformer and at the customer.

Fuses and circuit breakers trip out quicker due to over-current when there is a good return path to the source. Phase-to-phase short circuits on both wye and delta circuits trip out a fuse quickly because of the good return path to the source through the other phases. Phase-to-ground faults have a good return path to the source on wye circuits through the multigrounded neutral. A tree contact would blow a fuse much quicker on a wye system than on a delta system.

4.5 Three-Phase Power

The Combined Power in Three Phases

4.5.1 When there are voltage and current in each phase of a three-phase circuit, there is also power being delivered. The power delivered by each phase will be 120 degrees out of phase with the other phases. In other words, the power delivered by one phase could be at its peak when the power delivered by another phase is one-third farther into the cycle and the power delivered by the third phase is two-thirds farther into the cycle.

Therefore, the total power in a balanced three-phase system is not simply three times the power of one phase. The square root of three, which is 1.73, times the power in one phase will give the power in three phases. When the load is unbalanced, the load in each phase needs to be measured individually and the average value is put into the equation. The power factor must be known to calculate the true power delivered by a three-phase system.

Field Calculations

4.5.2 Converting amperes in a circuit to kilovolt-amperes (kVA) and converting kVA into amperes per phase are sometimes valuable tools in the field. Field calculations are valuable when balancing three-phase circuits. The ampere loading of a phase should be converted to kVA to determine which transformers to transfer over to another phase.

Before working live line on a transmission circuit where jumpers are going to be installed, it is valuable to know the expected load on a phase. The controlling station can give the total load on the circuit, which can be converted to amperes per phase using field calculations.

Apparent power (kVA) is used in the field instead of true power (kilowatts) because apparent power is what a feeder carries. True power (kilowatts) is measured by revenue meters. To determine the total load on a transformer, field calculations can convert amperes per phase to kVA.

Calculations Involving Power

4.5.3 Table 4–1 gives formulas used to calculate power in three-phase and single-phase circuits. Calculations for field applications use the kVA formulas.

Calculations involving three-phase power use line-to-line (phase-to-phase) voltage. The value of the square root of three is normally 1.73.

Calculations with "Handy Numbers"

4.5.4 To be able to do the power calculations quickly in the field, a "Handy Number" can be used for approximations (Table 4–2).

Approximate kVA = amperes × "Handy Number"
Approximate amperes per phase = kVA ÷ "Handy Number"

Examples Using "Handy Numbers"

4.5.5 ***Question*** One phase of an 8.3/4.8-kilovolt feeder has 80 amperes more load than the other two phases. To balance the feeder, how many kVA should be transferred to the other two phases?

Answer Using a "Handy Number," which is 5 for a 4.8-kilovolt single-phase line, × 80 amperes = 400 kVA. Therefore, 200 kVA should be transferred to each of the other two phases.

Question Load on each of the three phases of a 120/208 50-kVA pad mount transformer bank are 150 amperes, 170 amperes, and 140 amperes. Is the transformer overloaded?

TABLE 4–1 Power Formulas

To Find	Direct Current	Single-Phase AC	Three-Phase Wye AC
Kilowatts (kW)	$\frac{I \times E}{1000}$	$\frac{I \times E \times PF}{1000}$	$\frac{I \times E \times 1.73 \times PF}{1000}$
Kilovolt-Amperes (kVA)		$\frac{I \times E}{1000}$	$\frac{I \times E \times 1.73}{1000}$
Amperes (when kW are known)	$kW \times \frac{1000}{E}$	$kW \times \frac{1000}{E \times PF}$	$\frac{kW \times 1000}{1.73 \times E \times PF}$
Amperes (when kVA are known)		$\frac{kVA \times 1000}{E}$	$\frac{kVA \times 1000}{1.73 \times E}$

*PF = power factor

TABLE 4–2 Examples of "Handy Numbers" for Some Voltage Systems

Calculating Approximate Loads with "Handy Numbers"				
	Three-Phase kV	*Handy #*	*Single-Phase kV*	*Handy #*
	230	400		
kVA = Amperes ×	115	200		
"Handy Number"	69	120		
Or	46	80		
Amperes per Phase =	25	40	14.4	14
kVA	12.5	22	7.2	7
"Handy Number"	8.32	14	4.8	5
	0.208	0.36	0.12	0.12

To calculate the "Handy Number" for other voltage systems:
For three-phase lines, the *"Handy Number"* = *L – L voltage* × 1.73 ÷ 1000
For single-phase lines, the *"Handy Number"* = *L – N voltage* ÷ 1000
where *L* = Line and *N* = Neutral.

Answer The average load on the three phases is 150 + 170 + 140 ÷ 3 = 153 amperes. Using the "Handy Number," which is 0.36 for 120/208-volt service, × 153 = 55 kVA. The transformer is only slightly above its rating, but the load should be balanced more between the three phases to prevent one transformer winding from being overloaded.

Review Questions

1. What can be done at a three-phase transformer bank if a customer's three-phase motor is running backward?
2. When a feeder is fed from a substation transformer bank with a delta-connected secondary, what type of circuit would the feeder be?
3. What is the voltage-to-ground on a 4160-volt delta circuit?
4. Why does a source breaker see a phase-to-ground fault on a delta circuit?
5. If the phase-to-phase voltage on a wye feeder is 25 kilovolts, what is the phase-to-ground voltage?
6. What condition would put a very high current on the neutral and also raise the voltage between the neutral and ground?
7. Name four advantages for converting a delta circuit to a wye circuit.
8. A control room operator tells you a 115-kilovolt circuit is carrying 70 megavolt-amperes. How much current would a barehand crew be working with on a phase?

CHAPTER 5

Circuit Protection

5.1 Introduction

Purpose of Circuit Protection

5.1.1 An abnormal voltage or current is an indication of a problem somewhere in the circuit. Protective equipment is installed to detect and clear these abnormal voltages and currents.

- Circuit protection limits the time people are exposed to hazardous voltage and current due to situations such as a fallen conductor.
- Circuit protection limits the time equipment is exposed to damaging voltage and current.
- Circuit protection minimizes the number of customer outages during adverse conditions by automatically isolating and removing a faulted circuit from the system. Service to other customers is maintained on the rest of the system.

A common task for a powerline worker is tracing problems due to overcurrent, abnormal voltage, or poor system grounding. Troubleshooting will be easier with an understanding of how protective equipment is specified to operate.

5.2 High-Voltage System Protection

Two Types of Switchgear

5.2.1 An electrical system has two types of switchgear, each of which has a purpose:

- *Protective switchgear,* such as a circuit breaker or a fuse, protects a circuit by providing automatic isolation of the circuit when it is exposed to damaging faults. This type of switchgear is specified to interrupt extremely high currents that occur during a short circuit.
- *Isolation switchgear,* such as a disconnect switch, an air-break switch, or a load interrupter, does not operate automatically during a fault but provides operating capability to isolate, sectionalize, or transfer loads at strategic locations. This switchgear cannot interrupt a high current during a fault. Some types can interrupt a load current while other types do not have the capability to interrupt any load current.

Circuit Breakers in an Electrical System

5.2.2 Circuit breakers are the backbone of protection in an electrical system. They are installed to protect generators, transformers, transmission lines, subtransmission lines, and, in many cases, distribution lines (Figure 5–1). They provide sophisticated circuit protection and switching capability.

Types of Circuit Breakers

5.2.3 Circuit breakers are named according to their arc-quenching medium. There are oil circuit breakers, air-blast circuit breakers, vacuum circuit breakers, or sulfur-hexafluoride (SF6) circuit breakers. Oil circuit breakers (Figure 5–2) are the most common circuit breakers and have been used for the longest time. The circuit-breaker contacts are immersed in an insulating oil, which quenches the arc. The oil will eventually become contaminated by electrical-arc products.

Air can be used as an insulation medium to quench an arc if the air is prevented from being ionized between the contacts. Compressed air is piped to air-blast breakers and will "blast" the arc away from the contacts. When these breakers are opened, the shotgun-type blast is extremely loud.

A vacuum provides a good insulation for quenching an arc. The contacts for a circuit breaker can be in a vacuum and the remaining part of the breaker immersed in oil. A separate vacuum container for the contacts prevents the oil in the rest of the circuit breaker from becoming contaminated.

SF6, an inert nonflammable gas with high insulating properties is a common insulation for circuit breakers (Figure 5–3). Breakers using this gas are much smaller and can be used indoors. The operation is quieter and is ideal for substations in cities. SF6 is not toxic, but under arcing, toxic by-products are produced that require special handling during maintenance.

Fault Detection

5.2.4 Other than a lightning surge, the two main types of faults in an electrical system are short circuits and open circuits. Relays can only detect a fault if mea-

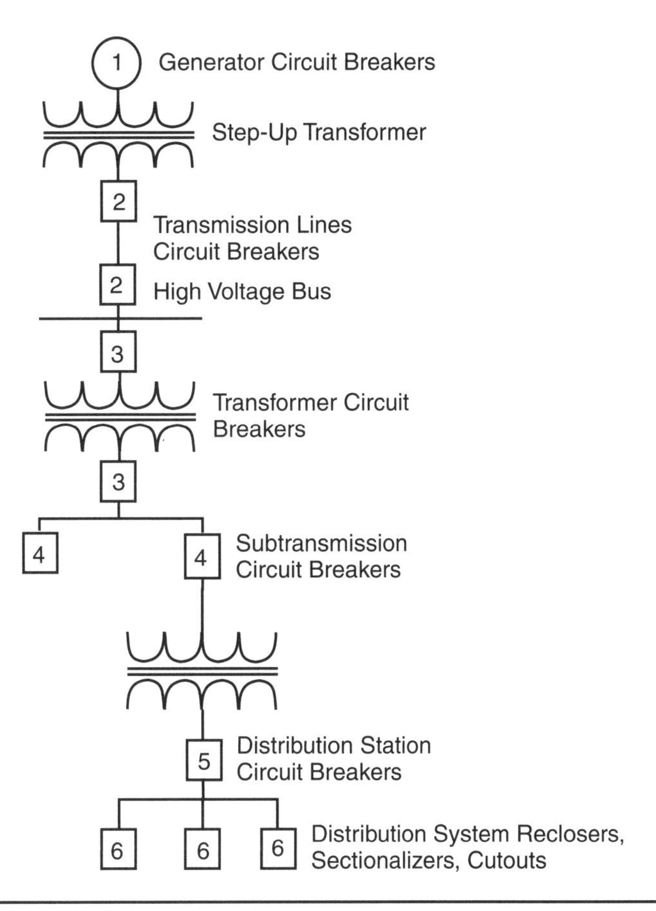

Figure 5–1 Circuit breakers in an electrical system.

surable conditions show up at the terminals of the circuit. Three types of faults, or combinations of faults, can show up and are measured at a terminal:

1. The current flow in one or more phases becomes abnormally high.
2. The current in the three phases becomes abnormally unequal.
3. The voltage in one or more phases becomes abnormally low.

Protective Relaying

5.2.5 A relay is a low-voltage switch that is normally in either an open or a closed position. Depending on the design, the relay switch will operate when electrical current passes through it or electrical current stops passing through it. The relay can be electromechanical or solid-state (electronic). Each relay is designed to operate when a certain amount of current is reached.

Potential transformers and current transformers installed on high-voltage circuits send representative low voltage and low current to the relays. Relays are

Figure 5–2 46kV Oil circuit breaker.

Figure 5–3 46kV SF6 circuit breaker.

designed to operate when the voltage and/or current is beyond the range of the specified relay settings.

A relay designed to detect overcurrent would operate if the representative current from the current transformer was beyond the specified setting. The relay switch would operate and send a low-voltage signal to the circuit-breaker control mechanism and cause it to open.

Relays Controlling Circuit-Breaker Operation

5.2.6 An electrical system can become unstable with adverse conditions ranging from a low-current fault, such as a tree contact, up to a *geomagnetically induced current* (GIC), due to an increased activity of sunbursts on the sun.

Relays can be designed to detect various conditions on a circuit as long as the voltage and current representing the condition of the circuit can be brought into the relay. Examples of some of the relays that can be installed to protect a circuit are listed:

- An overcurrent relay detects current when it exceeds predetermined limits.
- An undercurrent or underpower relay detects current when it has decreased beyond a predetermined limit.
- An over- or under-voltage relay detects voltage change beyond predetermined limits.
- A differential relay detects current entering a protected zone that does not equal the current leaving the zone.
- A current- or voltage-balance relay can detect a predetermined difference between two circuits or between phases on a circuit.
- An under- or over-frequency relay detects frequency when it changes abnormally.
- A thermal relay detects an abnormal rise in the temperature of a generator or transformer.
- A directional-power relay detects a change in the direction the power is flowing on a circuit within the grid.
- A power-factor relay detects changes to the reactance in a circuit beyond predetermined limits.

Circuit-Breaker Operation

5.2.7 Circuit breakers will open and reclose circuits in a sequence determined by relay settings. A circuit breaker must open within a speed range that will protect equipment and still coordinate with other switchgear on the system so that only the faulted section will be isolated.

The tripping time for a breaker is very fast. The speed is normally expressed in cycles. For example, a breaker can open on as few as three cycles, which means that a circuit breaker on a 60-hertz line can operate in 0.133 seconds.

Zone Protection

5.2.8 The protection scheme of the high-voltage system is divided into protective zones. When trouble occurs in a zone, relays sense the problem and send a signal to the appropriate circuit breakers to disconnect the zone from the system.

The usual protective zones (Figure 5–4) are generator, transformer, bus, and lines. Due to its length, the line zone has the greatest exposure to faults and is the most frequent zone to be disconnected from the system.

The Blackout of 2003

5.2.9 The blackout, on August 14, 2003, is an example of how a protection system can fail.

The complete cause of the blackout is complex; the system was already stressed with a high customer demand, voltage swings, and a need for more reactive power (capacitance) in the grid.

Under these unstable conditions, a transmission line tripped out due to a short circuit because of a brush fire under the line. Hot gases from a fire ionized the air

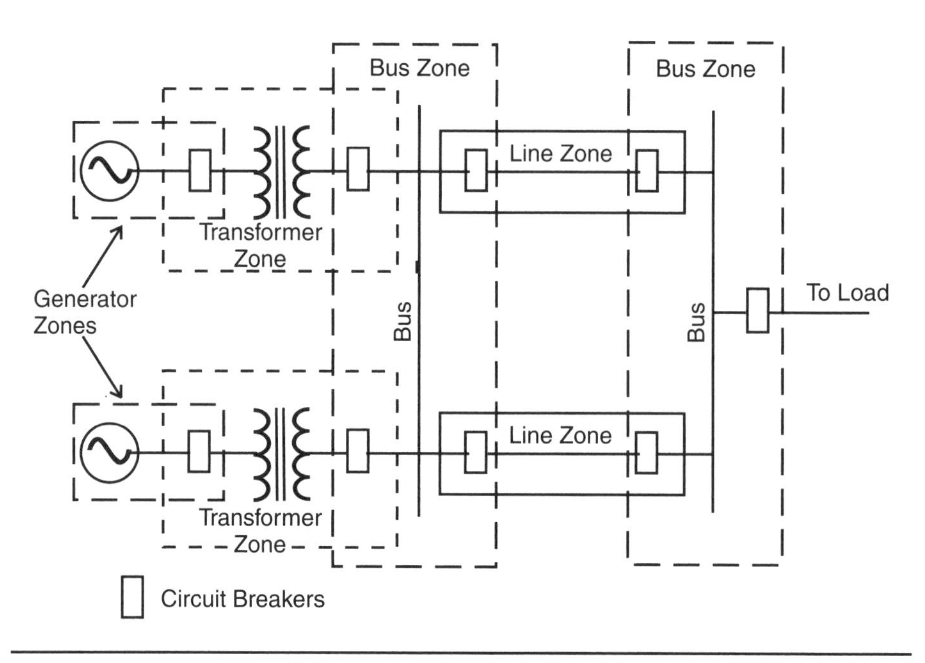

Figure 5–4 Typical protection zones.

and caused the air under the line to become conductive. Normally within a grid, other lines pick up the load dropped by the fault line. Extra load on another circuit caused the power conductors to warm up and sag into a tree and trip out. A generation plant tripped out when the transmission lines were no longer available. The resultant overload, voltage collapse, and current swings in the rest of the grid caused instability in the power system and circuits tripped out, one after the other.

System operators and the protective relays are supposed to isolate a fault to protect equipment from damage and prevent the fault from cascading into neighboring systems. However, unless there is intervention, the route power will flow in an electrical grid is controlled by the laws of physics. The flows will go over many different lines choosing a path of least resistance to get to the load center; meanwhile, relays will sense any overload, a voltage collapse, low frequency, reactive power overload, and other factors and cause signal breakers to open.

System operators in different jurisdictions can re-route power to a different transmission line by turning down the generation in one location and turning up generation at a different location. This assumes that there is reserve generation capacity available and that there is excellent coordination and communication between jurisdictions. Meanwhile, under normal conditions, a system operator (market player, generator, or trader) considers the price of power from different generation sources. A system operator is supposed to ignore economics and markets to relieve a transmission line overload, but it could still be a complicating factor for making quick decisions.

5.3 Distribution Protection

Distribution-Protection Requirements

5.3.1 A distribution system is exposed to overcurrent, overvoltage, and open-circuit conditions. Protective equipment is strategically placed to limit and isolate the faults so that the remainder of the system is not affected.

In many cases, especially in urban areas, a circuit breaker in a substation can detect faults in the full length of the circuit. Sensing problems such as open circuits, abnormal ground current, or abnormal phase unbalance can be obtained by using circuit breakers or electronic reclosers and their associated relays.

Downstream equipment, such as fuses or hydraulic reclosers, will trip out and sectionalize a circuit automatically, without the need for relays or other power sources. Only an overcurrent condition will trip out these devices.

Protection in a Distribution Substation

5.3.2 At a distribution substation, there are normally high-voltage fuses or a circuit breaker on the high-voltage side of the substation transformer and circuit breakers or reclosers on each distribution feeder on the low-voltage side.

Figure 5–5 shows a distribution substation with high-voltage (HV) fuses on the primary side of the transformer. A fault in the substation should trip out the high-voltage fuses before the circuit breakers at the source of the incoming subtransmission line trip out. The low-voltage (LV) reclosers should trip out a faulted feeder before there is any damaging current through the substation transformer. Figure 5–6 is a view of a small distribution substation.

Protection Using Home Wiring as an Example

5.3.3 The protection for the electric wiring in a home (Figure 5–7) is not unlike the protection of a distribution system. Home wiring for a 120/240-volt service is fed to a main breaker (or main fuses). Many individual circuits, each with its own breaker (or fuse), go from the panel to various loads in the home.

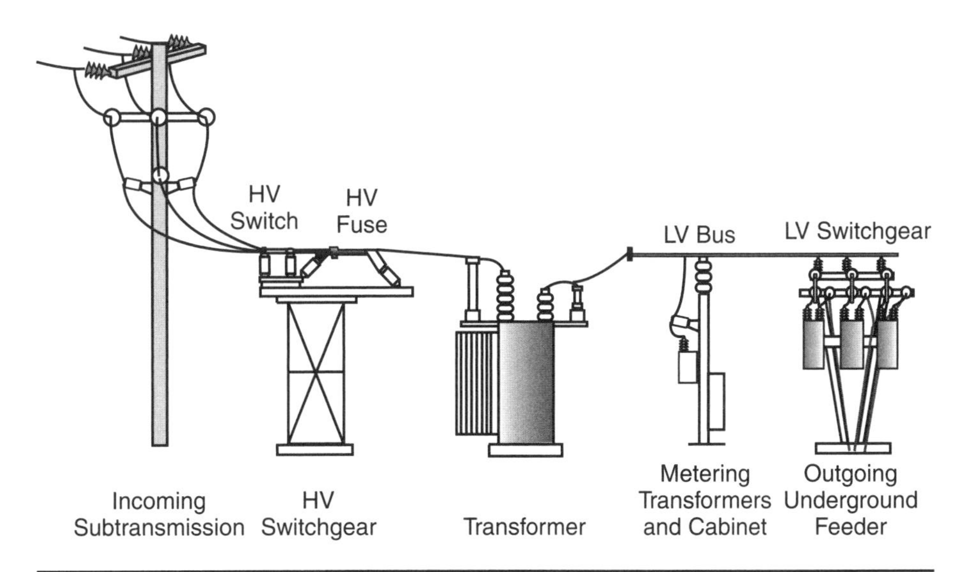

Figure 5–5 Distribution substation protection.

Figure 5–6 Distribution substation.

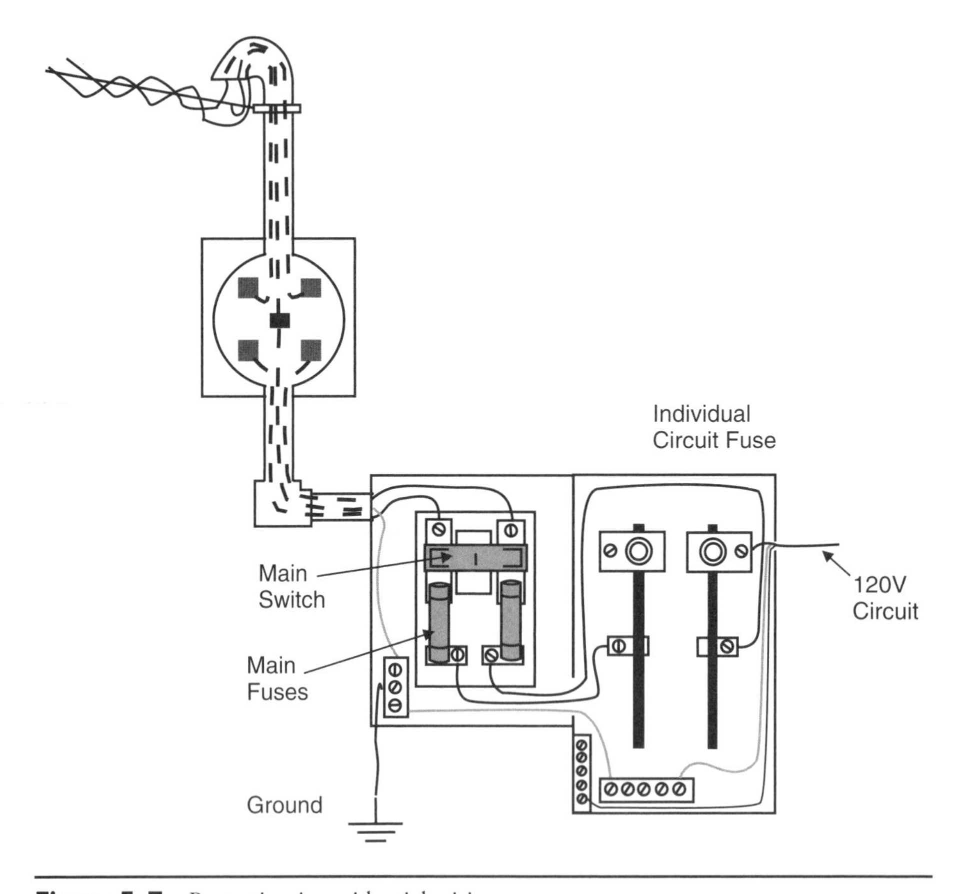

Figure 5–7 Protection in residential wiring.

The breaker (or fuse) on an individual circuit trips out when it is exposed to overcurrent, such as an overload or a short circuit. The rating of the main breaker is coordinated so that the individual circuit breaker will trip before the main breaker. This protection scheme is similar to the protective setup with a distribution substation and its feeders.

Circuits feeding a bathroom or the outdoors need more sophisticated protection to protect people in these well-grounded locations. Protection that will open a circuit at a current level below the threshold of a person's sensation is available. A ground-fault circuit interrupter (GFCI or GFI) is used on these circuits where it is desirable for the circuit to trip out quickly for even a minor fault.

The protection of a household circuit using a GFI can be compared to the ground-fault relay protection available on transmission and distribution circuits. A GFI monitors the current flowing to the load and the current returning to the source on the neutral. The live and neutral wires pass through a coil. Under normal conditions, the current through each wire would be equal but flow in opposite directions. No magnetic field would be induced into the coil because of the canceling effect of the two wires of opposite polarity. If some of the current flow returned to the source through ground, there would be more current flowing in the hot wire than in the neutral wire, and there would be a small magnetic field induced in the coil. This primary coil induces a voltage into a secondary coil whose output is amplified by an electronic amplifier. The output from the amplifier is applied to a relay coil, which opens the circuit.

A typical GFI will trip out within 30 milliseconds when there is a current imbalance of only 5 milliamperes. Faster and slower GFIs are available. The speed and low level of current that will trip a GFI is quick enough to prevent a person from feeling any electrical sensation.

The Need for Overcurrent Protection

5.3.4 A fault such as a phase-to-phase contact, a phase-to-ground short, or a follow-through current after a lightning strike shows up in a circuit as an overcurrent condition.

If the circuit stays in service during an overcurrent condition, *then:*

- Equipment such as transformers, regulators, and conductors would overheat and be damaged.
- An overcurrent condition would cause the voltage on the system to collapse, which, if not corrected, can damage customer equipment.
- There would be a greater risk to the public when they are involved with a conductor falling to the ground or when they have an accidental powerline contact with a ladder, antenna, or tree.

Just as it is essential to have a fuse or breaker panel to protect household circuits, utility circuits also need protection. Having no protection is similar to a householder putting a coin behind a fuse to bypass the fuse protection.

Transient or Permanent Faults

5.3.5 Depending on the protection scheme and location, about 90 percent of overhead-circuit faults are transient. A transient fault such as a lightning strike or a conductor contact with a tree in a windstorm can cause an initial trip of the circuit, but the circuit is automatically reclosed after the fault is gone. A lines crew is needed to repair and reenergize the circuit for the 10 percent of faults that are permanent.

Transient faults are rare on underground circuits. Most faults are permanent and require a lines crew to make repairs. Underground faults are not as obvious as overhead faults because faults on underground systems tend to be high-resistance faults. The protection opens the circuit before the fault becomes a dead short, therefore making the location of a fault more difficult to find.

Causes of Overcurrent

5.3.6 A circuit is faulted anytime a live conductor makes contact with another element that is at a different potential. When the resistance or impedance between a live conductor and another element is low enough, the fault becomes a short circuit.

- Lightning is the most frequent cause of transient faults. At the flashover point, a high-voltage arc establishes a path of ionized air to ground. A high follow-through current is established through the ionized air and causes a fault current to flow in the circuit.
- A phase-to-ground or a phase-to-neutral fault is the cause of about 70 percent of permanent overcurrent faults. A phase-to-ground fault could be caused by a broken insulator, a contaminated insulator, a tree contact, a broken conductor, an animal contact, a car accident, or an accidental public contact with a crane, sailboat, ladder, or antenna.
- A phase-to-ground fault on an underground cable can occur due to an insulation breakdown, a dig-in, or a driven fence post.
- A phase-to-phase contact is fairly rare. It occurs when conductors slap together in long, slack spans or when ice-covered conductors gallop. Phase-to-phase contact can also occur due to external forces such as a car accident or a falling tree.
- Overcurrent due to an overload occurs when the customer demand exceeds the specified setting of the circuit protection. Circuit protection does not differentiate between overcurrent due to an overload or a fault. Protection settings become difficult when the load current approaches the value of the available fault current.

Magnitude of Fault Current

5.3.7 Fault current, like any electrical current, needs a complete circuit in order to flow. The same current level flows into the fault, returns to the source, and returns through the protective switchgear to the fault. When a fault occurs, the amount of current that flows through the complete circuit depends on the capacity of the electrical system, the conductivity of the circuit, and the type of fault.

The capacity of the electrical system feeding the fault starts with the size of the transformer at the source of the feeder. The larger the transformer feeding the circuit, the greater the ability of the system to supply a high-fault current.

A short distance and a large conductor provide a low-resistance (impedance) path for a current to flow. A low-resistance path provides a greater capacity to carry a high-fault current.

A poor return path to the source reduces the current in the complete circuit and reduces the current through the protective switchgear. A neutral or other phases provide the best return path for a fault current.

The resistance of a fault affects the amount of fault current flowing in the circuit. A phase-to-phase contact or a phase-to-neutral contact is a dead short and has virtually no resistance to impede current flow. A conductor lying on the ground or contacting a tree has a relatively high resistance and limits the amount of fault current generated. A high-resistance fault is really just another load on the circuit if the current generated in the circuit is not high enough to trip any overcurrent protection switchgear. Likewise, a customer load is just like a high-resistance fault.

Impedance to Fault Current

5.3.8 A fault can be a dead short or a partial short circuit depending on the impedance or resistance between a live conductor and the object causing the short. A dry tree limb or a broken conductor lying on dry or frozen ground is a high-impedance fault and does not provide a good path for current to flow into the earth. A high-voltage feeder is more likely to overcome any impedance and generate a higher current flow back to the source.

A phase-to-phase or a phase-to-neutral fault is a low-impedance fault. The circuit conductors provide a good return path to the source. The complete circuit has a high-fault current flowing through it and trips the protective switchgear quickly.

Distribution-Protection Scheme

5.3.9 A protection scheme for a distribution feeder has a protective device at the source and protective devices downstream at junctions, lateral taps, and transformers.

Feeder Protection at the Substation

The source of a feeder can be protected by fuses, reclosers, or circuit breakers. A fuse is a one-shot device and is sometimes used in small, older stations as feeder protection. A recloser and a circuit breaker are multishot devices that provide a number of tripping and reclosing operations to give transient faults time to clear.

Downstream Multishot Protection

A substation recloser or circuit breaker is often able to provide multishot protection to the end of the feeder, especially on the shorter urban feeders or on high-voltage distribution feeders. Long rural feeders require downstream reclosers to maintain multishot protection to the end of the line.

Downstream Sectionalizers

A sectionalizer is a slave device to an upstream multishot device. A fault downstream from a sectionalizer will cause an upstream multishot device to trip out and reclose the circuit for a specified number of times. Depending on how the protection is specified, a sectionalizer will isolate a permanent fault during one of the intervals while the circuit is deenergized by the multishot device. A sectionalizer can interrupt a load current, but it does not have the capacity to interrupt a fault current.

Downstream Fuses

A fuse can be specified to isolate a permanent fault after the upstream multishot device has operated a specified number of times. Generally, the speed of the specified fuse will be set to melt before the third or fourth operation of the multishot device.

SCADA Systems

5.3.10 A *supervisory control and data acquisition* (SCADA) system is a distribution automation system that brings needed information and remote control of switchgear into a central operating control room. It brings the operation of a distribution system to a level similar to a transmission system. A communication system, such as telephone lines, fiber-optic cable, radio, or coaxial cable, brings information from *remote terminal units* (RTUs) in substations and at remote switchgear to a *central processing unit* (CPU) in the control room. The SCADA system gives an operator in a control room information such as system voltage data, feeder loading data, and the status of switchgear as open or closed.

Remote-controlled switchgear needs a remote-controlled motor that can open and close the circuit breaker or switch. The motor is set in motion by sending a signal to a motor-operating mechanism. The motor-operating mechanism has a low-voltage supply, a battery, RTU circuitry, and a communication connection to the control room.

Maintaining and troubleshooting a SCADA-controlled switch includes bypassing the remote control and operating the switch manually. A lines crew can be asked to go to a remote unit and operate it from the site when there is a breakdown.

5.4 Distribution Switchgear and Their Operation

Two Types of Switching in Distribution

5.4.1 There are two main categories of switchgear in distribution systems:

- Switchgear such as a disconnect switch, a load interrupter switch, or a live-line clamp is installed to provide operating capability. Some of this switchgear can interrupt load current, some need a load-break tool to interrupt load current, and some are not capable of interrupting any load current. None of this type of switchgear can interrupt a fault current.
- Switchgear such as a circuit breaker, a recloser, a sectionalizer, or a fused cutout is installed to provide circuit protection. Protective switchgear can be automatic where the device itself senses an overcurrent and automatically trips out the circuit, or it can be a relay-controlled device where the relay senses a fault and signals the protective device to trip out the circuit.

Reasons to Operate a Switch

5.4.2 It is important to know why a switch is being opened or closed. Incorrect switching can cause damage and injury. When it is known ahead of time that a switch is to drop load or will only drop a section of line, it can be determined if the switch being operated is capable of doing the job. See Table 5–1.

TABLE 5–1 Reasons to Operate a Switch

If	Then
A length of line with a load on it is to be isolated or interrupted.	A load current will be interrupted, and an arc will need to be extinguished. The switchgear must be designed to be able to interrupt its rated load current. Switchgear, such as a circuit breaker, a recloser, or a load-interrupter disconnect switch, is capable of dropping load. A disconnect switch without a load interrupter is not designed to interrupt any load. It may have a rating of 600 amperes, but that refers to the load it can carry through the switch continuously.
A length of line, with no load on it, is to be isolated.	It can be isolated with an ordinary disconnect switch. The amount of line that can be dropped with a disconnect switch depends on the voltage and length of the circuit. For example, an air-break switch that does not have a load interrupter can, with permission, safely drop 3 miles (5km) of 230kV or 16 miles (26km) of 50kV.
Switchgear is opened to break parallel in a circuit that is being fed from two directions.	Some current and voltage will be interrupted. If the switch breaks parallel between different stations or two unequal lengths of line, there will be a voltage difference (recovery voltage) and a current flow between the two feeds, immediately upon opening the switch. A control-room operator has the needed information to calculate the amount of current to be interrupted when parallel feeds are opened. Switchgear capable of interrupting load current can break parallel easier than an ordinary disconnect switch.
Switchgear is closed to make parallel between two sources.	There can be a voltage difference and current flow created when the switch is closed. Unless there is a restrike, all switches should be capable of picking up their rated load current.

(continued)

Table 5–1 *Continued*

If	Then
	Closing a switch between two circuits fed from two different substations that are fed from two different transmission systems can be dangerous because there may be a large voltage difference and it can create a large current flow through the switch. A system-control operator should have control of such a switch.
A loop must be opened or closed. (Opening a circuit fed from two directions but fed from the same circuit breaker is breaking a loop.)	Unless a loop is excessively long, it can be closed or opened with an ordinary disconnect switch or live-line clamps. An example of a loop is where a new line is constructed next to the old one, such as at a road-widening project. The new circuit can be energized in parallel with the old circuit by tying it in at both ends forming a loop. Load on the old circuit would need to be opened before breaking the loop and isolating the old circuit.
A switch is to be closed as a test to determine if a line is still faulted.	Switchgear can be closed without an arc. When closed in on a fault, the hot metal fragments blowing out of an expulsion cutout are a hazard to the operator. A major arc occurs if a switch is closed on a fault and the powerline worker immediately reopens the switch. It is important to let the normal protective switchgear trip out the circuit automatically.

Arc Length

5.4.3 The main concern to a powerline worker when opening a switch is the possibility of an excessive arc. Many types of switchgear are designed to extinguish an arc, but there are many disconnect switches with no ability to extinguish an arc.

When an arc is established, the heat from the arc causes the air to become ionized. The *ionization of air* means the air has become charged with atoms, and the air is changed from being an insulator to being a conductor. Once the air between the switch contacts becomes conductive, a follow-through current will flow across the switch gap.

There is more probability of an arc when opening a switch than when closing a switch. Normally, a circuit can be closed without concern about an arc unless the device did not close the first time and a restrike occurs. It is the interruption of current that causes an arc.

Arc-Length Formula

5.4.4 As an item of interest, a potential arc length can be calculated. However, a lineworker should not be asked to open a device unless:

- The switchgear is designed to interrupt load current.
- The switchgear is capable of accepting a load-bust tool.
- The switchgear will not be interrupting any load.

An arc length is dependent on the circuit voltage *plus* the current being interrupted. Formulas that can be used as estimates for the expected arc length when opening a switch follow for when the current is less than 100 amperes:

$$\textit{Arc length in cm} = 0.5 \times kV \times I$$

and when the current is more than 100 amperes:

$$\textit{Arc length in cm} = 50 \times kV$$

Designs to Extinguish Arcs

5.4.5 The design of switchgear includes the requirement to quickly extinguish any arc that is created when a switch is opened. There are many methods used to extinguish arcs:

- When contacts are immersed in an insulating medium such as oil, vacuum, or SF6, an arc is extinguished quickly.
- When air is the insulating medium, an arc ionizes the air, which causes the air to become conductive. An air-blast breaker will blast high-pressure air at any potential ionized air and blow it away.
- An arc can be transferred from the switch contacts to a spring-loaded horn gap or whip, which separates the arc.
- A spring-loaded fuse link separates the two parts quickly when the fuse melts.
- A special coating inside the fuse holder reacts with the arc and produces a gas at high pressure, which expels the ionized gas out of the tube.

Zero-Awaiting and Zero-Forcing Interrupters

5.4.6 Most switchgear takes advantage of the fact that in an AC circuit there is no current flow twice during each cycle. A zero-awaiting interrupter is a switching device that requires AC passing through zero before it can interrupt a load current. Examples of zero-awaiting switchgear are cutout fuse links, Bay-O-Net links, and cartridge fuses.

A zero-forcing interrupter is a device that forces the current to zero. It forces the system current to zero by inserting a large resistance into the circuit. An example of a zero-forcing device is a current-limiting fuse.

Disconnect Switches

5.4.7 A disconnect switch is any solid-blade switch installed to provide a means to isolate or sectionalize a circuit. Unless a disconnect switch has a load interrupter, it has no capability to interrupt load current. Disconnect switches have no capability to open automatically under fault conditions.

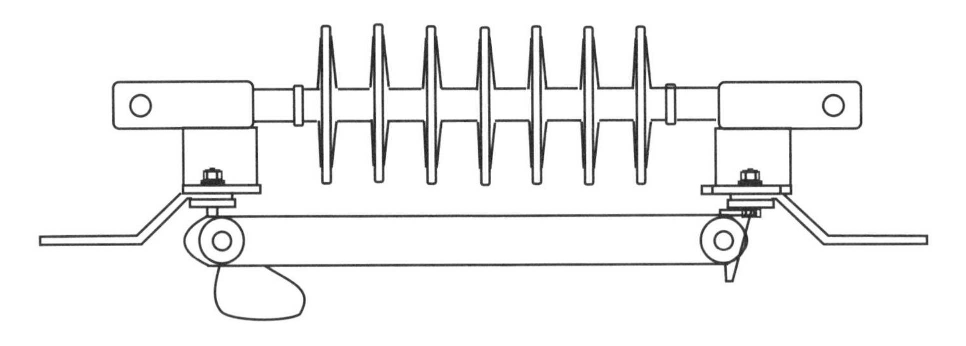

Figure 5–8 Single-phase inspan disconnect switch.

Figure 5–9 Subtransmission gang-operated disconnect switch.

Disconnect switches come in all voltages and can be single-phase bracket mounted, single-phase inspan (Figure 5–8), or three-phase gang-operated (Figure 5–9). Depending on the type, disconnect switches are operated by hotstick, operating handle, or remotely by way of a low-voltage motor.

Load-Interrupter Switches

5.4.8 Disconnect switches with load interrupters come in all voltages. They can be three-phase gang-operated switches or single-phase switches. Depending on the type, load interrupters are operated by hotstick, operating handle, or remotely by way of a low-voltage motor. A load-interrupter switch has no capability to open automatically under fault conditions. Load-interrupter switches are normally rated at 600 amperes or less.

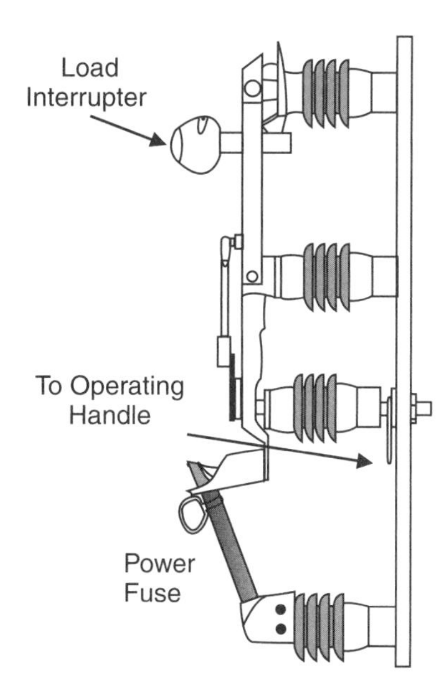

Figure 5–10 One pole of a load-interrupter switch.

Figure 5–11 Three-phase load interrupter and power fuses.

The most common load interrupter extinguishes the arc within an interrupter housing by deionizing gases that are exhausted flamelessly through a muffling device (Figure 5–10). This type of load-interrupter switch is commonly used in substations (Figure 5–11). The power fuse provides overcurrent protection. The load-interrupter switch is used to drop load before opening the power fuses manually.

Operating a Disconnect Switch

5.4.9 Many disconnect switches are three-phase gang-operated switches with operating handles extending to ground level. Occasionally, an insulator breaks or other breakage occurs while operating the switch, causing live leads to contact the switch frame. It is critical, therefore, to stand on a ground-gradient mat or ground-gradient network that is bonded to the operating handle (Figure 5–12). When the mat and operating handle are bonded together, the two will stay at the same voltage no matter how high the voltage becomes. If there is no voltage difference between the operating handle and the mat, there will be no voltage difference between the operator's hands and feet. If there is no voltage difference, there is no current flow.

A non-load-break disconnect switch should be operated by using the *inching method.* Because the switch should not be used to drop load, there should be no arc developing as the switch is opened slowly. If there is an unknown load and an arc starts to develop, the switch should be closed immediately. At a distribution voltage, single-phase disconnect switches are normally equipped with hooks to allow a portable load-break tool to be used to safely provide load-break capability.

Operating a Live-Line Clamp

5.4.10 Live-line clamps are not switches, but they are used as a convenient way to disconnect and connect drop leads between equipment and circuits. There is, of course, no arc-extinguishing capability with live-line clamps. Depending on the design of the live-line clamp, it also has a limited ability to carry any substantial load current. Live-line clamps can be used to drop a limited length of line, but, to

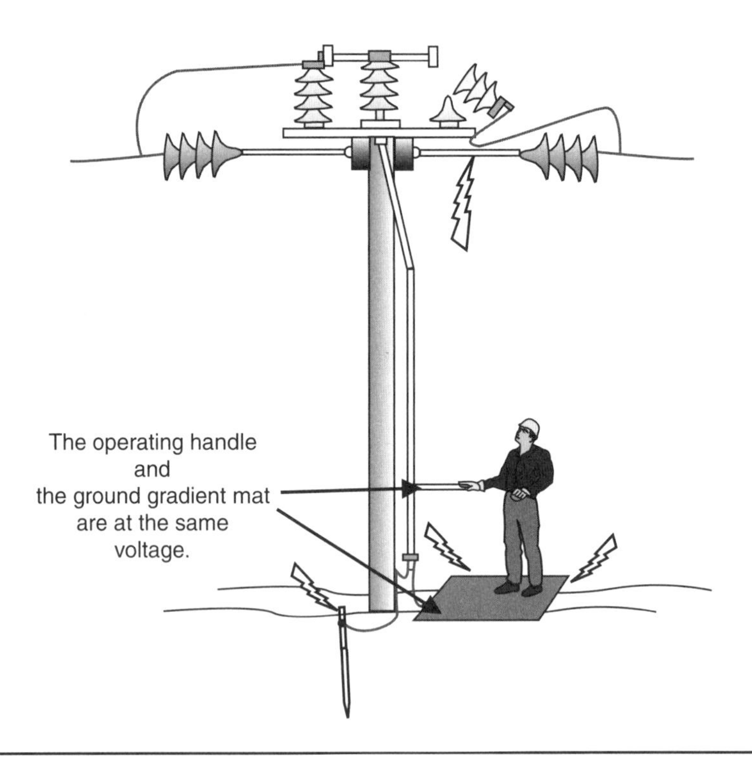

Figure 5–12 Use of ground gradient mat.

reduce the risk of an unmanageable arc, they should not be used to drop any load or even an unloaded transformer winding larger than 10kVA.

Operating a Distribution Fused Cutout

5.4.11 A fused cutout is a one-shot device that will interrupt a high-fault current due to a fault. When the fuse in the fuse holder melts, it breaks a mechanical link, which causes the fuse holder to drop open. A fused cutout provides overcurrent protection and will isolate faulty equipment and circuits automatically.

When a cutout is opened with a switch stick, the cutout is operated as a regular disconnect switch with no arc-extinguishing equipment other than the air gap between the switch terminals. A cutout can only interrupt about 15 amperes safely when opened with a switch stick. Normally, a load-break tool should be used when interrupting load current.

A *link-break* fused cutout (Figure 5–13) is a cutout that can interrupt load. A link-break hook is pulled down with a switch stick to break the fuse link and safely extinguish any arcing inside the fuse chamber.

A load-break tool can be used to drop load when the distribution cutout is equipped with hooks to accommodate a load-break tool as shown in Figure 5–13. A load-break tool is a bypass for the load current. The opening of the circuit and any resultant arc occurs inside the load-break tool.

If a cutout is closed in on a faulted circuit during troubleshooting, it is important not to immediately open the cutout. The design of the fuse and fuse holder should be allowed to extinguish the arc.

Use of a Load Break Tool

5.4.12 A load break tool can be hooked up to a fuse cutout or disconnect switch to provide a parallel path for load current through the tube of the load-break tool as shown in Figure 5–13. The initial downward pull on the tool charges an internal spring. At a certain point in the downward pull, the spring is released resulting in a high-speed separation of the contacts. Any arc is extinguished inside the chamber

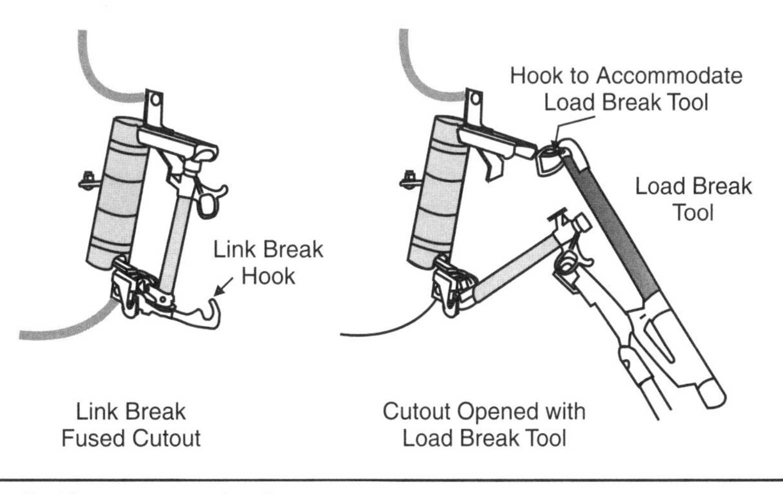

Figure 5–13 Operating fused cutouts.

by the fast elongation of the arc and the release of deionizing gases formed from the surrounding chamber material.

Fuses

5.4.13 A fuse is an overcurrent protective device that opens a circuit when it becomes overheated and melts. Fuses used in substations tend to be called power fuses, and fuses used on distribution feeders are distribution fuses.

A fuse carries load current without deteriorating, carries some overload without immediate rupture, and must be able to interrupt a very high fault current. The higher the current flowing through the fuse, the quicker it will blow. The amount of current a fuse is able to interrupt is based on its ability to extinguish an arc:

- An arc is extinguished by providing a fast separation of two parts of a melted fuse. One way to separate a blown fuse quickly is to have it spring-loaded so that the two parts of the blown fuse will separate quickly.
- An arc inside a fuse chamber forms gases when it acts on a special coating on the inside walls of the chamber. The formation of gas forces the arc products out of the expulsion chamber and extinguishes the arc.
- An increase to the resistance of an arc path will extinguish an arc. Having a fuse immersed in oil or other insulation will cool and increase the resistance of the arc.

The melting and clearing time of a fuse should coordinate with upstream and downstream protective devices. Different speeds of fuses are available; for example, a K-link fuse is fast, and a T-link fuse is slower. The K-link fuses and the T-link fuses have specific time-current characteristics recognized by all manufacturers.

Examples of fuses used in electrical utilities are:

Expulsion-cutout fuse links
Under-oil expulsion fuses (bayonet style)
Solid-material-filled power fuses
Nonexpulsion (NX) current-limiting fuses
Under-oil backup current-limiting fuses

Expulsion Cutout Fuse Links

5.4.14 An expulsion-cutout fuse link is the most common and economical fuse used in a distribution system. When the spring-loaded fuse melts, the fuse chamber drops open to provide a visible open switch. When a fuse element melts, the current continues to flow in the form of an arc through the particles of the vaporized fuse element and ionized gases. The heat from the arcing burns back the remaining element, and the heat generates the release of a large amount of gas from the inside wall of the fuse chamber. The resultant high-pressure gas and arc products are expelled from the tube.

Solid-Material Fuses

5.4.15 A solid-material fuse has a fusible element encased in a heat-absorbing and arc-quenching material such as silica sand or borax. During a fault, the fuse vaporizes and the solid material cools the arc. It can interrupt a higher fault current than an expulsion-cutout fuse link and is used in high-fault current or heavy-load current locations. An indicator at the end of the fuse unit projects out to show when a fuse is blown. When this type of fuse is used in a cutout, the fuse-unit projection will trip the latch and cause the fuse chamber to drop out.

Current-Limiting Fuses

5.4.16 An ordinary expulsion fuse is not current limiting. It will limit the duration of an arc but not the magnitude. A current-limiting fuse will limit the magnitude of the current flow by introducing a high resistance after the fuse element melts. A current-limiting fuse is used in locations where there is a very high fault current available on the electrical system. It is needed to protect transformers, underground cable, or other equipment from damaging fault current.

A current-limiting fuse, as shown in Figure 5–14, consists of a silver-ribbon element wound around in an insulated tube of silica sand. The silver ribbon is perforated with holes for its full length. During a fault, the silver-ribbon element vaporizes along its length, starting at the perforations. The vaporized element is blown into the surrounding sand. The resulting arc heats the sand and turns it into a glass-like material, which increases the resistance to current flow and chokes off the arc.

Types of Current-Limiting Fuses

5.4.17 The two main types of current-limiting fuses are partial range and full range. A *partial-range,* or backup, current-limiting fuse (Figure 5–15) is designed to limit only high-fault currents. It is used in series with a fused cutout to protect equipment from the high energy levels available in a high-fault-current location. It is available in different sizes and speeds to coordinate with a cutout fuse link. For

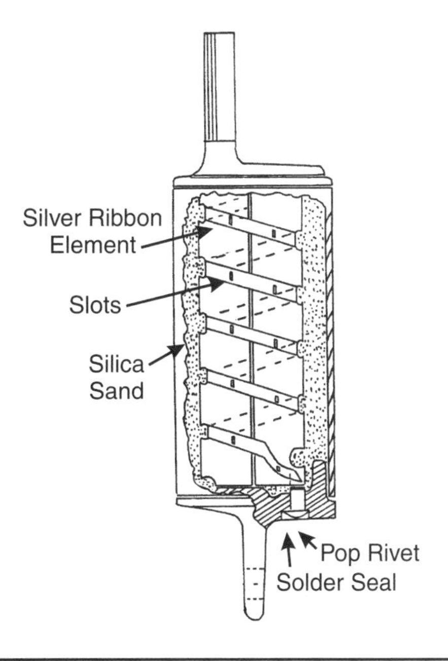

Figure 5–14 Current-limiting fuse.

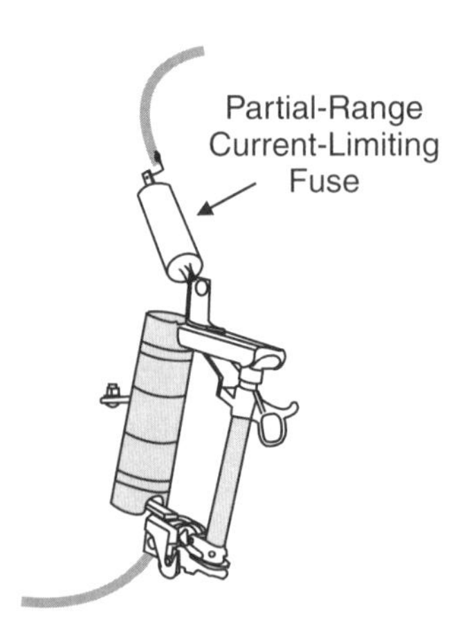

Figure 5–15 Application of partial-range current-limiting fuse.

example, a 25K current-limiting fuse coordinates with a 25K cutout fuse link. An air-insulated, partial-range current-limiting fuse is installed right on top of a fused cutout as shown in Figure 5–15. An under-oil submersible partial-range current-limiting fuse is used in series with low-current protective devices such as a bayonet-style under-oil expulsion fuse, cartridge fuse, or fuse link.

A *full-range,* or general-purpose, current-limiting fuse is not used in series with another fuse but is the only fuse needed to protect equipment. It prevents a high-fault current from going through equipment, and it isolates defective equipment from the circuit. It is most often used to protect underground or metal-clad equipment. A full-range current-limiting fuse can be air-insulated and used for high-amperage-rated applications in live-front switchgear or as an under-oil submersible current-limiting fuse (bayonet-style mounted fuse) used with dead-front equipment.

Reclosers

5.4.18 A recloser, like a circuit breaker, is a multishot device with the ability to automatically interrupt a circuit for a fault and reclose. Reclosers are designed for installations on distribution feeders, whereas circuit breakers are almost always found in substations where they can be provided with a low-voltage supply from voltage transformer (VT) and current transformer (CT).

A recloser can be set to open and reclose a number of times before locking out on a permanent fault. On a transient fault, the recloser will stay closed once the fault has cleared from the circuit. Reclosers can have their contacts enclosed in vacuum bottles, SF6, or oil.

A recloser can be set to open at various current levels. For example, a recloser is often set to trip out when the current going through the recloser coil is twice its rating. Therefore, it takes a minimum of 200 amperes to open a 100-ampere recloser. This ensures a trip-out is probably for a fault and not a temporary overload.

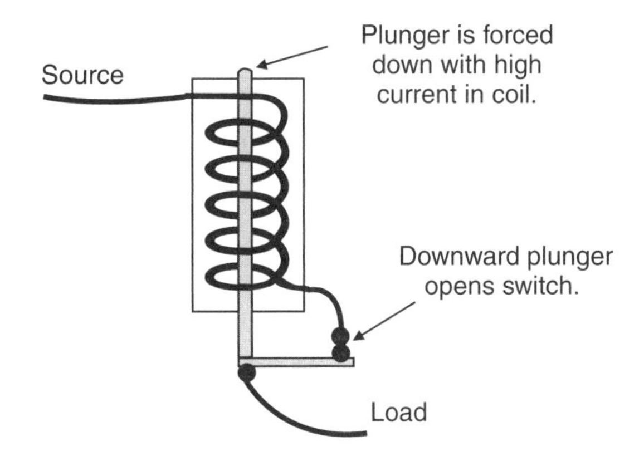

Figure 5–16 Basic solenoid coil.

Hydraulic Reclosers

5.4.19 A hydraulic recloser is an oil-filled recloser that automatically opens when an overcurrent flows through its trip coil. A solenoid coil (Figure 5–16) in the recloser is the fault sensor. When the current flowing through a coil is higher than the current rating of the coil, the electromagnetic action on a movable plunger inside the coil opens the contacts.

The recloser will close and reopen a specified number of times and at a specified speed. If the fault has cleared, the recloser will close and reset itself for the next time a fault occurs. On a permanent fault, the recloser will go through the full sequence of specified operations before staying open.

Electronic Reclosers

5.4.20 Electronic reclosers are relay-controlled. Relays can be programmed for a range of minimum trip values, a number of operations to lock out, and a range of minimum response times to coordinate with upstream and downstream protective switchgear. The size of the plug in the resistors in the control box (Figure 5–17) determines the trip-current level.

Operating an Electronic Recloser

5.4.21 An electronic recloser has an open-and-close handle and a non-reclose handle under the sleet hood, the same as a hydraulic recloser. The recloser can be *opened* with the handle, but some electronic reclosers must be *closed* from the control box. The handle must be in the *up* position, or the control cannot close the recloser.

An electronic recloser has a low-voltage supply, which allows the recloser to be operated from a control box. A low-voltage supply can be from a regular distribution transformer on the supply side of the recloser or from a bushing-current transformer mounted in the recloser.

The main operating switch in a control box can open and close the recloser. The switch can be closed and held in the closed position to provide a cold-load pick-up capability. Holding the switch blocks the instantaneous trip function. In other electronic reclosers, the cold-load pick-up feature is programmed to prevent the

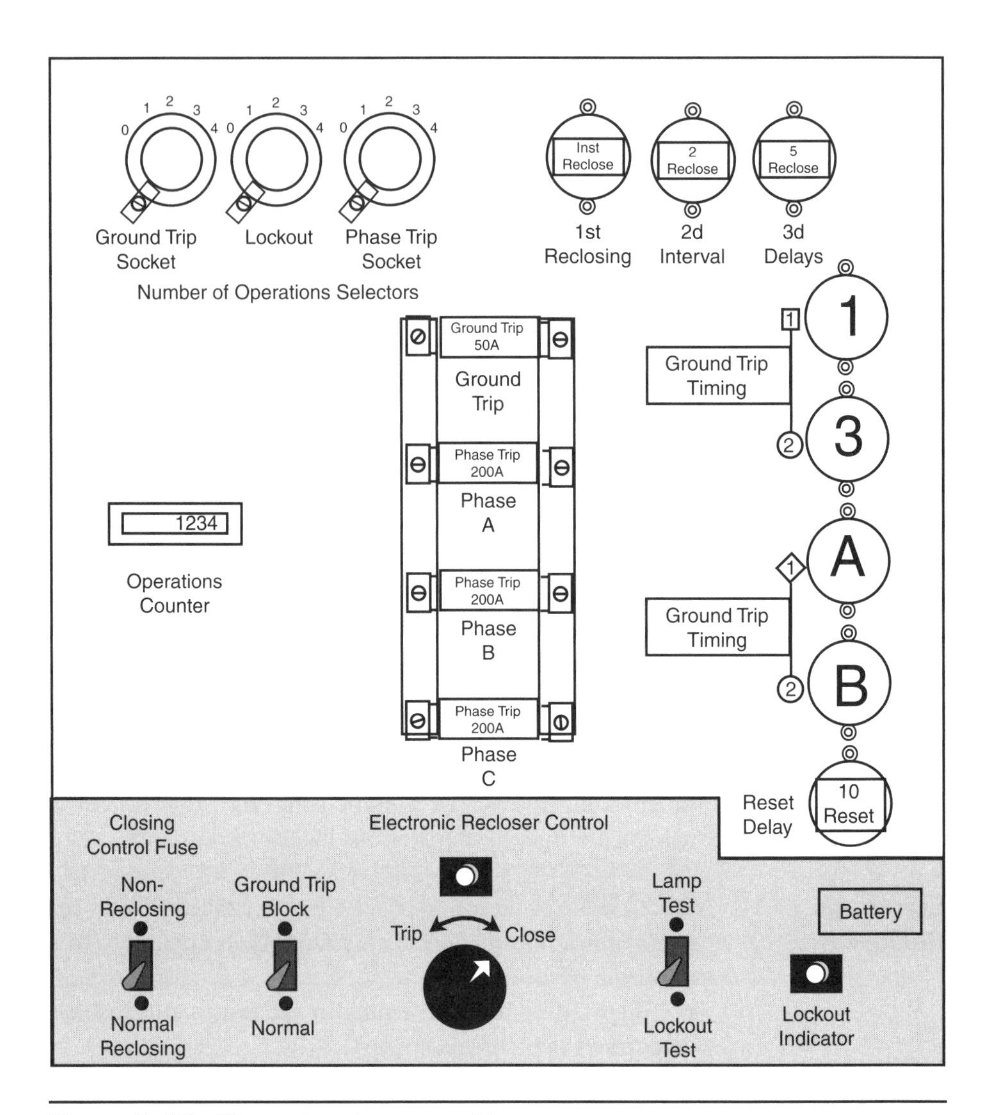

Figure 5–17 Electronic recloser control box.

instantaneous trip when closing it after a lockout. The delayed trip will still open the circuit if an overcurrent situation lasts long enough.

An electronic recloser can detect a fairly small phase-to-ground fault. This sensitivity can prevent reenergizing at the recloser because the ground-fault relay can interpret an unbalanced load as a phase-to-ground fault. Temporarily switching the *ground trip* toggle switch to the block position will prevent a small ground current from tripping out the circuit.

A ground-gradient mat bonded to the control box should be used when operating the main switch. The control cable from the control box is bonded to the recloser tank. The recloser can become alive if the recloser malfunctions or the bushings flash over.

Non-Reclose Feature on Reclosers

5.4.22 Reclosers can be set in a non-reclose position to prevent a circuit from being reenergized automatically. Putting a recloser in a non-reclose position gives a crew working on a circuit the assurance that the circuit will not be reenergized automatically if an incident on the job trips out the circuit.

Placing a recloser in a non-reclose position does not guarantee that a circuit will trip out during an accidental contact; it only ensures that a circuit will stay out of service once it is tripped out. Placing a tag at the recloser will prevent other utility personnel from closing the recloser without first checking with the crew working on the circuit.

Typical Recloser Ratings

5.4.23 Reclosers vary depending on the voltage rating, the interrupting capacity, and their mode of operation. Table 5–2 shows a variety of recloser frame sizes and the voltage rating, maximum current rating, and maximum interruption rating for each. Each manufacturer would have some kind of catalog designation to identify each type of recloser. In this table, the designation of the recloser types have been made up to avoid using a particular manufacturer's designation. Identifying the types of reclosers as H_1, EL_1, ET_1, etc., simplifies reference back to this table for the protection examples shown later. The table does not give all the needed information about a recloser rating. For example, the interrupting capacity of the H_3

TABLE 5–2 Typical Recloser Ratings

Types of Reclosers	Types	Maximum Rating (kV)	Maximum Current (A)	Interruption Rating (A)
The ratings of five typical sizes of *hydraulic* units are shown. The contacts on hydraulic reclosers open when overcurrent flows through a solenoid coil, which creates a magnetic force on a plunger and opens the contacts.	H_1 H_2 H_3 H_4 H_5	14.4 14.4 14.4 24.9 24.9	50 100 280 100 280	1250 3000 6000 2000 4000
The two typical *electric* reclosers shown require a low-voltage supply to operate their heavy contacts.	EL_1 EL_2	14.4 34.5	560 560	12,000 8000
The ratings of four typical three-phase relay-controlled *electronic* reclosers are shown. The relays can be set to detect specified overcurrent and ground-current conditions.	ET_1 ET_2 ET_3 ET_4	14.4 14.4 34.5 34.5	400 560 400 560	6000 10,000 6000 8000

recloser is shown as 6000 amperes, but its actual interruption rating is 6000 amperes at 4.8 kilovolts and only 4000 amperes at 14.4 kilovolts.

Sectionalizers

5.4.24 A *sectionalizer* is a device that will isolate a faulted section of line when coordinated with an upstream multishot recloser or circuit breaker. A sectionalizer is a slave device that cannot interrupt a fault current and opens the circuit during a specified second or third interval while the upstream device has the circuit deenergized.

A *hydraulic sectionalizer* is an oil switch with a mechanism that will open automatically after fault current goes through the coil a specified number of times. A hydraulic sectionalizer is not capable of interrupting fault current. It opens automatically for a downstream fault during an interval while the upstream multishot device has the line isolated.

As with reclosers, sectionalizers come with various voltage ratings and current ratings. A lines crew can use a sectionalizer to interrupt load current, as it can with any oil switch.

An *electronic sectionalizer* is a device that fits into a cutout in the same manner as a fuse chamber. It is made up of a copper tube with a bronze casting in each end. Current flows through the copper tube when the cutout is closed. A current transformer is mounted on the copper tube. The secondary of the current transformer is wired to an actuator on the bottom of the chamber that, when activated, drops the chamber. An electronic sectionalizer is a slave device that opens during an interval when the upstream multishot device has isolated the circuit.

An electronic sectionalizer is like any solid-blade switch when it is to be operated manually. It requires a portable load-break tool to interrupt load current.

Operating Three-Pole Switches

5.4.25 If only one phase of a three-phase circuit trips out, the protective switchgear and relays see a large unbalanced load with two phases loaded and one not loaded. The current and the voltage on the whole system would become unbalanced. The unbalance due to one phase being out of service is not as severe to distribution circuits as it is to transmission circuits. Three-phase motors heat up when one phase is out, but the thermal protection prevents them from being damaged.

Much of the switchgear in a distribution substation and on distribution feeders is gang-operated. All three phases will trip out, even if only one phase is faulted.

Bypassing Defective Switchgear

5.4.26 When defective protective switchgear is temporarily bypassed with a solid jumper, the circuit beyond the device will have less protection. Any fallen conductor or other fault will probably not cause the upstream protective device to trip out the circuit. This temporary fix should be corrected as soon as is practical.

The design of the recloser installation in Figure 5–18 and in Figure 5–19 allows the recloser to be bypassed for maintenance or when it becomes defective. The bypass should be fused to give protection downstream.

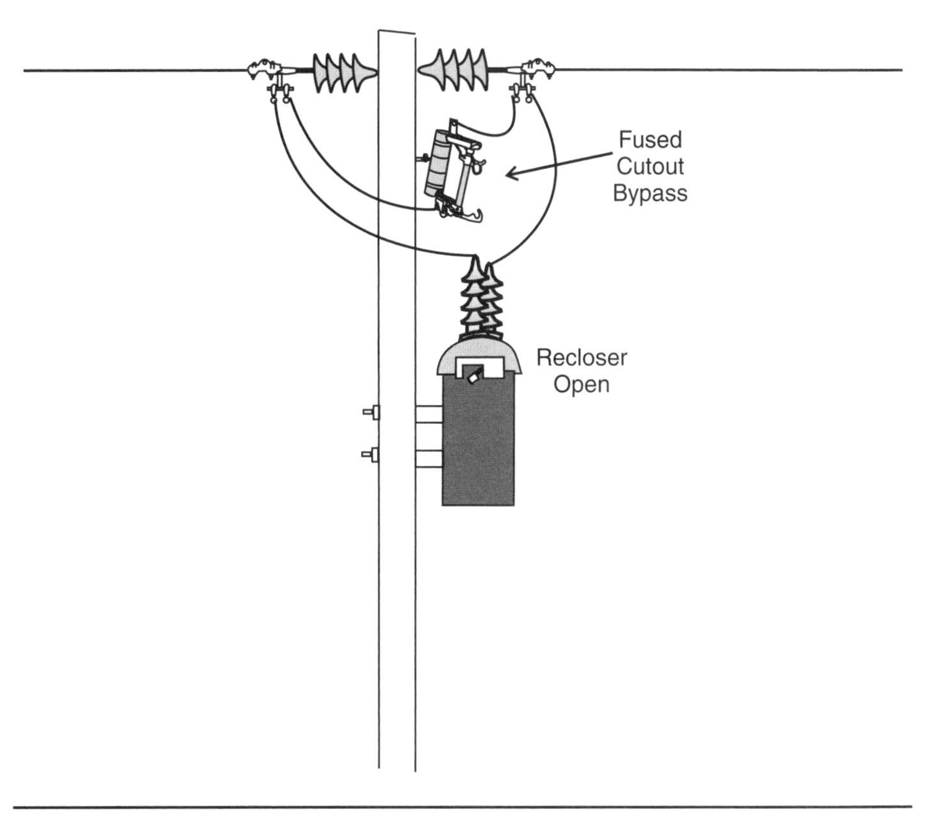

Figure 5–18 Recloser bypass.

5.5 Underground Distribution Switchgear

Switchgear in Underground Distribution Systems

5.5.1 Switchgear in underground distribution systems is in confined areas where any arc could easily spill over to grounded cabinets or other phases. Therefore, underground switchgear is designed to extinguish or limit the arc.

Switching distribution underground involves operating load-break-separable connectors (elbows), arc-strangler switches, bayonet-style under-oil expulsion fuses, vacuum switchgear, SF6 switchgear, and other devices with arc-quenching capability. Fused cutouts, with no load-break capability, are used in some vaults and with some equipment. A portable load-break tool should be used to open these cutouts.

The original design of some types of underground switchgear may have allowed its operation under energized conditions, but field experience, higher fault-current levels, and higher distribution-voltage levels have changed this. Usually, an accident or incident investigation will restrict the manner in which some types of switchgear are operated.

Some switching is carried out in a vault or manhole where there is the additional hazard of creating an arc in a confined space. A line crew always checks for the oxygen level and for the existence of flammable or toxic gas before and during switching inside a confined space.

Figure 5–19 Single-phase hydraulic recloser.

Figure 5–20 Switching cabinet using elbows.

Cable-Separable Connectors (Elbows)

5.5.2 Separable connectors are the live-line clamps of an underground system. They are used to isolate equipment or to sectionalize segments of cable (Figure 5–20). They are found at the inputs and outputs of transformers and at other locations where it is desirable to be able to separate a cable from equipment under live conditions.

There are non-load-break elbows (Figure 5–21) that should not be operated alive. A non-load-break elbow can have a current rating as high as 600 amperes and is often used to terminate a main feeder into a switching cabinet.

A load-break elbow can be pulled from the load bushing with a live-line tool to interrupt a 200-ampere load current. The elbow needs to be pulled off in one quick motion, especially on high-voltage distribution. If the lubricant between the elbow and the bushing has dried, an elbow does not always come all the way off with the first pull. An arc between the load break pin and the bushing contacts can spill over to the outside of the bushing to a grounded tank or cabinet. An elbow-pulling tool or some restrictive procedures are needed to reduce the risk of flashover incidents.

The capacitive test point (Figure 5–22) on an elbow can be used to determine whether or not a circuit is energized. The cap over the test point is removed with

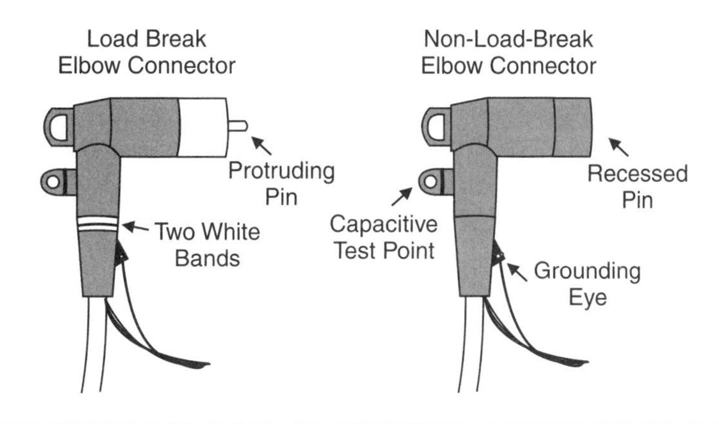

Figure 5–21 Load break and non-load-break elbows.

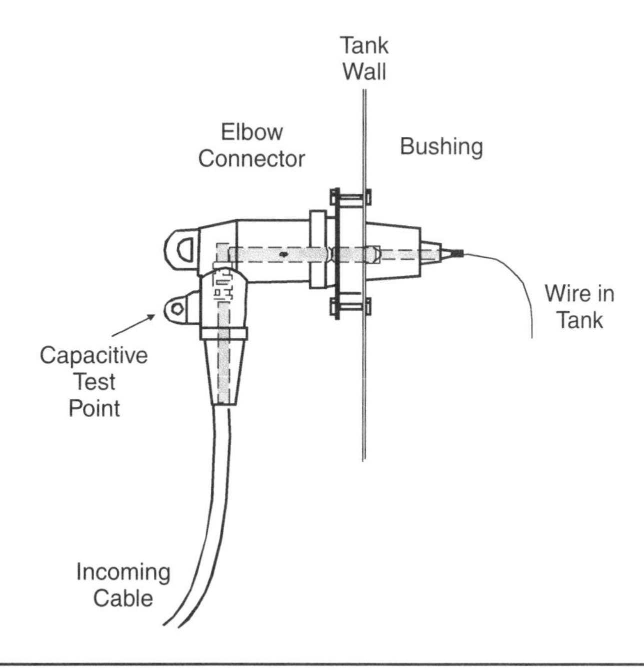

Figure 5–22 Elbow plugged into bushing.

a live-line tool, and a potential tester designed to be able to pick up the capacitive voltage is used at the exposed test point. The test point is not a direct connection to the live conductor inside the elbow but has a potential because of capacitance. An ordinary voltmeter will not pick up a potential at the capacitive test point.

Metal-Clad Switchgear

5.5.3 Urban distribution substations with underground feeders have the high-voltage and low-voltage switchgear in metal-enclosed construction. Subtransmission switchgear can be a circuit breaker or a load-interrupter switch to prevent an arc from spilling over to the grounded-metal enclosure. Any subtransmission power fuses are

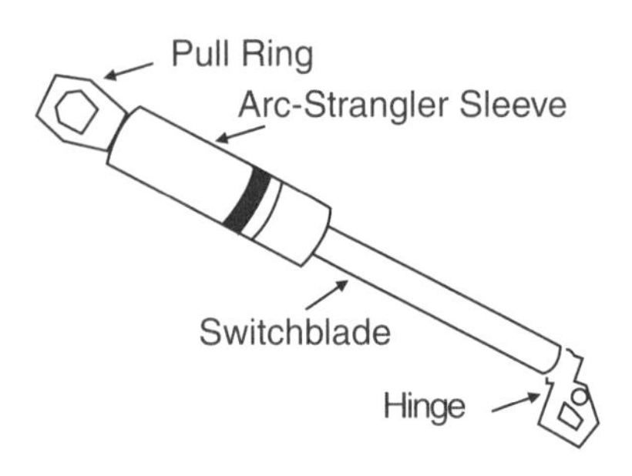

Figure 5–23 Arc-strangler switch.

fitted with flame suppressors, current-limiting fuses, or solid-material fuses that do not exhaust hot gases when blowing.

Low-voltage distribution-voltage switchgear consists of circuit breakers or reclosers. The circuit breakers have *plug in* input and output contacts, allowing the breakers to be removed and installed for maintenance.

NX-Type Switches

5.5.4 NX switchgear is found in live-front switching cabinets and transformer vaults. It is operated with a live-line tool in the same manner as a fused cutout. The NX switchblade (Figure 5–23) is an arc-quenching load-break device that is available as a current-limiting device or as a solid-blade device. The load-break device works similarly to a portable load-break tool and must be cocked to operate. To cock the barrel, the arc-strangler sleeve is pulled down and held in that position by a latch spring. When the switch is closed, the latch spring is depressed and the sleeve is held down by the switch itself. When the switch is opened, the arc-strangler sleeve is released and snaps up to activate the load-break feature of the switch barrel.

5.6 Specifying Protection for a Distribution Feeder

Planning a Protection Scheme

5.6.1 Planning a protection scheme is normally outside the scope of a powerline worker's duties, but it is useful and interesting to have an understanding of the process. While there is more than one way to prepare a protection scheme for a distribution feeder, an example of a common method is described here. There are four major steps in planning a protection scheme for a distribution feeder:

1. Gather the data needed to calculate a feeder profile.
2. Calculate the voltage, load current, and fault current on the feeder.
3. Interpret the results of the study.
4. Specify the protective devices needed on the feeder.

Gather Feeder Data

5.6.2 A planning engineer prepares a map or a schematic drawing of a feeder, as shown in Figure 5–24, and breaks the feeder into segments. Each line segment is assigned a numbered node or point.

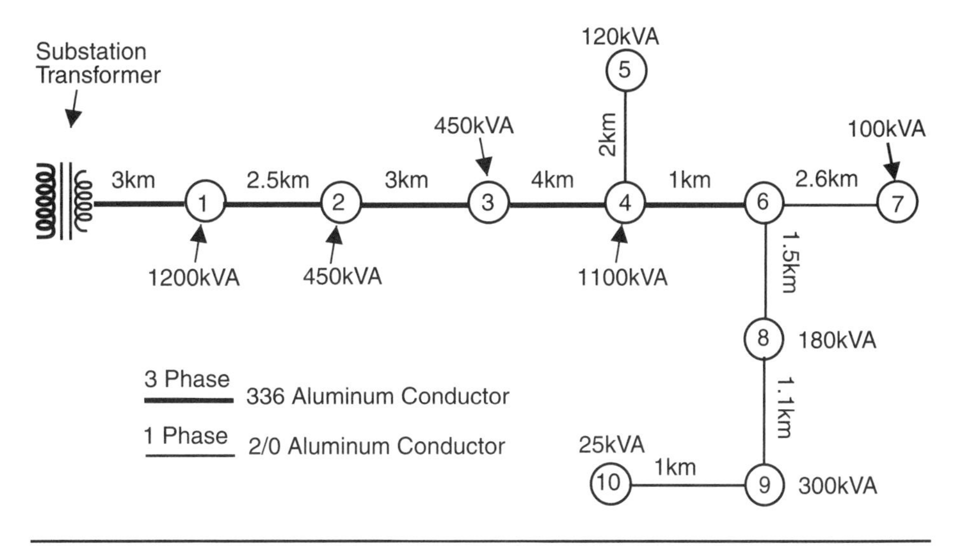

Figure 5–24 Distribution feeder data collection.

- Nodes are assigned to each junction, switch, conductor-size change, and load center.
- The length and conductor size in each line segment is recorded.
- The transformer load from node to node is added up and recorded. Transformers on a feeder are not always loaded to 100 percent capacity, so a diversity factor is used to approximate the actual loading. In the sample feeder shown in Figure 5–24, a balanced load is assumed between three phases. In reality, many utilities would add up the load on each individual phase.
- The available fault current at the source of the feeder is needed to do the feeder calculations. This information is based on the impedance and size of the distribution substation transformer and the voltage, conductor, size, and length of the circuit feeding the substation. In the sample feeder shown in the figure, the available phase-to-phase fault current at the secondary of the substation transformer will be given as 5000 amperes.

Calculating Feeder Information

5.6.3 Converting the field data to fault current, voltage, and peak-load current at each node is not normally done manually. The feeder data is input into a computer program, and the output shows the fault current, the voltage, and the peak-load current at each assigned node.

Calculations for fault current are based on a dead short, such as a phase-to-phase fault or a phase-to-neutral fault at each node. Most faults are, of course, not dead shorts, and this is taken into consideration when protective switchgear is specified. Figure 5–25 shows the results of the calculations.

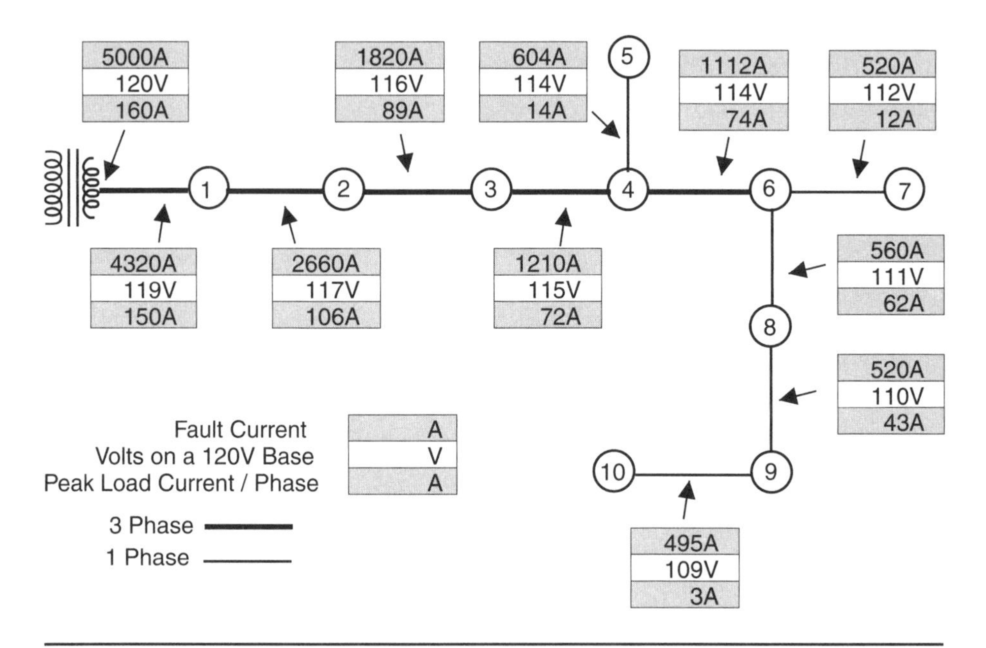

Figure 5–25 Results of calculations.

To keep the sample-feeder drawing uncomplicated, only the phase-to-phase fault current is shown in the three-phase sections. At node six, for example, the phase-to-phase fault current is shown as 1112 amperes, but the phase-to-ground fault current at this location is 960 amperes. Notice how quickly the fault-current values drop off with distance from the source.

The peak-load current shown at each node in the figure was based on the kVA loading in each line segment between nodes. A diversity factor is put into the calculations to change the transformer kVA loading to a more realistic value. A planning engineer would base a diversity factor on field data such as a feeder voltage/current survey and a recording of ammeters at the distribution substation.

The voltage at each node is also calculated by the computer program. The voltage in the sample feeder is shown to drop in proportion to the feeder load and the distance from the source. In this example, the voltage was put on a 120-volt base. The voltage on the single-phase line farthest from the station is shown to be too low, and the planning engineer would probably specify a voltage regulator.

Protection-Scheme Philosophy

5.6.4 Protection schemes will vary depending on utility preferences and whether the feeder is urban, rural, underground, or overhead. A protective coordination scheme provides automatic isolation of faulted circuits from the system while leaving the rest of the system in service. Where two protective switchgear devices are in series with each other, the protect*ed* switching device is on the source side and the protect*ing* switching device is on the load side. If a fault develops downstream from the protecting device, it should trip out before the protected device. For example, a transformer fuse should have a size and speed that will clear a faulted transformer from a circuit before a protected upstream-line fuse is damaged.

Most distribution feeders have a recloser or a circuit breaker that provides multishot protection to the feeder. A multishot protective device can be set for a sequence of a specified speed and number of trips and reenergizations.

A *fuse-saving* philosophy has the first trip-out operation of a source-multishot device happen so fast that a downstream fuse does not even begin to melt. For lightning and other transient faults, this protection scheme reduces outages and trouble calls. Everyone on the feeder experiences at least one trip-out for each transient, which starts their digital clocks blinking. A permanent fault on a lateral will cause a multishot device to trip out and reenergize the circuit for the complete specified number of times. The downstream fuse will blow on the second or third reenergization depending on how the protection is set up.

A *time-delayed-instant* philosophy has the first trip-out operation of the source-multishot device delayed long enough for a downstream fuse to blow first. This prevents everyone on the feeder from being exposed to a momentary trip. This philosophy is often used with fuses protecting underground feeders, where most faults are permanent. The time-delayed-instant philosophy keeps a feeder exposed to a fault longer and increases the likelihood of damage, such as a conductor burning down.

Interpreting the Results of a Feeder Study

5.6.5 The results of a feeder study, similar to the sample, gives a planning engineer all the information needed to specify, prepare, and justify betterment specifications for the feeder.

- The maximum and minimum fault current available along the feeder gives the necessary information to prepare a protective coordination scheme. For example, the sample feeder study shows that the reclosers in the distribution substation need to be big enough to interrupt 5000 amperes of fault current and carry a load current of approximately 200 amperes continuously.
- Normal data collection and calculations are done for each individual phase. The results show the load current on each phase and whether there is a need for some phase-balancing work.
- The study results show where there is a need for voltage correction. A voltage profile of the feeder is available from node to node. Voltage regulators or capacitors can be located near the node where the equipment will provide maximum benefit.
- Justification for major betterment work is based on a feeder study. The cost of line losses can be worked out and compared to the cost of reconductoring. A feeder study will show whether conversion of a single-phase circuit to three phases will solve phase-balancing, voltage, or protection problems.

Specifying a Hydraulic-Recloser Frame Size

5.6.6 When a recloser is chosen for a specific location, it must have a suitable frame size, which means a unit must have its contacts and its trip mechanism big enough to interrupt the highest fault current that could occur at that location. In the sample feeder, a fault immediately in front of the substation recloser could generate 5000 amperes. The recloser frame must have its contacts and a spring mechanism able to safely carry and interrupt 5000 amperes.

Table 5–2 in section 5.4.23 "Typical Recloser Ratings" shows that the smallest recloser for a 12.5/7.2-kilovolt system that can interrupt 5000 amperes would be an H_3 recloser. The table shows that this recloser can interrupt 6000 amperes. However, this interruption rating applies to 8.32/4.8-kilovolts. The interrupting capacity for the H_3 recloser on a 12.5/7.2-kilovolt system is 5000 amperes.

Specifying a Trip-Coil Size

5.6.7 An overcurrent situation will cause a recloser or circuit breaker to trip out a circuit. Relay-controlled devices, such as a circuit breaker or an electronic recloser, will trip out when a relay senses the overcurrent and sends a signal to trip out the circuit. A hydraulic recloser opens when overcurrent flows through its trip coil.

A trip coil installed in a recloser must be able to carry the expected peak-load current. For example, the sample feeder shown in Figure 5–25 has a projected peak load of 160 amperes per phase. A 200-ampere trip coil would probably be specified to allow for load growth and unbalanced load.

A 200-ampere recloser will carry 200 amperes continuously and will carry some overload without damage. In general practice, an overcurrent device will not trip for a small or temporary overload. A trip coil in a hydraulic recloser is normally designed to trip a circuit when the current flowing through it is twice the rating of the coil; therefore, when 400 amperes flow through a 200-ampere coil, the circuit will trip out. An overcurrent of double the recloser setting would normally be a fault.

A Recloser Protection Zone

5.6.8 A recloser will provide multishot protection downstream to any location where the circuit has the capacity to generate the needed fault current to trip it out. For example, the lowest fault current available in the sample circuit shown in Figure 5–25 is 495 amperes phase to ground at node 10. A dead short between phase and neutral at node 10 could generate 495 amperes, which would be enough to trip a 200-ampere recloser at the substation. A more common higher impedance fault, such as a broken conductor lying on dry ground or a tree contact, will not generate the 400 amperes needed to trip the circuit.

In the sample feeder, the reclosers at the substation will likely "see" most faults just beyond node 6 where the circuit can generate a phase-to-phase fault current of 1112 amperes and a phase-to-ground fault current of 960 amperes. At node 6, even a high-impedance fault is likely to generate the minimum 400 amperes needed to trip a 200-ampere recloser.

Ideally, a safety factor is needed to provide a greater likelihood of a circuit tripping out for a high-impedance fault. A safety factor of two would require a minimum of an 800-ampere fault current to be available at the end of a zone protected by a 200-ampere recloser.

Specifying Recloser Speed and Operating Sequence

5.6.9 A planning engineer chooses a recloser with a sequence of opening and closing speeds based on coordination with upstream and downstream protective devices. When a fault is beyond a downstream fuse, the planning engineer would choose a fuse size and speed to blow at the desired time; for example, at the second or third closing of the recloser. The timing sequence of a recloser is coded:

$A = \textit{instantaneous}$
$B = \textit{retarded}$
$C = \textit{extra retarded}$
$D = \textit{steep retarded}$

A typical speed and sequence may be A1B3, which means the recloser opens instantaneously on sensing a fault and recloses. If the fault persists on the line, the recloser opens and recloses after a retarded delay. For a permanent fault, the recloser will go through the complete sequence of one instantaneous trip and three retarded trips before remaining open. The speed and sequence chosen by the planning engineer depends on the protection scheme philosophy and on the need to solve particular protection problems.

Typical TCC Curve for a Recloser

5.6.10 When specifying a protective device, the speed of operation must be known in order to be able to coordinate the device with other protective equipment in the circuit. A *time current characteristic* (TCC) curve is available for each size of fuse, sectionalizer, recloser, or relay. A TCC curve is plotted on graph paper and shows the response time, minimum damage time, and total clearing time.

Note that the TCC for a 200-ampere recloser in Figure 5–26 shows that the minimum fault the recloser will see is 400 amperes. The A curve shows that 400 amperes will trip out the circuit in about 0.5 seconds. The recloser will clear a permanent fault of 400 amperes in about 15 seconds. A permanent 5000-ampere fault will clear in about 0.2 seconds.

Specifying a Downstream Recloser

5.6.11 In the sample feeder shown in Figure 5–25, a downstream recloser is needed to provide multishot protection to the end of the feeder. A higher impedance fault would not be "seen" by the source recloser.

Ideally there is multishot protection to the end of every feeder. For economic reasons, multishot protection may be unavailable for higher impedance faults near the end of a feeder. The end of the feeder can still be protected by a fuse, but without multishot protection a transient fault will blow a fuse and cause an easy callout for a line crew. To have multishot protection for a higher impedance fault on the single-phase line past node 6, a downstream recloser should be installed.

As shown in Figure 5–27 a 100-ampere H_2 recloser on the single-phase line at node 6 will carry the 62 amperes of load current and should see most faults at node 10, which has 495 amperes of available fault current. To provide node 7 with multishot protection, another recloser would need to be installed in that line.

Specifying Fuse Size and Speed

5.6.12 A line or a station fuse is chosen based on its current rating and speed. The speed of a fuse is indicated by the letter on the fuse link (e.g., K-link or T-link). In

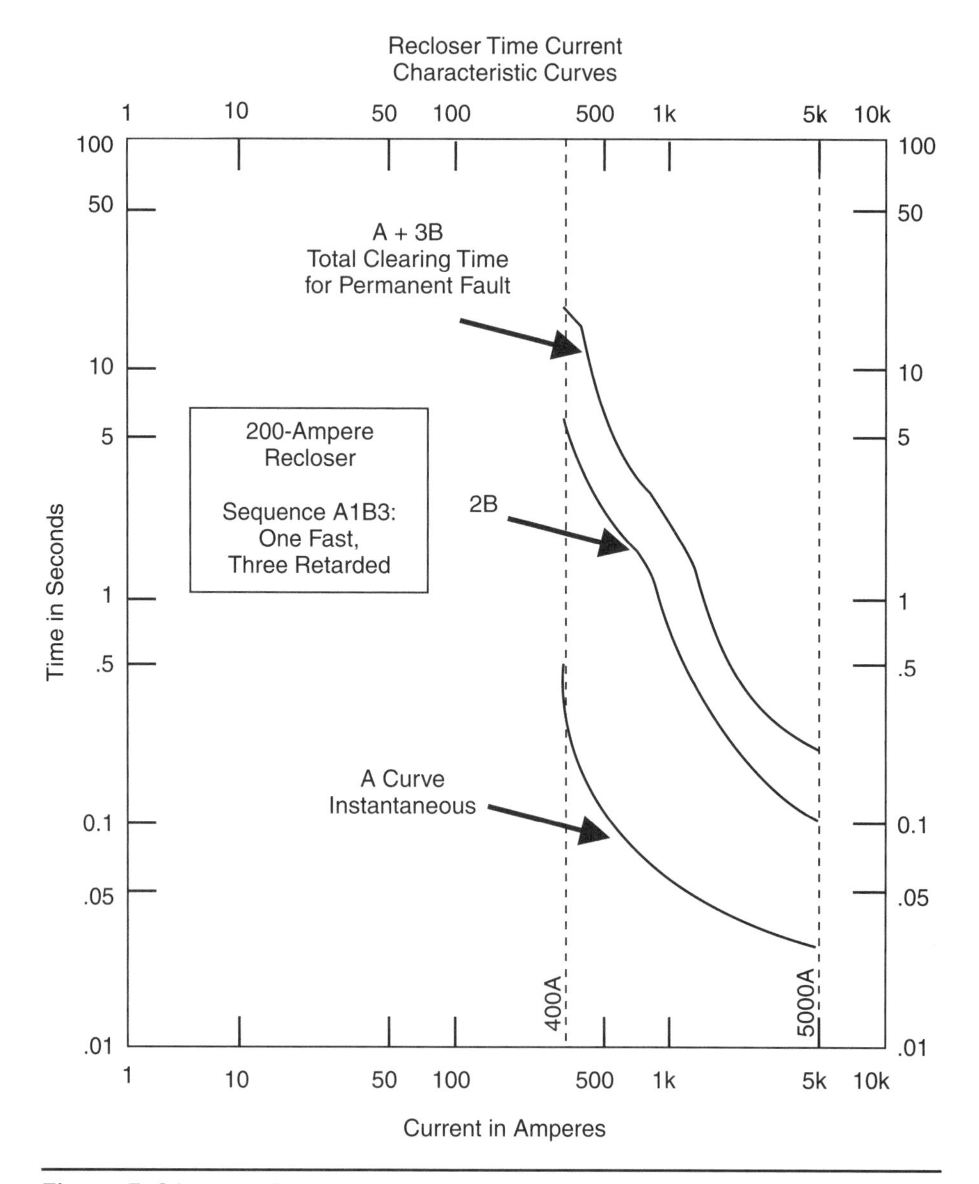

Figure 5–26 TCC for a 200A recloser.

order to coordinate the speed of the fuse with upstream and downstream protective switchgear, the minimum damage time, minimum melting time, and total clearing time must be known.

The TCC curve for a 100-ampere K-link fuse (Figure 5–28) shows how quickly the fuse melts as the current going through it is increased. In this example, 200 amperes will melt the fuse link after about 100 seconds of exposure, and 1000 amperes will melt the fuse in about 0.2 seconds.

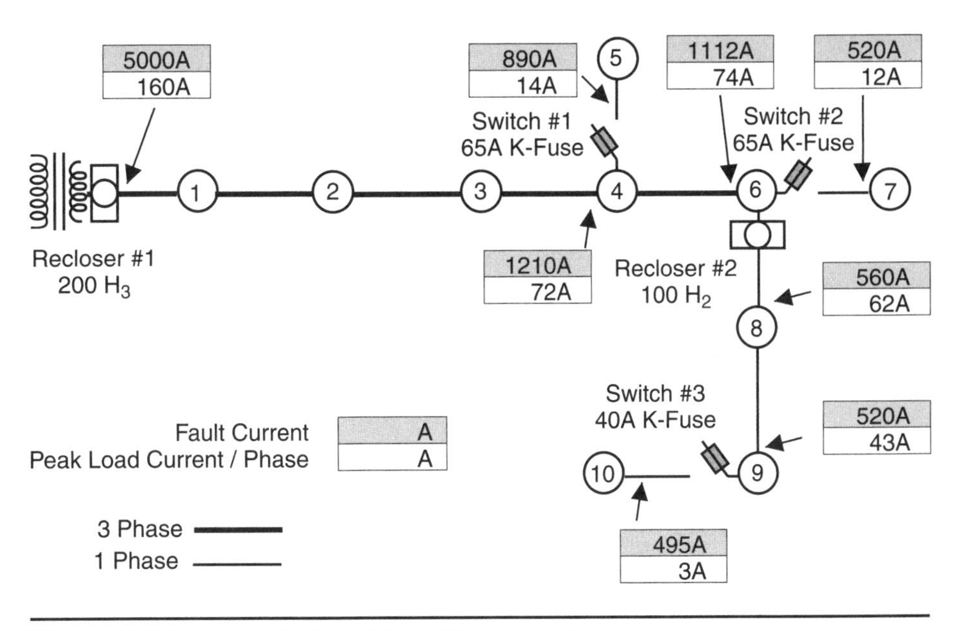

Figure 5–27 Sample feeder with specified protection.

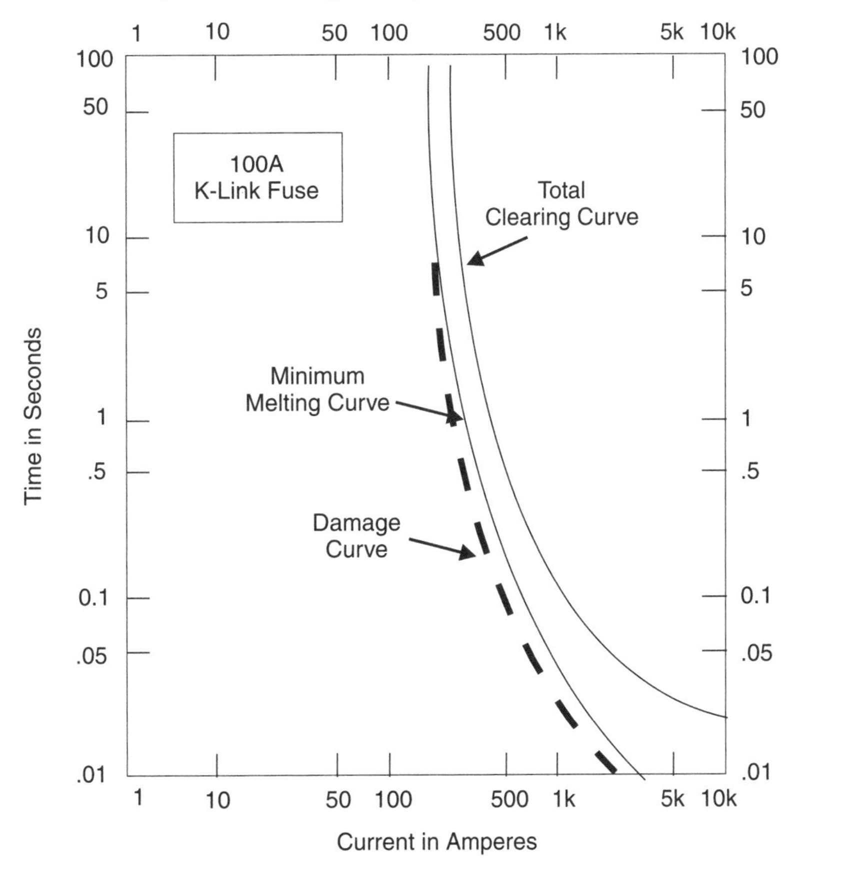

Figure 5–28 TCC curve for 100A K-link fuse.

Specifying Downstream Fuses

5.6.13 A fuse size and speed can be chosen so that it will melt in any of the energized intervals of an upstream multishot device as it goes through its sequence during a permanent fault. In the sample feeder shown in Figure 5–27, the lateral protected by a fuse in switch #1 is specified to coordinate with the 200-ampere substation reclosers. The lateral has a peak-load current of only 14 amperes. A 25-ampere fuse would protect this lateral, but it would melt before the upstream recloser could go through the instantaneous trip. A 65-ampere fuse would stay intact as the substation recloser went through its instantaneous trip for a fault on the lateral. The 200-ampere reclosers can see to the end of the lateral because the available fault current is 890 amperes.

A TCC curve for a 65-ampere fuse superimposed on the TCC curve for a 200-ampere H_3 recloser can show that the two devices coordinate with each other. A quicker method is the use of a recloser-to-fuse coordination table. Table 5–3 is a table that can be used with a 100-ampere H_2 recloser.

In the sample feeder shown in Figure 5–27, the lateral tap protected by a fuse in switch #3 is specified to coordinate with recloser #2, which is a 100-ampere H_2 recloser. The lateral is lightly loaded with a peak load of 3 amperes, but if the circuit is exposed to transients from lightning or trees, it would save a callout to have the fuse in switch #3 coordinate with recloser #2. The fault current at the end of the lateral is 495 amperes, which means that the 100-ampere recloser should see most faults. Figure 5–27 shows that the lateral at switch #3 has a minimum of 495 amperes and a maximum of 520 amperes fault current. Table 5–3 shows, therefore, that a 40-ampere fuse is needed to coordinate with a 100-ampere recloser.

In the sample feeder, the lateral protected by the fuse in switch #2 has multishot protection for low-impedance faults from recloser #1. The available fault current of 620 amperes at the end of the line may not generate enough current for a higher impedance fault such as a tree limb. A judgment has to be made as to whether to:

TABLE 5–3 Recloser-to-Fuse Coordination Table: A 100-Ampere H_2 Recloser Set at Sequence: One Fast, Three Retarded (A1B3) Type K Fuse Links

	Recloser-to-Fuse Coordination Range in Amperes		Maximum Fuse-to-Fuse Coordination Current within the Recloser Zone of Protection					
				Protected Link				
Protecting Fuse Link	*Min*	*Max*	*Protecting Fuse Link*	*50*	*65*	*80*	*100*	*140*
40	200	690	40	600	690	690	690	
50	200	1020	50		510	1020	1020	1020
65	200	1400	65			600	1400	1400
80	280	1880	80				1280	1880
100	560	2560	100					2560
140	1970	3000						

1. Install a recloser at switch #2.
2. Install a fuse that coordinates with recloser #1.
3. Install a small fuse that would blow quickly for any fault.

Option #2 is a compromise wherein recloser #1 sees most faults except for higher impedance faults near the end of the line. A 200-ampere recloser table similar to Table 5–3 shows that a 65-ampere fuse coordinates with recloser #2.

Specifying Sectionalizers

5.6.14 Sectionalizers are specified similar to the way fuses are specified. TCC curves or tables determine the most appropriate size and speed sectionalizer to coordinate with an upstream multishot device. A sectionalizer must coordinate with a multishot device because it isolates or opens a faulted circuit in an interval when the multishot device has tripped out the circuit.

A sectionalizer is more reliable than a fuse for specifying at which interval of the upstream multishot-device sequence it will open during a permanent fault. A trip coil for a sectionalizer is specified for its current rating and has a two-shot or three-shot sequence.

Overcurrent Trouble Calls

5.6.15 During an overcurrent situation, a line crew is called out after circuit breakers, reclosers, sectionalizers, or fuses have gone through their sequences and have isolated any circuit with a permanent fault (Table 5–4).

TABLE 5–4 Overcurrent Trouble Calls

If	Then
There has been wind, ice, wet snow, or lightning.	The problem is often the traditional broken conductors, broken poles, or fallen trees and branches.
Protection tripped out on a nice clear day.	The fault is sometimes due to public contact, such as a car accident, tree cutting, or a boom contact. An outage on a clear day suggests that a patrol should be carried out before energizing the circuit.
An overload is the likely cause of an outage.	The most common fix is phase balancing. The planning engineer often assumes that the load on the feeder is balanced among the three phases. When the load is not balanced, one phase of the three-phase system is carrying more than its share of the load and the protective device will trip out.
The circuit has been out for a while, especially during peak-load periods.	There is a high initial inrush current when the switch is closed, and the line trips out again. A heavily loaded circuit may have to be picked up a section at a time.

High-Voltage Distribution Conversion

5.6.16 Many utilities have voltage conversion programs to convert their system from typical primary voltages such as 8.3/4.8kV and 12.5/7.2kV to higher distribution voltages such as 25/14.4kV and 34.5/19.9kV. The advantages of a higher-voltage distribution system are:

- The substation breaker or recloser can see out much farther because the available fault current level remains high much farther downstream. There is often no need for downstream reclosers or fusing in order to maintain protection at the end of the line. Sectionalizing devices are still installed for operating purposes.
- There are fewer substations needed and fewer feeders needed from the substation.
- For a given load, a higher-voltage feeder would have less load current and, therefore, there is less voltage drop along the feeder. Downstream voltage regulators are rare except for a very long, or very heavily loaded, feeder. There is a rule of thumb that a distribution feeder with an average customer load can feed about a mile (1.6 km) per 1kV (phase-to neutral) before needing voltage regulation. Based on this rule of thumb, an 8.3/4.8kV feeder would need voltage regulation about five miles from the substation and a 34.5/19.9kV feeder would need voltage regulation at about twenty miles from the station.
- A higher-voltage feeder can feed much more load. For example, an 8.3/4.8kV feeder typically can be considered heavily loaded when supplying about 3MW while a 34.5/19.9kV feeder is considered heavily loaded when it is supplying about 15MW.
- A utility would normally not be able to feed an individual customer that has a requirement for a 750kW service on an 8.3/4.8kV system. A subtransmission feed would be required along with all the associated costs. Meanwhile, if a 25/14.4kV or 34.5/19.9kV feeder was available, a 750kW load could be supplied on that system with standard distribution transformers and not create a problem for other customers on the feeder.

5.7 Overvoltage Protection

Causes of Overvoltage

5.7.1 Other than lightning, there are not many trouble calls due to overvoltage. Lightning is the most common source of overvoltage and is the source of most transient faults on an electrical system. The high voltage and current generated by lightning usually cause a momentary outage to customers because the automatic circuit protections open and reclose after the strike.

Switching operations can cause a transient overvoltage. The magnitude of the surge is much lower than lightning and normally is significant only on circuits of 230 kilovolts and above. An overvoltage situation can occur when a voltage regulator is stuck in a maximum-boost position, a switched capacitor is left in service during an off-peak time period, a higher-voltage wire falls onto or makes contact with a lower-voltage wire, or, in rare cases, a transformer with a partially shorted coil results in a high secondary voltage.

Convection and Frontal Thunderstorms

5.7.2 There are two types of thunderstorms—a convection storm and a frontal storm. A *convection* thunderstorm is the most common thunderstorm, especially on a hot summer day. It is a localized storm that occurs when the hot air near the earth rises and meets the cold air at a higher altitude. A convection thunderstorm does not last long because it is usually accompanied by rain. The rain cools the earth, which removes the energy source for the storm.

A *frontal* thunderstorm occurs when a cold front meets a front of warm, moist air. The storm can stretch for hundreds of miles and last for hours. It can regenerate itself because air masses continue to move in as the fronts collide. A frontal storm is more severe than a convection storm.

Description of Lightning

5.7.3 Lightning is part of an overall electrical circuit between the earth and the atmosphere (Figure 5–29). Electrical charges are created in the atmosphere by the friction between particles of rapidly moving air.

The cloud normally associated with a frontal storm or a convection storm is the cumulonimbus cloud, also described as a thunderhead. During rain or hail, negative charges fall to the bottom of the cloud. When these negative charges travel toward positive charges, the air becomes electrically stressed and breaks down, resulting in a high voltage and high-current discharge (lightning) between

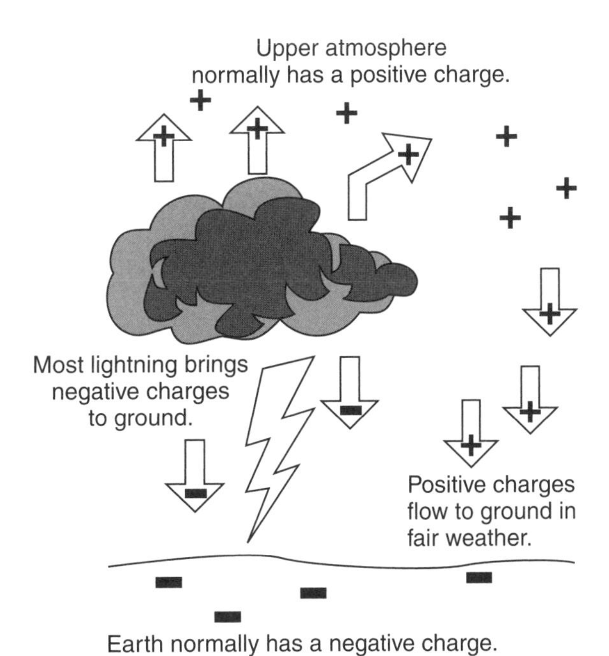

Figure 5–29 Electrical circuit between atmosphere and ground.

the positive and negative charges. The electrical discharge can be within the same cloud, to other clouds, or to earth.

The direction or polarity of the current flow due to lightning is not important as far as protection is concerned. Both positively charged and negatively charged lightning strikes the earth. Contrary to its appearance, the most common direction for the current flow is from cloud to earth. A rarely visible charged leader from the cloud reaches the earth, and, as it starts to neutralize at the earth, it becomes a visible arc from the earth to the cloud. Some characteristics of lightning are:

- Lightning can have a charge of up to 100 million volts.
- Lightning establishes an electrical path through which the follow-through current flow can be from 2000 to 200,000 amperes.
- Lightning can generate heat in a 2.5-cm (one inch) diameter path of 33,000°C (approximately 60,000°F).
- Lightning generates heat, which causes an explosive expansion of air and is heard as thunder.
- Lightning, on average, strikes the earth somewhere at about 100 times per second.

Estimating Distance from Lightning

5.7.4 The *flash-bang method* is a quick method to estimate the distance to lightning. Count the number of seconds between seeing the lightning and hearing the thunder:

1. The number of seconds divided by 3 = the number of kilometers.
2. The number of seconds divided by 5 = the number of miles.

Determining the direction in which a storm is moving is important for determining safe working conditions on distribution lines. If the storm is five miles away but heading toward the work area, it is time to get clear of the circuit. The next lightning flash will be closer. As soon as lightning is seen or thunder is heard, work on powerlines, especially transmission lines, should be suspended because some part of the line could reach into the storm area.

Effects of a Direct Stroke of Lightning

5.7.5 The effect of a direct stroke of lightning can be very dramatic. When lightning flashes, an electrical circuit is being completed. When lightning strikes the earth, an electrical circuit is completed through the closest and lowest-resistance path to ground.

A tall, pointed object is the most attractive object for lightning to strike. A 1000-foot (300-meter) structure may be struck four times a year while a 100-foot (30-meter) structure is struck once every 25 years. Tall towers and buildings with lots of steel in them are struck by lightning but can carry the current to ground without overheating and causing damage.

Concrete footings of transmission-line towers can be damaged because lightning will heat the moisture in the concrete and expand it explosively. Concrete footings are, therefore, usually bypassed with a heavy-gauge conductor to carry the current.

Trees and wood poles are more attractive to lightning than the earth is. Although they are a higher-resistance path than a steel building or tower, they are still a lower-resistance path than air. When lightning strikes these objects, the moisture in the wood is suddenly vaporized and the resultant explosion will cause the wood to splinter.

Ground-Gradient Effect

5.7.6 There are step and touch potentials up to 70 kilovolts per foot (200 kilovolts per meter) in the immediate area of a lightning strike. People standing near a tall object that is struck by lightning are injured because of the high step and touch potentials at the base. As seen in Figure 5–30, people standing on the ground can be hurt by touching an object struck by lightning as well as by standing near the object struck by lightning.

Even when the lightning discharges in the clouds and does not reach earth, there are voltage-gradient changes on the ground. Induction from the lightning discharge can cause gradients of 3 kilovolts per foot (10 kilovolts per meter) on the ground. The induced gradients are the cause of corona discharge from pointed, grounded objects and can cause hair to stand on end and skin to tingle. Sparks from induced gradients and corona discharge can cause fires around unprotected flammable liquids even without a lightning strike.

An object isolated from the ground can have an induced potential up to 100 kilovolts. When lightning is seen or thunder is heard, there can be an induced-potential buildup on isolated conductors.

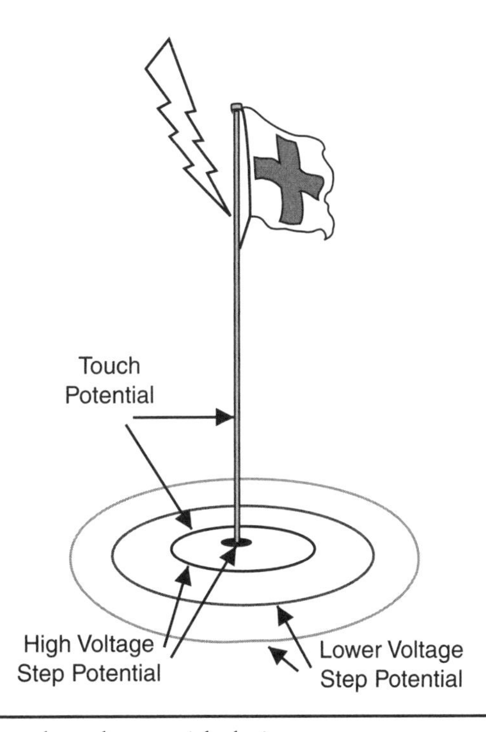

Figure 5–30 Step and touch potentials during a storm.

Safety Tips for Working Near Lightning

5.7.7 These safety tips for working near lightning can apply to on-the-job and off-the-job situations:

- The safest place to be during a thunderstorm is in a vehicle. If a vehicle is struck by lightning, all the metal in the vehicle will be at the same potential and anyone in the vehicle will not be exposed to any potential difference.
- When your hair stands on end and your skin starts to tingle, there is a potential difference building up between the earth you are standing on and the clouds above. Lightning is about to strike. Drop into a fetal position.
- Usually lightning strikes tall objects such as trees. Nearby people are injured due to the ground gradients of the lightning dissipating into earth. The ground gradients are like the rings formed when a stone is thrown into water. Each "ring" is a different voltage with the greatest potential differences being near the center. Staying away from tall objects reduces exposure to the highest voltage gradients. Keeping your feet together reduces exposure to the different potentials of the rings or to ground gradients. In the outdoors, with no nearby shelter, crouch low with feet together.
- In buildings, stay away from grounded objects. Grounded objects rise in potential when there is a nearby stroke of lightning.

Tracking Lightning Storms

5.7.8 Many electrical utilities have access to information from lightning-detection systems, which detect the location and intensity of lightning discharges. This information is valuable to determine which transmission lines are at risk of a lightning strike and therefore at risk of being tripped out.

When possible, the system operator reduces the load on vulnerable transmission lines and increases the load on other generators and lines to make up the difference. Putting a storm limit on a key extra-high-voltage (EHV) circuit is critical because an unexpected loss of a large source of power could destabilize the whole electrical system.

Lightning-detection systems can also provide an early warning to line crews working on transmission lines. A voltage surge due to lightning is a hazard when working on a live line as well as when working on an isolated and grounded line.

Effect of Lightning on a Circuit

5.7.9 A voltage surge from lightning travels like an ocean wave in all directions, except that it travels at the speed of light. The wave is attenuated (diminished) when it encounters a location where it will flash over, preferably at a surge arrester where a sparkover occurs and some or all of the energy is dissipated. There are other ways a voltage surge is attenuated.

A voltage surge causes a coupling effect between the phase conductor and the neutral or shield wire. The coupling effect reduces the voltage on the phase conductor while it raises the voltage on the neutral or shield wire. When the neutral and shield wire are well grounded, the voltage is quickly lowered and the coupling effect, in turn, lowers the voltage on the phase conductors.

A high-voltage surge will be attenuated by corona discharge because distribution conductors are too small for such high voltages. The relatively small conductor also limits the current-carrying capacity of a conductor because of skin effect. The current flow is subject to skin effect in the conductor even though the current resulting from lightning is DC. However, it is not a steady-state current; it reaches a peak value and then diminishes. The effect of the change in current means it has a similar effect on conductors as AC.

When an ocean wave hits a wall, it is reflected back double in size and then dissipates quickly. A voltage surge acts out similarly when it comes to a dead end. That is why transformers on dead-end poles are more vulnerable to lightning damage.

Switching Surges

5.7.10 Switching operations cause transient overvoltages. The magnitude of the voltage surge is much lower than lightning and is significant only on circuits 230 kilovolts and above. If a switch opens and is immediately reclosed, there is a brief overvoltage when energizing a capacitor. A length of parallel conductors on a transmission line is like a big capacitor. Arcing inside a circuit breaker can have the same effect as though the breaker is reclosing and there is a restrike in closing or opening. Arcing or restrike on a capacitor causes a switching surge in the line. When carrying out live-line work, the risk of a restrike or switching surge is reduced if the circuit breaker is put into a non-reclose position.

Effects of a Voltage Surge

5.7.11 A voltage surge can cause a flashover, a sparkover, or a puncture.

- A *flashover* refers to a disruptive discharge along a solid material such as an insulator string or a live-line tool.
- A *sparkover* refers to a disruptive discharge through the air.
- A *puncture* refers to a disruptive discharge through a solid material such as rubber cover-up, rubber gloves, or a fiber conductor cover.

To better understand the behavior of a voltage surge, it can be compared to the action of a wave of water as shown in Figure 5–31 and described in Table 5–5.

Circuit Protection from Overvoltage

5.7.12 The basics of overvoltage protection are to intercept the overvoltage and conduct it to the earth, where it is dissipated. Circuits are protected from overvoltage by surge arresters, insulation, shield wire, and good system grounding. When a transient overvoltage exceeds the sparkover value of a surge arrester, the arrester drains the energy from the system before insulators and equipment insulation flash over or are punctured.

Protection from overvoltage includes providing insulation on equipment and lines that can withstand a voltage considerably higher than the voltage rating of the circuit. Shield wire, which is overhead ground wire, protects circuits and stations from direct strokes of lightning. Shield wire strung high above a circuit or station is more attractive to lightning than the power conductors and equipment below. Shield wire is grounded at every structure and will discharge a voltage surge to

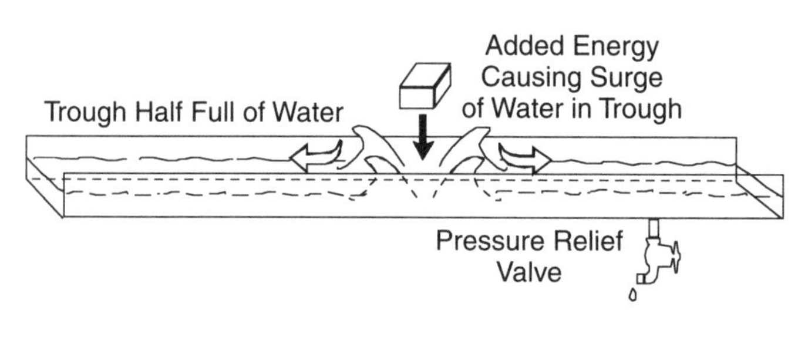

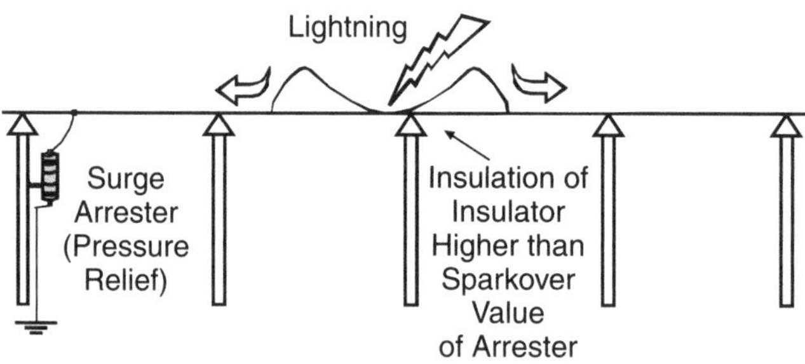

Figure 5–31 Voltage surge/wave action.

ground. Good system grounding provides a low-impedance path for a voltage surge that has gone to ground at an arrester to dissipate into the earth.

Surge Arresters

5.7.13 A *surge arrester* is a device that provides a path for the surge current associated with a voltage sparkover to discharge to ground. After the voltage surge is diverted to ground, the arrester chokes off the follow-through current, and the arrester is restored and ready for the next surge.

A normal voltage will not discharge to ground across a gap or many gaps inside the arrester. Surge arresters are designed to operate at a voltage higher than the circuit voltage but lower than the voltage that would cause an insulator to flash over or a transformer to be damaged. The arrester must operate in time to avoid damage to equipment or, in other words, coordinate with the withstand curves of equipment needing protection.

Types of Surge Arresters

5.7.14 An arrester provides a sparkover location for a voltage surge, provides a path for a high-surge current flow to ground, and stops the current flow to ground before it becomes a permanent ground fault. Some types of surge arresters are listed, but there are continuing improvements being made.

A *spark gap* or rod gap is installed between the line and ground to discharge a surge and protect insulators (Figure 5–32). This type of protection is found mostly on transmission lines. Once an arc is established, however, the ionized air becomes a conductor and the arc continues until the circuit trips out. This continuing arc is a fault current called the power-follow current. The spark gap will erode and be damaged by a prolonged arc.

Table 5–5 Comparison of Voltage Surge to Water Surge

Surge Activity	Water System	Electrical System
Initiation of Surge	An outside energy source such as a brick is thrown into the water.	An outside energy source such as lightning strikes a line.
Speed of Energy Wave	The speed of a wave of water can be calculated with a relatively complicated formula. The speed of a wave of water is relatively slow.	The speed of a voltage surge through a powerline is at the speed of light.
Surge or Wave Impeded	A wave of water is impeded by: • The size and shape of the trough or conduit holding the water • The frictional resistance of material making up the trough	A large conductor offers less impedance than a small conductor.
Surge or Wave Action at Dead End	The wave is reflected off the dead end and is doubled in magnitude.	The wave is reflected off the dead end and is doubled in magnitude. It diminishes in a relatively short distance.
Uncontrolled Surge Discharge	Water spills over the edge of the trough.	Voltage surge spills over or sparks over at location with least insulation.
Controlled Surge Discharge	A pressure-relief valve reduces the pressure of a surge in a water system. The amount of discharge is dependent on the spring pressure of the pressure-relief valve and the size of the discharge pipe.	A surge arrester with a voltage rating lower than the system insulation provides a path for a voltage surge to be dissipated into the ground. The amount of discharge depends on the resistance of the path to ground and the resistance of the ground.
Dissipation of Surge	Wave action on the edge of the trough reduces the size of the main wave.	A high-voltage surge on a small conductor causes some dissipation of the curve due to corona and skin effect. A high-voltage surge dissipates when it couples with a nearby phase, neutral, or ground.

Figure 5–32 Spark gap surge protection.

A *valve type arrester* is designed to limit a power-follow current after passing the voltage surge to ground. A valve-type arrester is made up of air gaps and a special resistor called a *valve element.* The valve element in the arrester will pass a lightning surge to ground, but it has a high resistance to 60-hertz power and stops any 60-hertz power-follow current.

In a *metal-oxide varistor,* the gaps are not air but layers of highly resistive metal-oxide additives. They do not need air gaps to prevent a normal voltage from going to ground. This material is able to pass a voltage surge without damage. The material is nonlinear, meaning that it sets up a reactance that prevents the voltage and current from being at their peak value at the same time, therefore limiting the energy being dissipated through the arrester.

A *deadfront-type arrester* is an arrester assembled in a shielded housing, such as an elbow, and installed for the protection of underground and pad-mounted distribution equipment and circuits.

Classes of Surge Arresters

5.7.15 Valve-type surge arresters come in four classes:

1. A *station-class* arrester provides the highest degree of protection. It is used in transmission substations because it is able to withstand a very high surge current.
2. An *intermediate-class* arrester is used in distribution substations and on subtransmission circuits where the high quality and cost of a station-class arrester are not justified. The difference between the circuit voltage and the sparkover voltage of the intermediate-class arrester is narrower than it is with the distribution-class arrester. This provides better protection for the equipment the arrester is protecting.
3. A *distribution-class* arrester is used on distribution transformers and other line equipment to provide a reasonable balance between protection and cost. Distribution-class arresters can be:
   - *Heavy-duty class,* used to protect overhead distribution systems exposed to severe lightning currents
   - *Light-duty class,* used to protect underground distribution systems where the major portion of the lightning-stroke current is discharged by an arrester located at the riser pole
   - *Normal-duty class,* used to protect overhead distribution systems exposed to typical lightning currents
4. *Secondary-class* arresters are used for the protection of secondary services.

Insulation Coordination

5.7.16 The design of an electrical system includes using insulation that meets the standard insulator *basic-impulse level* (BIL) for the voltage level on the system. The voltage level that can flash over the leakage distance of an insulator is the BIL. A straight-line distance through an insulator is much shorter than the distance over the outside. The longer distance over the outside is the leakage distance.

Overvoltage protection involves ensuring that a voltage surge will not exceed the BIL rating of the equipment or line it is protecting. For example, in a transmission station, the BIL for a 230-kilovolt system could be 900 kilovolts. Every insulator, breaker, and transformer is coordinated to have a BIL of 900 kilovolts. The 230-kilovolt portion of the station can, therefore, withstand up to a 900-kilovolt surge without a flashover. A surge arrestor would have the lowest BIL in the station; therefore, it controls the location where a high-voltage surge would flash over and be dissipated safely to ground.

Insulation and surge arresters on distribution feeders are specified in the same way. A 25/14.4-kilovolt system with a BIL of 95 kilovolts means that all the insulation and equipment should withstand 95 kilovolts. A surge arrester would have a voltage rating of less than 95 kilovolts but more than 14.4 kilovolts. On a 25/14.4-kilovolt system, an arrester would have a voltage rating of 18 kilovolts.

Equipment Voltage-Time Curves

5.7.17 Equipment such as a transformer has a BIL voltage-time curve, which represents the amount of voltage and time a transformer can withstand before it is damaged. Equipment is manufactured to meet a BIL voltage-time curve so that a standard surge arrester can be used to protect it. Figure 5–33 shows the BIL voltage-time curve for a transformer and how the curve of a voltage surge, shunted to ground by a surge arrester, is well below the level that would damage the transformer.

Shorting Insulators for Live-Line Work

5.7.18 When working on a live transmission line, the BIL level at an insulator string will be reduced by tools, insulated strain links, and live-line rope used by the line crew. On a compact-designed transmission structure with smaller spacings between conductor and ground, live-line tools have an even greater likelihood of lowering the BIL. A location with a lower BIL is more vulnerable to a flashover if there is a voltage surge. Although a voltage surge due to switching or lightning is extremely unlikely to happen while live-line work is carried out, protection from a possible flashover can be taken.

Prevention of voltage surges at the job site is improved by having a lightning warning system in place. Getting clear of the line when there are thunderstorms in

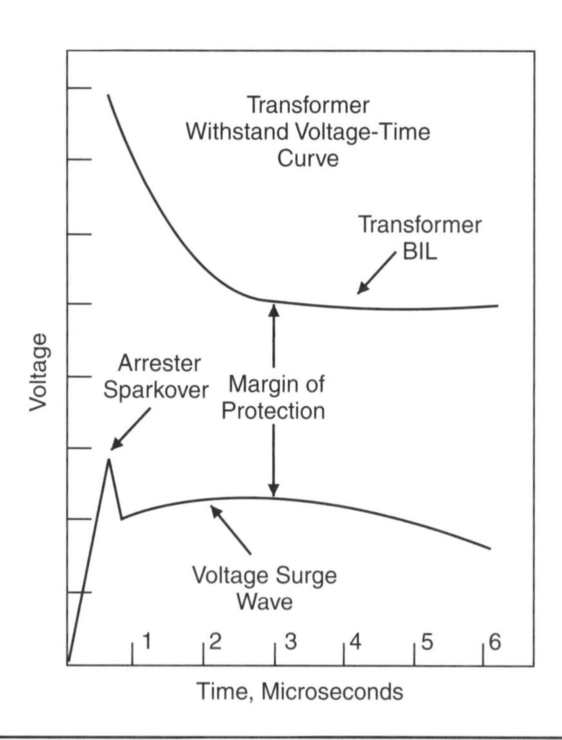

Figure 5–33 Transformer withstand voltage-time curve.

the area eliminates exposure to the greatest cause of voltage surges. To avoid a switching surge, the reclose feature of the circuit breaker is blocked. A blocked recloser avoids the switching surges involved with reenergizing the circuit if it is tripped out for some reason.

To reduce the likelihood of a flashover at a work site, the BIL of an insulator string at an adjacent structure can be lowered. Any voltage surge that occurs while working on the circuit would then flash over at the adjacent structure and not at the work location.

Lowering the BIL of an insulator string at an adjacent structure can be accomplished by shorting out some insulators or by installing a *portable protective gap* (PPG). A PPG is a portable device that can be set to provide a spark gap for a given value of voltage surge and shunt it to ground.

Surge Protection at Problem Locations

5.7.19 Some circuits and locations are exposed to more lightning than others. Additional steps can be taken to reduce outages to circuits prone to lightning strikes:

- On transmission lines where there is already a shield wire, improvements to grounding of the shield wire dissipate the effects of lightning quicker.
- On problem subtransmission and lower-voltage transmission lines, lightweight, polymer, metal-oxide arresters are installed on all three phases at the most vulnerable locations. In some cases the arresters are installed on every pole or tower. One arrester on each of the three phases is better than a single arrester on the center phase because, when lightning strikes one phase, the three arresters allow all three phases to go to the same voltage, preventing flashing (backflash) from one phase to another. These types of arresters have a hotline clamp on the top and a grounded isolator on the bottom and can be installed live line.
- Stringing a shield wire in a problem-plagued location will give additional voltage surge protection, although it is more expensive than installing arresters on each structure and phase.
- Problem distribution circuits can have surge arresters installed in more locations.
- Improving the grounding of the system neutral reduces a voltage surge quicker because of the coupling effect that occurs between the neutral and the phase conductors during a surge.
- On individual equipment such as a transformer, surge arresters should be installed as close as possible to the equipment. The ground wire from the arrester should go as directly as possible to ground.
- A transformer on a dead-end pole is more vulnerable to lightning damage because of the doubling of voltage when the lightning comes back from the dead end. Stringing the line one span farther is sometimes done in problem locations.

5.8 System Grounding for Protection

Importance of System Grounding

5.8.1 System grounding a power system is electrically connecting the neutrals of every wye-connected transformer or generator to earth. The importance of grounding an electrical system is not always obvious. Under normal conditions, an electrical system delivers power without good grounding. It is during a voltage surge, an unbalanced-load condition, a short circuit, or an equipment failure that a dangerous voltage rise occurs on the grounding system. The voltage rise occurs on the neutral and everything connected to it. Good grounding is necessary to protect the public and workers from a dangerous voltage rise at places where the system is grounded, such as at a downground on a pole.

Purposes of System Grounding

5.8.2 Good grounding provides people protection and circuit protection.

People Protection

Grounding an electrical system limits the rise in potential on the neutral, metal structures, non-current-carrying electrical equipment, and everything electrically connected to ground. When equipment is connected to a low-resistance ground, a voltage surge is kept to a lower level and will dissipate more quickly.

Circuit Protection

Good grounding improves the likelihood of a circuit breaker, recloser, or fuse tripping out the circuit for a phase-to-ground fault. Most faults on an electrical power system are line-to-ground faults. A well-grounded electrical system provides a better return path for the fault current to flow back to the source to complete the circuit for the fault current.

System Grounding for People Protection

5.8.3 When a line or substation is built, the installation of a grounding and bonding network is essential for safe operation. The design of electrical installations includes system grounding so that a person touching any equipment during a fault condition is not subjected to a dangerous current or potential.

- The installation of a ground-gradient control grid reduces step and touch potentials around equipment such as a pad-mount transformer or a switching kiosk.
- The use of crushed rock around equipment, especially in a substation, increases electrical resistance under the feet of workers.
- Standing on a ground-gradient control mat that is bonded to a switch handle protects the switch operator from electric shock if the handle becomes alive during a switch failure. With the mat bonded to the switch handle, the operator's feet and hands are at the same potential, which means that there will be no current flow.
- Grounding the system neutral at frequent intervals provides a good return path for current from a line-to-ground fault. The current from a fault would have a relatively short path back up to the neutral.

- During a fault condition, there is a voltage rise in the ground grid and everything attached to it. To limit the travel of this potential rise, station fences are isolated from the ground grid. Any railway spur that enters a station has one section of rail removed.
- To ground electrical equipment means to connect transformer tanks, metal-clad equipment, and so on, to earth. The main purpose is to reduce shock hazards by limiting the potential difference between the grounded equipment and earth. It is also important to electrically bond all tanks of electrical equipment together to prevent someone from getting in between and becoming a path from one piece of equipment to the other.

Ground versus Neutral Connections

5.8.4 On an equipment installation, ground wires and their connections should not be confused with neutral wires and their connections. A *neutral* is part of the electrical circuit and is a current-carrying conductor that provides a path for current to flow back to the source. If a neutral connection on energized equipment is opened, there is a voltage equal to the circuit voltage across the open point.

A *ground wire* grounds the equipment tank and surge arrester lead-to-earth potential. During a fault or a voltage surge, a good ground reduces the potential rise on the equipment and prevents equipment failure or flashover.

An equipment installation may appear to have a ground wire and a neutral connection doing the same job, but skipping one of the connections can cause a dangerous potential rise or service problem. Figure 5–34 shows a typical transformer-grounding installation with one major oversight. The primary neutral of the transformer was not connected to the system neutral. A person working on the ground was electrocuted when he tried to repair a damaged ground-rod connection. The down ground was part of the primary circuit, and, when it was opened, a primary-voltage difference appeared across the opening.

The Neutral as Part of the Grounded System

5.8.5 For a circuit to be complete, all the current leaving the source must return to the source. In a balanced three-phase circuit, the return flow occurs in the phase conductors. Unbalanced three-phase and single-phase wye-distribution circuits use the neutral for current to find its way back to the source.

The neutral also provides a return path for a fault that goes to ground. On a well-grounded, multigrounded neutral, about two-thirds of the current returns to the source through the neutral and the remaining current returns through ground.

Neutral Potential

5.8.6 From Ohm's Law ($E = I \times R$), it can be seen that if there is a current flow, there is a voltage. There is always some current on a distribution neutral. A neutral potential, which is measured between the neutral and a remote ground, should be kept at less than 10 volts.

A neutral potential will be lowered if the current flowing through it can be reduced. A neutral is the return path for any unbalanced current between phases. The current on the neutral can be reduced by converting heavily loaded single-phase circuits to three phases and balancing load between the three phases. A

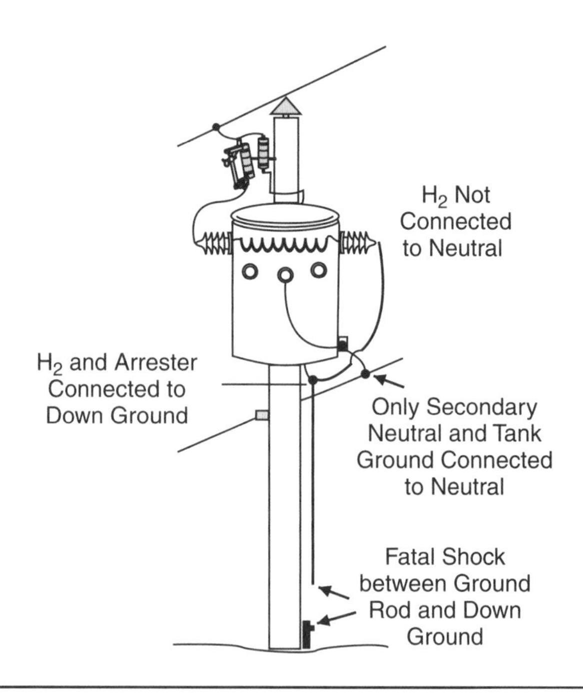

Figure 5–34 Transformer grounding accident.

neutral potential can also be lowered by installing more grounds in good earth, in many locations along its length.

If the resistance of a neutral is lowered, the potential is lowered. Ohm's Law shows that less resistance means less voltage. Neutral resistance can be reduced by stringing a larger conductor and ensuring that all neutral connections are in good condition.

Note: Under a line-to-ground fault condition, the voltage on the neutral can rise to many kilovolts.

Grounding for Overcurrent Protection

5.8.7 A line-to-ground fault is by far the most frequent type of fault. When a line-to-ground fault occurs, the protective switchgear trips out the circuit if the current feeding the fault is high enough. To draw a fault current high enough to trip the circuit, a return path to the source must also be able to carry the same amount of current. The return path is the neutral and earth.

Frequent grounding of the neutral in good earth aids the fault current to find its way back to the neutral and the source. When a conductor falls in a location that is not well grounded, such as rocky ground, sandy soil, or dry snow, there is not enough fault current generated to trip the circuit because the resistance of the return path through the earth is too high. Every powerline worker, however, has seen circuits fail to trip out when they should have. In these cases, the culprit is poor grounding or an outdated protection scheme.

Inserting an Impedance in the Ground Return

5.8.8 At some substation transformers, the neutral point is not grounded directly to earth. A resistor or reactor is placed in series with the connection to earth. The fault-current levels on high-voltage distribution feeders from some substations can be very high and very damaging. Inserting a resistor or reactor into the return circuit from ground has the effect of reducing the fault generated from a line-to-ground fault in the whole circuit.

Surge Protection

5.8.9 The dissipation of a voltage surge is greatly improved through good grounding. Circuits in a location with a poor ground are more vulnerable to outages due to lightning.

When a voltage surge occurs on a phase conductor, the neutral conductor acts as a coupling wire. The voltage on the neutral rises along with the voltage on the phase conductor. This coupling lowers the potential difference across insulators and equipment. The coupling of a well-grounded neutral or shield wire also helps lower the voltage of the surge more quickly. A good grounding design dissipates a voltage surge.

- The ground resistance of a driven ground rod should be low, preferably below 25 ohms.
- A grounding wire should be short because there is a big voltage drop during the high current due to a fault condition. The added resistance of a long lead will add substantially to the voltage drop.
- A ground lead should be as straight and direct as possible because a high-voltage surge will jump across sharp bends.
- A surge arrester should be well grounded and as close as possible to the equipment being protected.

Good Ground

5.8.10 The resistivity of earth is measured in ohms per meter. An ohm per meter is the resistance between the opposite faces of a cubic meter of soil. The resistance of earth varies with the type of earth, moisture content, and temperature. Table 5–6 shows the great variance found in soil resistance.

Soil with no moisture content would be an insulator. The mineral salts of the earth, dissolved in water, give the soil low resistance. A ground rod should be driven straight down, where there is more likelihood of moisture.

The temperature of the earth affects its resistivity. When the temperature decreases, the resistivity of the earth increases. When the moisture in the earth freezes, the resistance increases dramatically. A ground rod must be driven below the frost line to be effective in the winter. Frozen ground around part of a ground rod can double or triple the resistance.

Measuring Ground-Rod Resistance

5.8.11 Measuring the resistance of a ground rod can confirm that an installation meets design requirements. For example, an acceptable resistance for a ground rod at a transformer installation is typically 25 ohms. Ground-resistance measurements are also carried out when investigating calls such as tingle voltage, high-neutral voltage, excessive vulnerability to lightning, or failure of protective switchgear to operate.

An earth-resistance tester (Figure 5–35) is a specialized ohmmeter used to measure the resistance of a specific ground rod. A relatively high voltage is impressed on the ground rod, and any potential and current is measured at a potential probe and at a current probe far enough away to represent remote ground.

TABLE 5–6 Typical Soil Resistance

Earth Type	Resistivity (Ohms per Meter)
Sand Saturated with Sea Water	1 to 2
Marsh	2 to 5
Loam	5 to 50
Clay	5 to 100
Sand/Gravel	50 to 1000
Sandstone	20 to 2000
Granite	1000 to 2000
Limestone	5 to 10,000

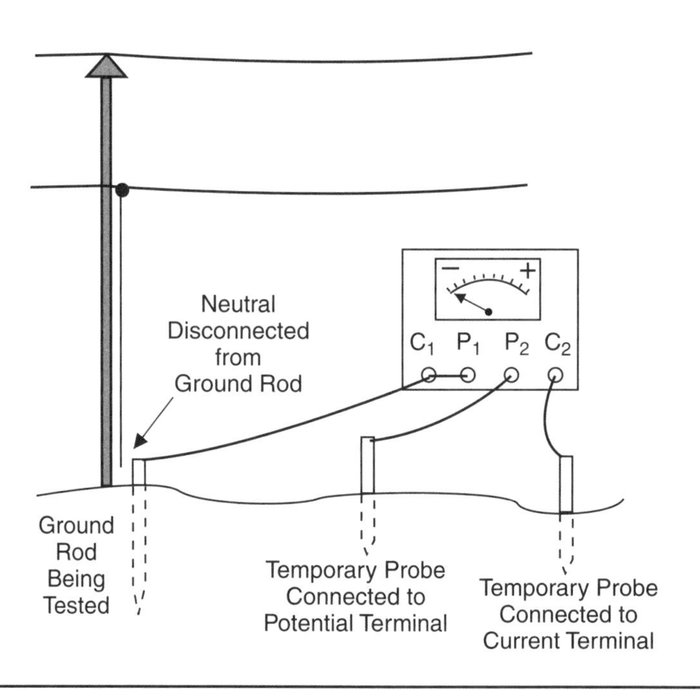

Figure 5–35 Measuring ground rod resistance.

Review Questions

1. What two main types of switchgear are used in an electrical system?
2. How does a relay receive a signal about a change in voltage and current on a high-voltage circuit?
3. Why is a good multigrounded neutral important for circuit protection on a wye system?
4. Why does a dry tree limb or a broken conductor lying on dry or frozen ground not always trip out the circuit protection?
5. Can a load-interrupter switch interrupt a fault current?
6. When a disconnect switch is operated, a breakage may cause live leads to contact the switch frame. What is the best protection for an operator, when standing on the ground, to open a three-phase gang-operated switch with an operating handle?
7. Why can a cutout not be opened with a switch stick to interrupt load?
8. May a sectionalizer be used by a lines crew to interrupt load?
9. Where is the safest place to be during a thunderstorm?
10. Name three ways to reduce the risk of a voltage surge on a transmission line while carrying out live-line work.
11. What is the difference between connections to a neutral and connections to ground?
12. How does good grounding of the neutral improve overcurrent protection on a circuit?

CHAPTER 6

Working in an Electrical Environment

6.1 Introduction

Working on or Near Electrical Circuits

6.1.1 When working on or around electrical circuits, people are exposed to electrical influences from sources that are not always obvious. The topics in this chapter apply electrical theory knowledge to the various tasks involved when working in an electrical environment. An increased recognition of a potential electrical hazard will allow for safer work in an electrical environment.

Making Contact with an Electrical Circuit

6.1.2 A person making contact with a live line becomes an electrical load that is connected in series with the circuit or connected in parallel with the circuit. The effect that voltage and current have on a load connected in parallel is different from a load connected in series.

6.2 Connecting a Load in Series with a Circuit

What Is a Series Circuit?

6.2.1 Two or more loads within a circuit are considered to be *in series* when there is one (common) current flowing through all the loads. When the current has only one path (Figure 6–1) to take through two or more electrical loads, then the loads are in series with each other. In other words, when a component is connected end to end into a conductor, it is connected in series.

Characteristics of a Series Circuit

6.2.2 In a series circuit, an equal amount of current runs through the complete circuit and all connected loads. An ammeter reading would show the same value any place in the circuit. If there is any change made to the current flow in any part of the circuit, the change to the current applies throughout the circuit.

$$I_{Total} = \textit{the current through any part of the circuit}$$

In a series circuit, there is a voltage drop across each load. The voltage drop across each load is dependent on the resistance of the load and can be calculated using Ohm's Law ($I \times R$). A large load or resistance would have a large voltage drop across it.

A break in the circuit would introduce an infinite resistance to current flow and would result in a total voltage drop. This means that the total circuit voltage (recovery voltage) is available across the break. It is, therefore, a basic lines-trade procedure to always jumper any conductor that is carrying current before it is cut.

The sum of all the individual voltage drops is equal to the applied voltage of the circuit. The voltage drops some as it goes through each load in the circuit and is "all used up" when it gets back to the source.

$$E_{Source} = IR_1 + IR_2 + IR_3 + IR_{etc.}$$

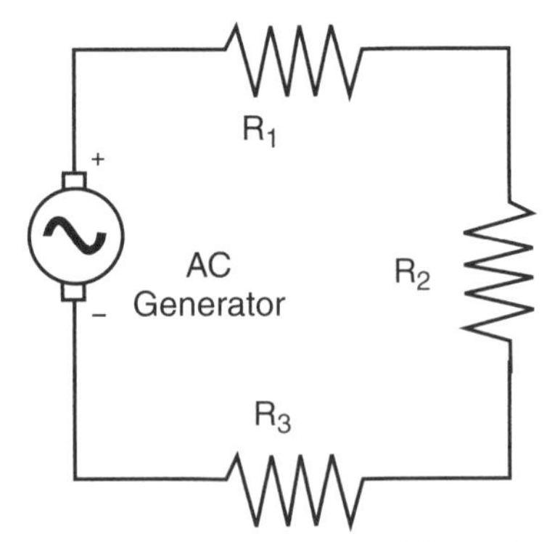

Figure 6–1 Loads connected in series.

The total resistance in the circuit is equal to the sum of all the individual resistances within the circuit.

$$R_{Total} = R_1 + R_2 + R_3 + R_{etc.}$$

Loads Placed in Series

6.2.3 There are few applications of series-connected loads in an electrical utility. There are still series street-lighting circuits in some utilities. The supply transformer keeps the current constant throughout the circuit, and the voltage varies depending on the number and size of the lights in the circuit.

There are series-connected loads at the utilization level, but they are mostly within electrical equipment such as appliances and electronic equipment. Some strings of Christmas-tree lights are connected in series. In a series-connected string of lights, when the element of one bulb fails, it opens the circuit and the lights go out in the whole string. The input voltage is equally divided over all the bulbs in the string. For example, there are strings with 10 bulbs, 25 bulbs, or 40 bulbs; the voltage rating of the bulbs from a 120-volt source is 120 divided by the number of bulbs in a string. Similarly, a person with a 240-volt service could connect two 120-volt lights in series to allow the use of standard 120-volt bulbs.

Making Series Connections

6.2.4 The line trade is involved in connecting equipment such as switches, reclosers, and voltage regulators in series within the circuit. Anytime a person is involved with cutting or joining a conductor, there is the hazard of a person being able to get into series with the circuit. When a power conductor is cut, all of the current flow is interrupted. The resistance across the open point is infinite; therefore, the voltage drop (IR) is at full-line voltage. This full-line voltage (recovery voltage) appears across the two ends of the cut conductor. Installing a jumper across an open point keeps the current flowing and keeps the potential across the cut at zero volts.

The Human Body in a Series Circuit

6.2.5 A person's body can complete a circuit when it bridges across an open point within the circuit. When a body completes a circuit, it is exposed to the full-line voltage of the circuit and all the current the voltage is able to drive through the resistance of the body.

Every time a person makes a cut or joint in a current-carrying conductor, there is danger of accidentally putting oneself into series with the circuit. Even on an isolated and grounded circuit, there can be enough current flow in the conductors to cause a lethal shock. Only body resistance, clothing, and gloves limit the current flow when a body is put into series with the circuit. If the voltage is high enough across the open point, this resistance will be overcome, a person's body will complete the circuit, and a current flow will be established through the body.

Accident Examples Involving Series Contact

6.2.6 Accidents occur when a person's body completes a circuit:

- A powerline worker received electrical burns when he opened the neutral of a live circuit without installing a jumper bypass across the open point. His body was in series with the neutral conductor. This occurred even though the neutral was multigrounded.

- A powerline worker received electrical burns when he removed a loop on a grounded transmission line. There were grounds installed on each side of the loop, but there was no bypass jumper installed.
- Attaching oneself to (getting in series with) a live secondary bus with one hand and an isolated service drop with the other hand caused a powerline worker to be "frozen" on. The voltage across the open point was equal to the service voltage.
- Forgetting to install a jumper before cutting a conductor while doing rubber-glove work or bare-hand work results in an electrical flash and exposure to a lethal voltage across the open point.

Protection from Series Contact

6.2.7 There would not be an open point in a circuit, and therefore no voltage across an open point, if a jumper was installed before cutting or joining a conductor. When a jumper is installed or removed using live-line techniques, a person avoids bridging across an open point.

It should be a normal habit for powerline workers to make or break a connection without having hands on each side of the open point. For example, when making a connection, the worker should have both hands on one wire until contact is made between the two wires. Even with a jumper in place, workers still should never put their hands across an open point, as a matter of habit, just in case some earlier step has been forgotten. Cover-up or an insulated aerial bucket provides protection from electric shock in cases of accidental contact. However, a bucket or cover-up will provide no protection from contact made by hands across an open point. Wearing rubber gloves provides protection, but there is no backup protection if the only thing preventing a current flow from hand to hand is the integrity of the rubber gloves.

6.3 Connecting a Load in Parallel with a Circuit

Making Parallel Contact

6.3.1 A powerline worker is frequently in a parallel path to ground when working on a wye circuit. For example, when working on a live conductor with rubber gloves, a person is in parallel with other loads on the circuit. The rubber gloves may prevent a dangerous current flow through the worker. On a delta circuit, a person would need to make phase-to-phase contact to make a parallel connection with other loads on the circuit.

What is a Parallel Circuit?

6.3.2 When there is more than one path for current to flow through (Figure 6–2), the loads in each path are connected in parallel with each other. The current divides into each branch of the circuit. The magnitude of the current in each branch depends on the resistance within that branch.

Characteristics of a Parallel Circuit

6.3.3 The same voltage appears across each path connected in parallel because each path is connected between the same two wires supplying the source voltage. Each branch is independent of the others. Each branch has the same voltage, but the current and resistance can change in a branch and it will not affect another branch.

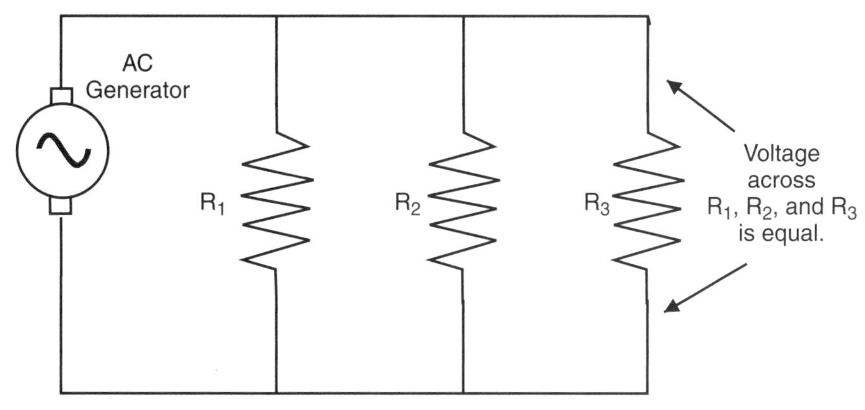

Figure 6–2 Loads connected in parallel.

$$E_{Source} = E_1 = E_2 = E_3 = E_{etc.}$$

The total current in a circuit with its loads connected in parallel, is equal to the sum of the currents flowing into each path.

$$I_{Total} = I_1 + I_2 + I_3 + I_{etc.}$$

The current within each path is dependent on the resistance or load in each path. The lower the resistance in a parallel path, the greater the load current will be in that path. Therefore, the path with the least resistance is also the path with the greatest load current. Most current takes the path of least resistance, but *each* path will have *some* current flow through it. The magnitude of the current through each path is inversely proportional to the resistance. The total current in the circuit is always more than the current in the branch with the least resistance.

The total resistance of a parallel circuit is less than the smallest resistance in the circuit. Expressing this in terms of load current, which is the inverse of resistance, the total load in a circuit will always be greater than the largest load.

A short-circuit fault is the lowest-resistance load in a circuit and, therefore, attracts the most current flow. However, every other element connected in parallel still takes a portion of the current.

Note: A load such as a transformer on an a-c circuit has both resistance and reactance that makes up the total impedance to a current flow. The terms resistance and impedance are often used interchangeably, although technically incorrect. The term resistance is used here even though the examples are showing a-c circuits, because powerline workers use the term resistance more often than impedance.

Calculating the Total Resistance of a Parallel Circuit

6.3.4 The total resistance of a parallel circuit is not the sum of the resistance as in a series circuit. The total resistance is calculated using the formula

$$\frac{1}{R_{Total}} = \frac{1}{R_1} + \frac{1}{R_2} + \frac{1}{R_3} + \frac{1}{R_{etc.}}$$

Example Calculation

In Figure 6–3, three transformers are connected in parallel. Calculate the equivalent or total resistance of the circuit, given a source of 5000 volts and three parallel branches with a resistance of 250, 500 and 100 ohms.

Substituting the resistance values into the formula

$$\frac{1}{R_{Total}} = \frac{1}{250} + \frac{1}{500} + \frac{1}{100}$$

$$\therefore \frac{1}{R_{Total}} = \frac{2}{500} + \frac{1}{500} + \frac{5}{500}$$

$$\therefore \frac{1}{R_{Total}} = \frac{8}{500}$$

$$\therefore R_{Total} = \frac{500}{8} = 62.5 \text{ ohms}$$

Note that the total resistance of this parellel circuit is smaller than the resistance in any one branch. The larger transformer has a larger load current but a smaller resistance.

Loads Connected in Parallel

6.3.5 Almost all loads are connected to an electrical system as parallel-connected loads. Loads on a distribution feeder are fed through transformers. All transform-

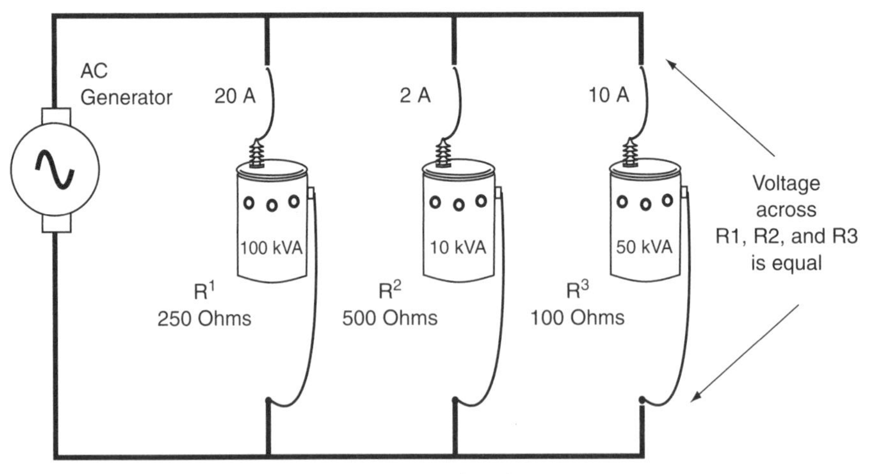

Figure 6–3 Calculating total resistance in a circuit.

ers on a feeder are connected in parallel to each other. All transformers on a feeder are connected to the same voltage, and the current flowing into each transformer is inversely proportional to the load.

Loads at the utilization level in homes and factories are connected in parallel. When an appliance is plugged into a socket, it is connected to the same voltage as other appliances plugged into that electrical service.

Leakage Current

6.3.6 In a wye circuit, anything that is in contact with both the ground and a live conductor is a parallel connection and has some current flow through it. Even high-resistance objects, such as insulators, have some current flow through or along them. This small leakage current is in inverse proportion to its resistance and is measured in microamperes.

Live-line tools, rubber gloves, and insulated booms are all subject to some leakage current. The care taken to keep these tools clean and in good condition is to keep the leakage current below a certain threshold.

Leakage Current Working Barehand

6.3.7 When working barehand from an aerial basket, the truck boom is a parallel path to ground. The resistance of the boom is very high, but there is still some current flow through the boom to ground. A boom contamination meter is used to confirm that the leakage current is at an acceptable level. The leakage current is never at zero. A typical acceptable leakage current for an insulated boom is less than one microampere for each 1000 volts to ground.

A person in an aerial basket is exposed to even less current because the metal grid below the feet is bonded to the conductor. In other words, the potential of the hands is the same as the potential below the feet. On high-voltage circuits, a conductive suit is worn to reduce the uncomfortable voltage gradients around the body and to keep all of the body at the same potential.

The Human Body in a Parallel Circuit

6.3.8 Anytime a person is in parallel with any load in a circuit, that person is exposed to some voltage and some current regardless of the amount of resistance being offered to the current. A person making contact with a live conductor is in parallel with other loads on the circuit. Even when working from an insulated basket, there is leakage current flowing down the rubber gloves and down the insulated boom. The current flow is very small, probably less than 10 microamperes, and it is below the threshold where one would feel anything.

If a person fails to wear rubber gloves or to reduce the resistance of the electrical path through the body in any way, then the body will take a greater share of the current flow. There is always some current flow when making parallel contact. The only defense from a lethal current is to keep the path through a human body at a high resistance.

Accident Examples Involving Parallel Contact

6.3.9 Accidents occur when a person's body is a parallel path to a current flow:

- A powerline worker received a lethal shock when he made an inadvertent contact with a live circuit while standing on a pole waiting for the circuit to be isolated. The 4800 volts in the line were high enough to overcome the resistance of his leather work gloves and the wooden pole. The amount of current going through him, as a parallel path to ground, was well above the usually fatal 100 milliamperes.

- An uninsulated vehicle boom made contact with the bottom of a transformer cutout while a powerline worker was getting material from the truck bin. The truck was grounded to a temporary ground probe but the circuit did not trip out. He was a parallel path to ground and received fatal electrical burns.
- While a powerline worker was opening a gang-operated switch, the operating handle became alive when an insulator broke and a live lead dropped onto the steel framework of the switch. The powerline worker was standing on a ground-gradient mat that was bonded to the operating handle and his back was in contact with some brush growing next to the ground-gradient mat. The brush was a parallel path to a remote ground, and the powerline worker received electrical burns on his back.

Protection from Parallel Contact

6.3.10 Because there is always some current flow when a parallel contact is made, a person is protected by keeping the current flow through the body to ground below a dangerous threshold. Current cannot flow unless there is a circuit. A person must prevent a circuit from being completed through his body. For example, a person in contact with a single live conductor, while working from an insulated aerial device, is part of an open circuit. Other than a small leakage current, the current has no place to flow to because there is no second point of contact. Removing or keeping away from a second point of contact is a key element in all live-line work procedures.

Current flow through a person is kept below dangerous values when resistance is increased between a live conductor and the second point of contact by the use of rubber gloves, live-line tools, cover-up, and insulated platforms. Current flow through a person's body is kept below dangerous values when there is no potential difference between any part of the body and another part of the body. The use of a ground-gradient control mat when switching or operating stringing equipment ensures that the hands will stay at the same potential. When there is no potential difference, there is no current flow.

6.4 Electrical and Magnetic Induction

Three Electrical Effects without Any Contact

6.4.1 There are three different ways for a person to be affected by electrical phenomena without being in contact with a live conductor. All people, not just utility workers, are exposed to *static electricity.* Static electricity is noticed when one touches a doorknob after walking across a carpet. The spark to the doorknob is a DC charge, which is quite different from the type of induction experienced working near live AC circuits.

An object or person near a live AC circuit is exposed to *electromagnetic induction.* Electromagnetic induction is comprised of *electric-field* induction and *magnetic-field* induction.

Static Electricity

6.4.2 Although static electricity is not the same as the induction experienced by electrical-utility personnel working near live lines, it is a phenomenon that can be a hazard in some work settings. Static electricity is a DC charge. It is called static because it is a stationary electric charge. This charge can be generated by rubbing two nonconducting substances together. When a static charge is discharged, there is a current flow between a charged object and the body that makes contact.

Static buildup occurs on fast-moving conveyor belts and when removing synthetic clothing, walking on carpets, and sliding out of a car seat and contacting the metal door. A spark from static electricity can cause a fire, especially when the spark occurs in the presence of flammable vapor, dust, or gas. Some ways to eliminate fires due to static include:

- The hoses at gasoline pumps are conductive to prevent a static discharge.
- When fueling a helicopter, a bond is installed between the fuel pump and the helicopter.

Electric-Field (Capacitive) Induction

6.4.3 An isolated, ungrounded conductor strung near a live conductor has an induced voltage on it. It is called *capacitive induction* because, as illustrated in Figure 6–4, the live conductor acts as one plate of a capacitor, the isolated conductor acts as the other plate of the capacitor, and the air between is the insulation between the two plates.

The isolated conductor has a voltage on it induced from the live conductor. The isolated conductor is not part of a circuit if the switches are open at each end, and, therefore, no current flows. If the isolated conductor is grounded to earth at one location, the voltage is drained from the conductor and there is a minor amount of current flowing down the ground. The magnitude of the induced voltage is higher when:

- The voltage on the live conductor is high.
- The distance between the two conductors is decreased.

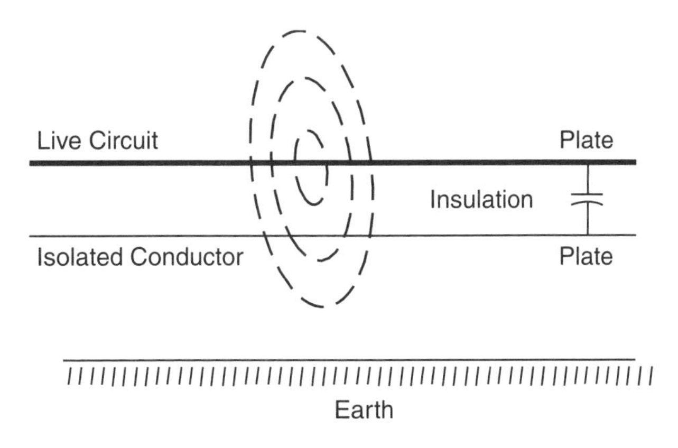

Figure 6–4 Capacitive effect between conductors.

The length of the parallel between the live conductor and the isolated conductor is not a big influence on the amount of voltage induced on the isolated conductor but is a big influence on the amount of current induced on the conductor.

Electric-Field Effect on an Isolated Conductor

6.4.4 The voltage induced on an ungrounded isolated conductor from a neighboring high-voltage circuit can be very high. Contact with such a conductor can result in a lethal steady-state shock that continues as long as contact is maintained. Installing a set of grounds on the conductor discharges the induced voltage effectively. Bonds between vehicles, ground probes, conductors, and so on, eliminate exposure to potential differences among them.

Electric-Field Effect Working in a High-Voltage Environment

6.4.5 When working in a high-voltage environment, such as climbing a transmission tower or working in a high-voltage substation, a person's body acts as the second plate of a capacitor even though the body is not necessarily well insulated from ground. An electric field induces a voltage on a body. The voltage becomes apparent every time a grounded object is touched and the voltage is discharged to ground. The resultant spark consists of a high-voltage but low-current discharge. One hazard from an electric field is an involuntary movement or a fall due to the surprise of a shock. Wearing conducting-sole boots reduces this hazard when working on transmission towers or working in high-voltage stations.

When approaching a live transmission line, the electric field gets more intense as a worker gets closer to a live conductor; in other words, the kilovolts per inch (cm) increase as a person gets closer to the conductor. When working bare hand, a line worker is shielded from the high intensity electric field with a conducting suit that includes a hood to fit over the hard hat, conducting sole boots, and conducting work gloves. This shield or conductive "blanket" around the worker keeps all parts of the body at the same potential and forms a "Farady Cage" so that any current from the electric field will flow through the conductive cover suit.

Similarly, a person rubber gloving on a high-voltage distribution line, such as 34.5kV, might feel a vibration or "bite" in the rubber gloves. The rubber glove acts as insulation between two conductive plates, the first being the 34.5kV and the second the worker.

Magnetic-Field Induction

6.4.6 An isolated and grounded conductor paralleling a live-current-carrying conductor has current induced on it. This occurs because each conductor acts like a coil in a transformer and the air in between acts like the core of a transformer. Current flowing through a live conductor acts as the primary coil. A nearby grounded conductor acts as the secondary coil.

Induced current can only flow when a circuit has been created in the isolated conductor. As illustrated in Figure 6–5, the installation of two sets of grounds (bracket grounds) on the isolated conductor creates a circuit with current flowing along the conductor, down one set of grounds through earth (or neutral), and up the other set of grounds. The magnitude of the induced current is higher when:

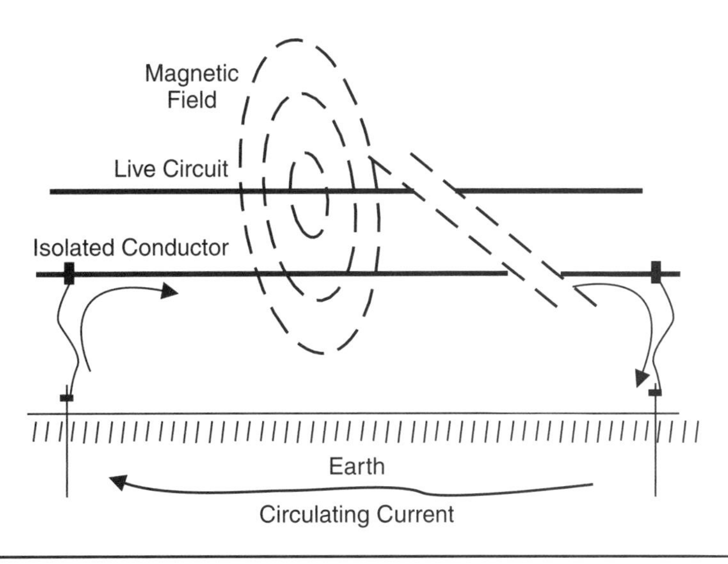

Figure 6–5 Magnetic field effect between conductors.

- The current on the live conductor is increased.
- The distance between the two conductors is decreased.
- The length of parallel between the two conductors is increased.

Because the amount of induction from a magnetic field depends on current and not voltage, this induction is also a hazard on lower-distribution voltages.

Working in an Environment with High Induction

6.4.7 A powerline worker is exposed to induction from a magnetic field when working on an isolated and grounded circuit that is parallel to a live circuit. When only one set of grounds is installed at the point of work, a circuit has not been created for the current to flow. When the conductor is grounded in two locations, a circuit is created and current flows through the conductor and through the grounds. If this circuit is interrupted by removing a ground or cutting a conductor, a high voltage is available across the open point. That is why it is important to install and remove grounds with a live-line tool.

Transpositions

6.4.8 Many transmission and subtransmission circuits constructed before 1955 had *transpositions* installed in them (Figure 6–6). A transposition involved changing the position of the three-phase conductors by crossing a conductor from one side of the structure to the other side. When the different phases crossed each other, the induced voltages on the phases canceled each other out. They were installed because large electromagnetic fields interfered with open-wire telephone circuits. In Figure 6–7, the A phase crosses under the other two phases to go from one side to the other. Three of these transpositions (one barrel) are needed to bring the three conductors back to the original ABC configuration.

Figure 6–6 Transposition on subtransmission circuit.

6.5 Voltage Gradients

Ground Faults

6.5.1 There is a hazard to anyone working at a location where a ground fault results in electrical energy flowing into the earth. A ground fault is a short circuit caused when an object in contact with the earth contacts a live conductor. Examples of ground faults follow:

- Electrical current flows into the earth when a utility pole becomes alive due to contact with a live conductor or because of a faulty insulator.
- Electrical current flows into the earth when a live conductor comes in contact with a tree or a truck boom.
- Electrical current flows into the earth when a portable ground is accidentally installed on a live conductor.
- Electrical current flows into the earth when an underground cable is dug up or punctured.
- Electrical current flows into the earth when a live conductor is lying on the ground.

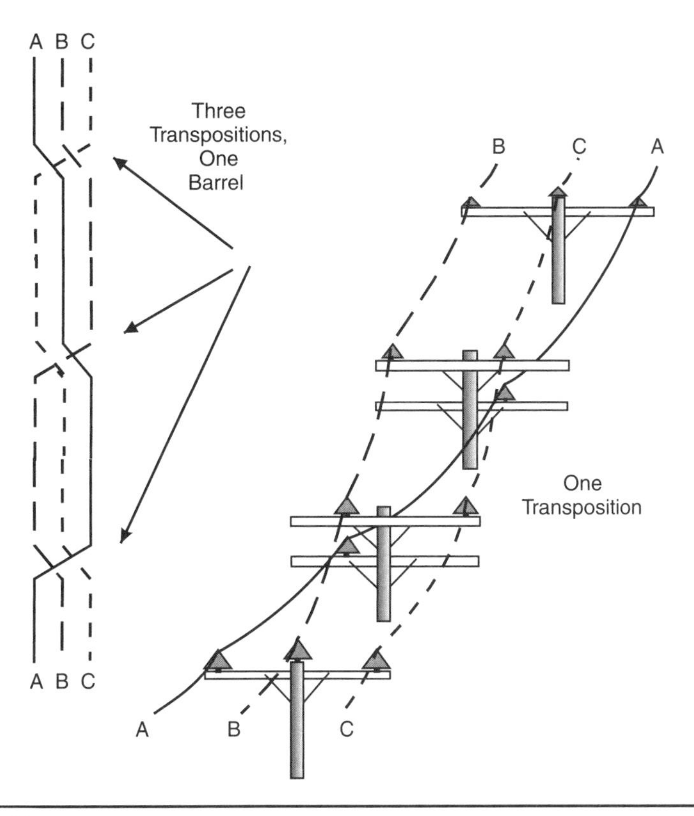

Figure 6–7 Transposition.

- Electrical current flows into the earth when the operating handle of a gang-operated switch becomes alive because of a switch-insulator failure.
- Electrical current flows into the earth at the reel stand or tensioning machine when the conductor being strung contacts a live circuit.

Voltage Gradients

6.5.2 At the point where the current enters the earth, the current breaks up and flows in many paths depending on the makeup and resistance of the earth. The voltage available is highest where the current enters the earth. The earth acts as a network of resistors, and the voltage drops as the current flows through these resistors. The voltage at the current-entry point is higher than the voltage at one or two paces away from the entry point. Therefore, there is a difference of potential in the earth around the current entry point.

It is easiest to visualize these voltage gradients as ripples in a pond from where a stone has entered. The ripples are strongest at the center and get weaker as they get farther from the center. The difference in the intensity of the ripples represents the difference in the voltage levels. Figure 6–8 shows the voltage between the

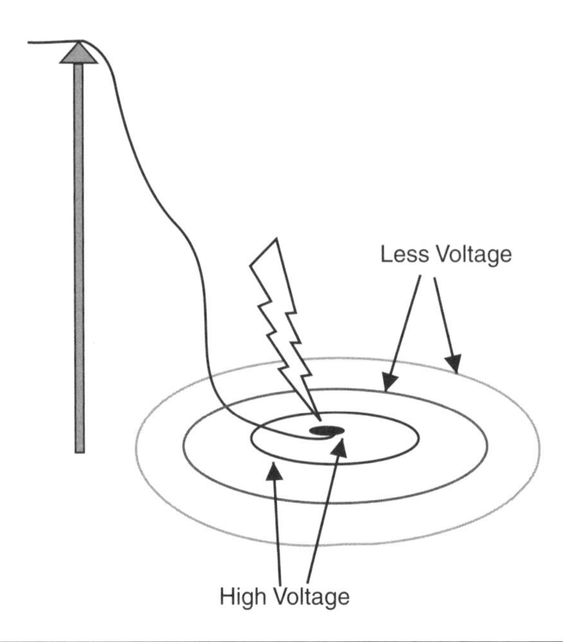

Figure 6–8 Voltage gradients.

ripples lessening as the distance from the contact point is increased. Voltage gradients are also known as ground gradients, potential gradients, and step potentials.

A Voltage-Gradient Example

6.5.3 The lines trade is exposed to a ground-gradient hazard when a pole is being installed in a live circuit. If a pole contacts a live circuit and also contacts the ground, then there is a voltage gradient set up at the base of the pole. The details involved with setting a pole in a live line are used in the remainder of this section to explain the specific hazards involving ground gradients.

Setting a Pole in a Live Distribution Line

6.5.4 A work procedure to set a pole in a live line requires insulated cover-up on the conductors and/or on the pole. With the use of cover-up and an observer, a modern vehicle can set poles in a live line without making contact with a bare live circuit. The procedure requires that the boom-equipped vehicle setting the pole be grounded so that the circuit trips quickly if the boom or its load makes accidental contact with the circuit. If the vehicle becomes alive, anyone near the truck is exposed to ground gradients. The boom operator stays on the operating platform or uses a ground-gradient mat. The circuit breaker or recloser can be put in a non-reclose position to ensure that the circuit remains isolated should it be tripped out due to an accidental contact.

A person controlling the butt of the pole wears rubber gloves to provide a barrier against touch potentials (Figure 6–9). Pole tongs or cant hooks are also used to keep a person away from the base of the pole, because the highest ground-gradient potentials are where the pole touches the ground. A person using rubber gloves to guide the pole into the hole could still have feet spanning across two dif-

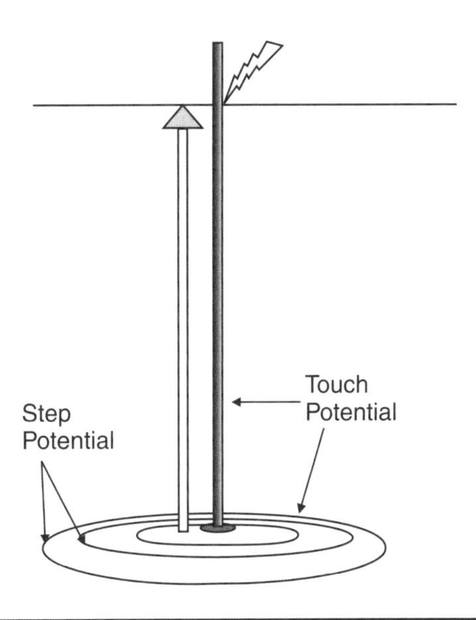

Figure 6–9 Step and touch potentials.

ferent voltage levels. At higher-voltage levels, guide ropes tied to the butt of the pole can be used to get even farther away from the current-entry point.

Step Potential and Touch Potential

6.5.5 Electrical current flows into the earth when an object such as a pole accidently makes contact with a live conductor. Where the pole touches the earth, there is a rise in voltage relative to any earth farther away from the base of the pole. The current takes many paths as it flows away from the pole. The voltage at the base of the pole is higher than the voltage farther away. The voltage or potential gradient around the base of the pole can be pictured as the ripples in a puddle after dropping a stone in the middle. There is a voltage difference between each pair of ripples, and the voltage decreases as the ripples move farther from the center.

A step potential refers to getting one foot on a high-voltage ring near the base of the pole and the other foot on a lower-voltage ring farther away from the pole. A *step potential* is defined as the voltage differential between two points on the ground separated by the distance of one pace or one meter.

A touch potential refers to touching the pole where the hands are at one potential while the feet on the ground are at a different potential. A *touch potential* is defined as the voltage differential between both feet on the ground and an object being touched by hand.

Magnitude of Step and Touch Potentials

6.5.6 The magnitude of the step and touch potentials depends on the voltage of the circuit, the conductivity of the pole, and the quality of the ground. The higher the voltage of the circuit, the easier it is for the current to overcome the resistance offered by the pole and the ground. A high voltage, therefore, is more likely to generate high touch and step potentials. The magnitude of the current flowing to ground depends on the type of earth at the base of the pole and can also be dependent on the contact with the grounded derrick handling the pole.

The resistance of a wood pole varies with the moisture, weather, and pole treatment. On low-distribution voltages, a wood-pole contact does not normally trip out the circuit, and the wood resistance may reduce the step and touch potentials at the base of the pole to a minimal hazard. A steel pole, a concrete pole, or a wood pole with a downground installed, on contact with a live conductor, has a touch potential at the base of the pole at almost the full line voltage. The current flow into the earth depends on the resistance of the ground that the base of the pole is touching.

The Size of the Step-Potential Gradient

6.5.7 Electricity needs a circuit before current can flow. Normally, current flows through a conductor to the load and returns to the source through another phase, neutral, and earth. When a fault to ground occurs, the current flows through the earth back toward the source through the easiest paths available. The return flow can be into the earth, back up any ground wires to the neutral, along fences or creek beds, and so on.

When setting a pole, the vehicle should be grounded. The return path for much of the current is back to the vehicle and up to the neutral. In other words, during a ground fault there is no way to know where and how far the ground gradients will travel. In good, moist earth there is less resistance, and, therefore, the voltage drops very quickly, before the current travels very far from the pole. When the voltage drops quickly, there is a high-voltage difference between the potential rings close to the base of the pole. On a high-resistance surface such as gravel, sand, rock, and dry snow, the potential gradients drop off more slowly and farther from the base of the pole. There is less voltage difference across each step.

Voltage Gradients on a Pole

6.5.8 Voltage gradients can also occur along a wood pole when a live conductor contacts a pole. For a person on a pole, the voltage at the contact point is higher than the voltage at a spot farther from the contact point. Anyone on the pole would have a potential difference between the hands and the feet. On a very conductive pole, such as a steel structure, there is less potential difference between the hands and feet because both are in contact with the same object at the same potential.

6.6 Working with Neutrals

Voltage and Current on a Neutral

6.6.1 A neutral conductor tends to be treated as a nonenergized conductor by the lines trade. This is probably because the voltage on a neutral is normally below the threshold of sensation. The neutral voltage is usually below 10 volts. There is, however, a current flow in the neutral, and when the current is interrupted there is a recovery voltage across the open point.

Sources of Neutral Current

6.6.2 In an electrical circuit, all the current must return to the source. On a single-phase wye circuit, all the current returns by way of the neutral and the earth. The neutral and the earth are parallel paths for current to flow back to the source.

Depending on the type of earth, about two-thirds of the current flows through the neutral and the remaining one-third flows through the earth. The more load there is on the single phase, the more current there is returning to the source.

On a *balanced* three-phase wye circuit, all of the current returns to the source on the other phases. However, the loads on distribution circuits are not perfectly balanced between the three phases. The neutral and the earth carry the unbalanced portion of the current back to the source. The more unequally the phases are loaded, the greater the current flow through the neutral. In an electrical system where feeders are tied between stations, the neutral can also carry current between substations.

The neutral on a secondary service also carries current. On a three-wire 120/240-volt service, the current returns to the source transformer on the opposite leg of the two 120-volt legs. When the load is not balanced equally between the two 120-volt legs of the service, all of the current will not return to the source through the opposite leg. The neutral and, to a lesser extent, the earth serve as the path for remaining current to flow back to the source.

Sources of Neutral Voltage

6.6.3 When there is a current flow, there also has to be some voltage pushing it. Therefore, if the current flow is reduced, the voltage is reduced. As the current gets higher, the voltage gets higher. Normally, on a neutral, the voltage is below 10 volts and not felt by a worker.

- The voltage on a neutral increases if the neutral resistance is increased. Based on Ohm's Law ($E = I \times R$), the voltage increases if the resistance or current is increased. A poor electrical connection in the neutral circuit raises the voltage.
- There can be a very high voltage on a neutral, in relation to earth, during a line-to-ground fault. A line-to-ground fault on one phase is like a large unbalanced load between the phases. A large portion of the return flow to the source is through the neutral and earth. A high-fault current will result in a high voltage on the neutral.
- There can be a very high voltage on a neutral when a phase is struck by lightning. The high voltage on the phase couples (is linked together electromagnetically) with other phases and the neutral. Electrical coupling with the neutral tends to bring the phase voltage down, but electrical coupling also raises the voltage on the neutral before the voltage is dissipated to ground on the multigrounded neutral.

An Open Neutral

6.6.4 An open or broken neutral is a major hazard in the lines trade. There is normally no apparent voltage on a neutral. A neutral is usually grounded on each side of a break at transformers or other equipment. There is, however, a high voltage across the break.

On a multigrounded neutral, the voltage is kept low. If the current is interrupted by opening or cutting the neutral, all the voltage pushing the current

Table 6–1 The Vehicle as an Electrical Hazard

If	Then
A person on the ground is in contact with a utility vehicle while the uninsulated portion of a boom accidently contacts a live circuit.	The person is a parallel path to ground and will take some share of the current flowing to ground.
A person on the ground is near a utility vehicle while the uninsulated portion of a boom accidently contacts a live circuit.	The person is exposed to ground gradients at each location that current is entering the earth.
An uninsulated boom is used to lift or handle an isolated and grounded conductor in a high-induction area.	Unless the boom is bonded to the grounded conductors, the boom can be at a different potential to the conductors being handled.

appears across the open point. A break in the neutral introduces an infinite resistance to current flow and results in a total voltage drop. Applying Ohm's Law shows that the total circuit voltage (recovery voltage) is available across the break. It is, therefore, a basic lines procedure to always jumper any neutral before it is cut.

Precautions for Working with a Neutral

6.6.5 A powerline worker needs to take special precautions that apply only to a neutral. The first precaution is to properly identify which conductor is the neutral. On older construction, the neutral position is not always standard. On some two-phase delta construction, one phase is in a position that would normally be the neutral position in a wye system. Placing a ground on the neutral is a positive identification method.

If a neutral is to be cut or joined, a bypass jumper must be installed. This avoids exposure to the voltage that is available across the open point. Stay away from all conductors, including the neutral, during a lightning storm.

6.7 Vehicle Grounding and Bonding

The Vehicle as an Electrical Hazard

6.7.1 Vehicles with booms are exposed to accidental or inadvertent contact when used in the vicinity of live circuits. There are two ways a vehicle can be become an electrical hazard to people on the ground at a job site (Table 6–1).

How Vehicles Can Become Alive

6.7.2 Most utilities have procedures to protect people on the ground in case a vehicle should become alive due to inadvertent contact with a live circuit. There are many types of mishaps that can cause a digger derrick or the uninsulated portion of an aerial device to become alive:

- A live conductor can fall onto a boom.
- A digger derrick can make contact while installing a pole or other equipment.
- The lower boom of an aerial device can inadvertently make contact with a tap running laterally from the circuit being worked on as shown in Figure 6–10.

Ground-Gradient Hazards near the Vehicle

6.7.3 When the vehicle becomes alive (Figure 6–11) there are ground gradients in the vicinity of the vehicle's outriggers, tires, attached trailers, and anything else in

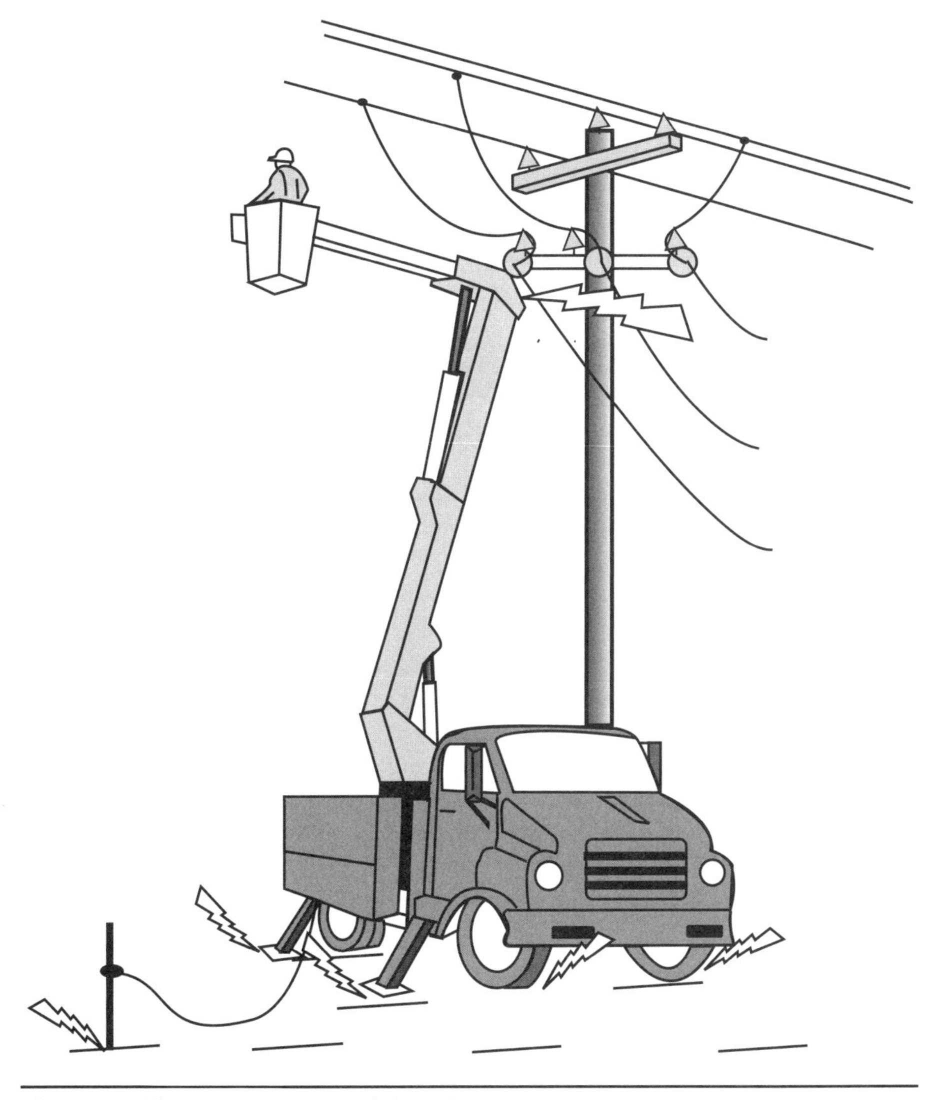

Figure 6–10 Boom contact with lateral tap.

Figure 6–11 Ground gradient hazards near a vehicle.

contact with the vehicle. If the vehicle is grounded to the system neutral, there is a voltage rise on the neutral and a ground gradient hazard at all the downground locations along the line. If a ground rod or pole downground is used to ground the vehicle, there is a voltage gradient around them as well.

Protection from Shock

6.7.4 The only real protection available to people working around a vehicle when a boom makes contact with a live conductor is to stay on the vehicle or stay away from the vehicle. Staying on the vehicle or on a ground-gradient mat bonded to the vehicle keeps a person at the same potential as the vehicle. As long as the person avoids contact with anything not connected to the vehicle, there will be no current flow through the person to another object. Keeping a safe distance from a vehicle that can become alive prevents contact with the vehicle and keeps one a safe distance away from high-ground gradients. Truck barricading promotes the need to stay away from the vehicle. High-resistance footwear provides additional protection by increasing the resistance of a person as a parallel path to ground. High-resistance footwear also provides protection from ground gradients.

Grounding the Vehicle

6.7.5 Anyone touching a utility vehicle when there is contact between its boom and a live conductor is a parallel path to ground and has some current going through him. Even when the vehicle is grounded to an excellent ground, like a neutral, the amount of current going through the body could still be fatal.

Grounding the vehicle to the neutral promotes a fast trip-out of the circuit should the vehicle become alive and also promotes the collapse of the voltage in the system. A fast trip-out reduces the exposure time to the hazard, and a collapse of the voltage reduces the voltage people are exposed to.

On lower-voltage wye distribution systems, the vehicle must be grounded to a neutral to trip the circuit. The resistance of a ground rod or anchor rod is normally not low enough to generate a fault current high enough to trip out the circuit. On a delta circuit, the vehicle should be grounded to a ground rod. Most delta circuits are protected by ground-fault relays that trip out a circuit when there is a line-to-ground fault. On higher-voltage distribution and subtransmission circuits, the voltage is normally high enough to overcome the resistance of a well-driven ground rod and cause the circuit to trip out.

Choice of Ground Electrodes

6.7.6 The purpose of grounding a vehicle is to promote a fast trip-out of a circuit if a boom contacts a live conductor. A vehicle should be grounded to the best ground electrode available at the work site so that any fault current has a good return path to the source. A complete circuit is needed for the current flow to be high enough to trip the protective switchgear. Typical ground electrodes, listed in order of priority, are:

- A permanent ground network such as a station ground, a neutral, or a steel tower.
- A ground rod or an anchor rod in earth.
- A temporarily driven ground rod.

A permanent ground network ensures that the fault current has a good path back to the source to complete the circuit. Ground rods and temporary ground rods place a relatively high resistance element in the circuit.

Typical Values during a Boom Contact

6.7.7 Table 6–2 gives some typical ampere values that can occur when a digger derrick makes contact with a 7200-volt conductor at a location where the circuit is capable of supplying a 6000-ampere fault current. In this example, a person with a resistance of 1000 ohms is touching the vehicle. When reading Table 6–2 remember that:

- A person's heart can go into fibrillation after 50 milliamperes or .05 amperes of current goes through the body for a very short time.
- A location with a 6000-ampere fault current available will probably be near the source of a feeder. Depending on the size of the recloser or the relay settings, the protection will probably need more than a 700-ampere fault current to operate.

TABLE 6–2 Boom Contact with Live 7.2-Kilovolt Conductor

Vehicle Ground	Fault Current Generated	Truck-to-Ground Voltage	Current through Person
Vehicle Not Grounded	200A	5500V	6A
Vehicle Connected to Ground Rod	700A	5000V	6A
Vehicle Connected to Neutral	5000A	200V	0.2A

Examples of Using a Ground Rod to Ground a Vehicle

6.7.8 A temporarily driven ground rod may not always trip out a circuit.

Transmission Lines Example

An accidental boom contact is made with a transmission line or subtransmission line. The vehicle is grounded to a temporary ground rod. The rod will probably provide a resistance low enough to trip out the circuit.

The resistance of a temporary ground rod varies with the soil conditions but would seldom be less than 25 ohms and frequently is 100 ohms. Using Ohm's Law and an example calculation of a 50-ohm ground rod for working on a 230-kilovolt circuit, the fault current generated would be:

$$I = \frac{E}{R} = \frac{133{,}000}{50} = 2660 \ amperes$$

where R = 50 ohms
$E = 230 \div 1.732 = 133$kV phase-to-ground

The fault current generated by the ground rod in this example is more than enough to trip out the circuit, especially a circuit protected by a breaker with relays that sense phase differential and ground currents.

Distribution Lines Example

An accidental boom contact is made with a distribution circuit. The vehicle is grounded to a temporary ground rod. There is a likelihood that the resistance of the rod is too high to trip the circuit quickly or at all. Using Ohm's Law and 50-ohm resistance for the temporary ground rod on a 4800-volt circuit the fault current generated would be:

$$I = \frac{E}{R} = \frac{4800}{50} = 96 \ amperes$$

In this example, a fault current of 96 amperes will not trip out a circuit protected by a fuse larger than 50 amperes because it takes 100 amperes to blow a 50-ampere fuse.

Boom in Contact with an Isolated and Grounded Conductor

6.7.9 A circuit is not *dead* when it has been isolated and grounded. There is often current flowing in the grounds, and there is often a voltage difference between the circuit and a remote ground. When a truck boom or crane is used on a job where a circuit is isolated and grounded, the boom should be bonded to the portable line grounds. On a right-of-way that has high induction from live neighboring circuits or in the rare case of accidental reenergization, bonding ensures that there will be no potential difference between the vehicle, boom, winch, and conductors.

Examples of Hazards Using a Boom around Isolated Conductors

6.7.10 The following examples of hazardous incidents are rare occurrences, but they have all happened:

- A person working on a conductor handles the winch of a crane (Figure 6–12) and bridges the grounded conductors and an unbonded crane. An unbonded vehicle is a remote ground in relation to the grounded conductors. In an area with high induction, there is a high voltage between the grounded conductors and the winch.

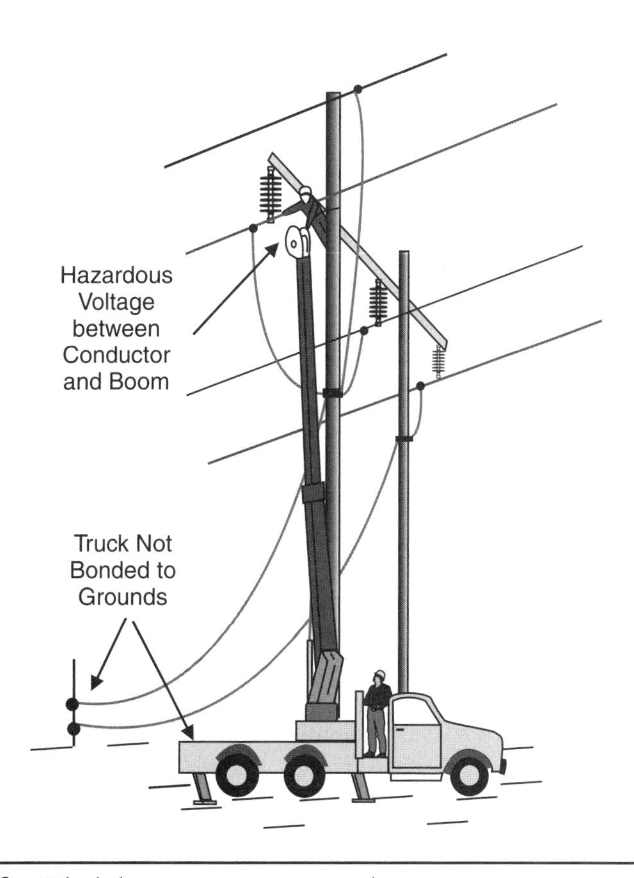

Figure 6–12 Vehicle boom as a remote ground.

- When a vehicle is bonded to grounded conductors, the vehicle and the grounded conductors are at the same voltage. In a high-induction area, the bond sets up a voltage between the vehicle and any remote ground. A person standing on the ground and touching the vehicle could receive a shock. Grounding and bonding the vehicle, as well as bonding, reduces the voltage difference between the ground and the vehicle.
- When using an aerial device to install grounds, especially on a multicircuit line, the vehicle is often grounded/bonded to the same ground electrode as the portable line grounds. If the grounds are mistakenly installed on a live circuit, the truck also becomes energized, endangering those on the ground near the vehicle. People on the ground should stand clear of the vehicle until the grounds are installed.

6.8 Electromagnetic Fields (EMF)

Why the Interest in EMFs?

6.8.1 Before the 1970s the only risks thought to be associated with electricity were electric shock or flash burns. There is now a question about the possible relationship between electromagnetic fields (EMFs) and some types of cancer. Most of the evidence comes from epidemiological studies. Epidemiology is a science that looks for statistical evidence of health patterns in people and the likely factors that may be responsible.

What Are EMFs?

6.8.2 Wherever there is electricity, magnetism appears, and vice versa. Both electricity and magnetism travel in the form of waves, in fields (in the same way that gravitation has a field) and spread everywhere. EMFs are energy waves with both an electric and a magnetic component. Energy is radiated out from a source and can travel without the need of any material to conduct it.

Electromagnetic waves can travel through a vacuum and at the speed of light. Electromagnetic waves have a wavelength, frequency, and amplitude (or field strength). A wavelength is the distance between one peak on the wave and the next peak. The frequency, measured in hertz, is the number of wave peaks that pass by in one second.

Electromagnetic waves from power frequencies of 50 or 60 hertz are just one portion of a large band or spectrum of electromagnetic waves. The 50- or 60-hertz waves from powerlines are very long waves (5000 kilometers) and are referred to as extremely low frequency (ELF) waves.

When EMFs are discussed in the media, the term usually refers to electric and magnetic fields in the power-frequency range that has a powerline as its source. The term EMF, however, is also used for electromagnetic fields with higher frequencies, which radiate much farther than power-frequency waves.

Electromagnetic Spectrum

6.8.3 All rays—including X-rays, ultraviolet light, visible light, infrared light, microwaves, radio waves, heat, and electrical powerlines—are electromagnetic waves. The properties that differentiate these various sources are the frequency and the wavelength.

TABLE 6–3 The Electromagnetic Spectrum

Source	Effects	Frequency (Hz)	Wavelength
ELF (Extremely Low Frequency) DC Is 0Hz Powerlines Are 50–60Hz	Low Energy No Thermal Effects	3 to 300	Powerlines are 6000km.
VLF (Very Low Frequency) AM Radio	No Proven Health Effects	3000 to 30,000	AM radio waves are about 300 meters.
VHF (Very High Frequency) FM and TV	Can Cause Heating High Induced Currents	10^7 to 10^9	Television waves are 0.3 to 5.5 meters.
Radar Microwave		10^9 to 10^{12}	Microwaves are about 12 cm.
Infrared Visible Light Ultraviolet	Energy Waves Can Be Seen Photochemical Effects	10^{12} to 10^{15}	Visible light waves are from 0.75 to .04 micrometers.
Ionizing Radiation Ultraviolet Radiation X-Rays Gamma Rays	High Energy Radiation Burns DNA Damage	10^{15} to 10^{22}	X-rays are about .03 to .00001 micrometers.

Table 6–3 shows some of the types of radiation that make up an electromagnetic spectrum. The lines dividing the different types of radiation overlap because the division between them is not accurately defined.

Sources of Power-Frequency EMF

6.8.4 All live electrical wires, equipment, and appliances have an *electric* field and a *magnetic* field around them. Combined, these two fields are referred to as EMF. EMFs should not be confused with emf, which refers to electromotive force. The electric and magnetic fields behave differently and are measured separately, as seen in Table 6–4.

Methods of Reducing the Strength of EMF

6.8.5 There are ways to reduce the strength of EMF coming from live power conductors:

1. Increasing the heights of poles and towers increases the distance from the ground to the powerlines.
2. A circuit converted to a higher voltage, given the same amount of load, would carry less current and, therefore, have a lower-strength electromagnetic field.

Table 6–4 Comparison of Electric and Magnetic Fields

Electric Fields	Magnetic Fields
The electric field is a voltage field. The higher the voltage source, the higher the electric field. An electric field is present around any live conductor and is independent of the current flow in the circuit.	The magnetic field is a field of magnetic lines of energy. The higher the current in a source conductor, the higher the magnetic field. If there is no load on the circuit, then there will be no magnetic field.
An electric field is measured in volts per meter (V/m).	A magnetic field is measured in microteslas (μT) or milligauss (mG).
The strength of the electric field drops off with the inverse of the distance from the source squared ($E = \frac{I}{r^2}$), where r is the radius from the source.	The greater the distance from a live source, the less the strength of an EMF. The strength of the magnetic field drops off with the inverse of the distance from the source ($H = \frac{I}{r}$), where r is the radius from the source.
Shielding from electric fields occurs if there are trees, walls, etc., blocking the way and draining the electric current to ground. The electric field from a powerline would not normally penetrate into a building.	Magnetic fields are only weakened, but not stopped, by barriers such as trees, etc. They are similar to the effect of a magnet where the lines of force can go through material. Magnetic fields from a powerline can penetrate a building.
When a person is within an electric field, there is a voltage induced on the body. Workers feel the existence of electric fields when they are working in a substation or on a high-voltage transmission line. Hair tends to stand up as a person gets close to a high-voltage source. A small arc can occur when the voltage on a person's body is discharged to a grounded object. An electric field can induce a weak electric current flow in the body by moving charges in the body. The redistribution of charges causes small currents, but these currents are typically much smaller than those produced naturally by the brain, nerves, and heart.	A magnetic field can induce a current flow in lengths of conducting material such as in a wire fence. A current, however, can flow only if there is a circuit. A fence kept insulated from the ground would have no current flow on it unless it was grounded in two places to form a circuit between the two grounded locations. An isolated fence would have a voltage on it. A magnetic field will pass through a body but, because a person's body is not normally in series with a circuit, there should be no current induced in it from magnetic waves.
An electric field can cause a fluorescent tube to light up when it is held under a high-voltage powerline.	A magnetic field can be induced on a circuit built near a high-voltage line. You could actually steal power if a circuit built near a powerline were connected to a load. You would have a small amount of current and no control over the voltage, which would make this option impractical.

3. There is a canceling effect from three-phase lines and multicircuit configurations. The EMF from the current returning on one phase cancels the EMF from the current traveling in the other direction on another phase. A closer spacing between conductors reduces the EMF. For example, a triplex service or house wiring has a reduced EMF due to the closeness and canceling effect of the returning current.
4. Underground powerlines have a reduced electric field because of the shielding from the cable sheath and the surrounding earth. There is also a reduction in magnetic fields for underground lines, not due to shielding, but because of the EMF-canceling effect of the neutral and other phases in the closer conductor spacings allowed in underground.
5. Transpositions in a powerline reduce the EMF. Transpositions were originally put into circuits to reduce the effect of EMF on open wire communications circuits. As a prudent avoidance measure, the EMF from electric blankets and waterbed heaters is reduced when they are manufactured with transpositions because the wires are continually crossing over each other.
6. There are methods available using metal shielding to reduce magnetic fields from powerlines, but they are considered an expensive method to shield large areas.

Measurement of EMF

6.8.6 To measure the strength of EMF, two instruments are needed, one for the electric field and another for the magnetic field. The strength of an electric field is measured as volts per meter (V/m) or in larger units of kilovolts per meter (kV/m). The strength of the electric field depends on the voltage of the source conductor and the distance away that the measurement is taken. The instrument needs to be held away from the body because the body acts as a shield and distorts the readings.

The strength of a magnetic field is measured in teslas or gauss. The gauss is common in the United States, and the tesla is used internationally. A tesla is a measurement of the magnetic-field intensity or the density of magnetic lines. The strength of the magnetic field depends on the amount of current flowing in the source conductor and the distance away that the measurement is being taken. Measurements must be done in different locations over a period of time, because the load current of the circuit changes during the day. Tesla units are too large to measure common exposure; therefore, the microtesla is used.

$$\begin{aligned} \text{One tesla} &= 1{,}000{,}000\ \mu T \\ \text{One tesla} &= 10{,}000\ \text{gauss} \\ \text{One gauss} &= 1000\ mG \\ \text{One}\ \mu T &= 10\ mG \end{aligned}$$

When taking measurements, it is important to understand the difference between an emission and exposure. A measurement taken at a conductor will give the emission, but exposure to the emission is normally much farther away.

Exposure needs to be measured at a location where people would normally be. The measurements will not mean much to the average person unless the numbers are used as a means to compare the EMF strength from various sources, as shown in Tables 6–5 and 6–6.

Dose

6.8.7 One problem in researching the health effects of EMF is to determine which *dose* to measure. The word *dose* means exposure that produces an effect. For example, if there is a risk due to EMF exposure, what type or dose would produce harmful effects?

- Is a weak field safer than a strong field?
- Is it exposure to the peak electric field or exposure to a constant electric field that is harmful?

TABLE 6–5 Sample Measurements of Electric-Field Strength

Location of Measurement	Typical Values (V/m)
Under a transmission line, the field strength depends on the voltage of the line and the height of the conductors.	One to 10kV/m, which is 1000 to 10,000V/m
On the edge of a transmission line corridor.	100 to 1,000V/m
Near an overhead distribution line.	2 to 20V/m
In homes.	200V/m close to an appliance to less than 2V/m in other locations in the home.
Exposure to powerline workers, cable splicers, and substation workers.	Typical 100 to 2000V/m with peaks as high as 5,000V/m.

TABLE 6–6 Sample Measurements of Magnetic Field Strength

Location of Measurement	Typical Values (μT)
Under a transmission line, the field strength depends on the line loading and the height of the conductors.	10μT
On the edge of a transmission line corridor.	0.1 to 1.0μT
Near an overhead distribution line.	0.2 to 1.0μT
In homes.	150μT near appliances to less than 0.02μT in other areas in the home
Exposure to powerline workers, cable splicers, and substation workers.	Average 0.5 to 4μT with as high as 100μT

- Is it the going into and coming out of an electric field?
- Is it exposure to the peak magnetic field or exposure to a constant magnetic field?
- Is it the going into and coming out of a magnetic field?

Natural Levels of Electric and Magnetic Fields

6.8.8 There is a natural existence of electric and magnetic fields on the earth. This EMF source is obviously not at 50 or 60 hertz, but is a DC source with a fluctuating voltage and current level. It is the fluctuations that cause the electric and magnetic fields to occur.

The earth's atmosphere has a naturally occurring electric field that fluctuates at around 130 volts per meter on the surface. Large electric current through the earth's core creates a magnetic field on the earth's surface that can range from 30 to 60 microteslas. The magnetic field is stronger at the North and South Poles than at the equator.

A thunderstorm is the ultimate demonstration of naturally occurring electric and magnetic fields in action. The electric-field strength is high enough to break down the insulation value of air and discharge as lightning. The high current within the electric arc generates a large EMF that can be noticed as interference on radio and television reception.

Health Risk

6.8.9 The health risk of EMF from 50- or 60-hertz sources has been the subject of more than 500 studies over a thirty-year period. There are biological effects produced by very high levels of electric and magnetic fields, but there has been no agreement on whether lower levels are any hazard to health. There has not been any definitive proof one way or the other that there is a health risk due to EMF. Experience shows that if there is a risk, it will be extremely small. Compared to smoking, diet, or sunlight as causal factors in cancer risk, the risk of cancer from EMF would be extremely small. Electric utilities will likely continue to monitor future studies and keep their staff informed of developments.

Limits of Exposure

6.8.10 The World Health Organization, the International Radiation Protection Organization, and other organizations have published threshold limits of exposure to EMF from power-frequency sources (50 and 60 hertz). There are limits for the general public and for workers. For example, one organization has a threshold limit for a workday exposure to EMF as 10 kilovolts per meter for electric fields and 5 gauss for magnetic fields. Utility workers should check for the limits their own utilities have adopted for exposure to EMF.

6.9 Minimum Approach Distance

Minimum Approach Distance as a Barrier

6.9.1 Maintaining a *minimum approach distance* to live exposed conductors is the most common barrier a powerline worker uses to avoid electrical contact. Government regulators and utilities have tables listing the minimum approach distance from various voltages for different levels of qualified people and equipment.

The minimum approach distance table (Table 6–7) applies to work in the vicinity of a live circuit, and it also applies to live-line work.

The flashover voltage for a live-line tool is the same as it is for air. A fiberglass live-line tool may be better insulation than air, but the distance needed on a tool is an air gap between the hands on a live-line tool and the live conductor.

Rubber-glove work and bare-hand work may appear to be exceptions to the minimum approach table, but the table still applies, in reverse. Rubber-glove and bare-hand procedures depend on the powerline worker being insulated from ground or other phases. The table should apply to the distance from the powerline worker to any exposed objects that are second points of contact. Unless they are covered with rubber or fiber barriers, a minimum approach distance is kept from other phases, the neutral, a grounded structure, or any other grounded objects.

The distances in Table 6–7 apply to a qualified, competent worker:

- working near a bare, exposed live circuit or equipment.
- keeping a length of clear live-line tool for the voltage being worked on.
- keeping any objects that are second points of contact at the distances listed while working with rubber gloves or bare hand.

Factors Considered for a Minimum Approach Distance

6.9.2 The minimum approach distance tables consider the hazard of an electric discharge from a live conductor to a person, and they consider an inadvertent movement by a worker in the vicinity of a live conductor. A minimum approach distance, therefore, consists of an *electrical-factor distance* and an *ergonomic* or *human-factor distance.*

TABLE 6–7 Typical Minimum Approach Distances and Phase-to-Ground Distances

Minimum Clearance from Live AC Parts	
Phase-to-Phase kV	*Phase-to-Ground Exposure*
1.1 to 15kV	2 ft. 1 in. (64 cm)
15.1 to 36kV	2 ft. 4 in. (72 cm)
36.1 to 46kV	2 ft. 7 in. (77 cm)
46.1 to 72.5kV	3 ft. (90 cm)
72.6 to 121kV	3 ft. 2 in. (95 cm)
138 to 145kV	3 ft. 7 in. (1.09 m)
161 to 169kV	4 ft. (1.22 m)
230 to 242kV	5 ft. 3 in. (1.59 m)
345 to 362kV	8 ft. 6 in. (2.59 m)
500 to 550kV	11 ft. 3 in. (3.42 m)
765 to 800kV	14 ft. 11 in. (4.53 m)

$$\textit{Minimum approach distance} = A + (F \times B)$$

where

- A is a human-factor distance that takes into account a momentary inadvertent reaching into the prohibited zone.
- F is the minimum electrical clearance that would prevent a flashover from the highest transient voltage from an internal source.
- B is a safety factor to ensure there is no chance there will be a flashover across the minimum electrical clearance.

The Electrical Factor (F × B) in a Minimum Approach Distance

6.9.3 Design engineers use standard electrical clearances when determining the distance allowed between a phase and a grounded object for a pole framing or transmission tower design. An electrical-clearance distance is considerably less than the minimum approach distance a person must maintain. A minimum electrical clearance, which is represented by F in the preceding formula, is just greater than the flashover distance from the highest possible internal transient voltage. An internal source for a transient overvoltage would be a switching surge. An external overvoltage surge, such as lightning, could flash over the minimum electrical-clearance distance; therefore, live-line work is not carried out with a lightning storm in the vicinity.

In the formula, the electrical-clearance distance F is multiplied by a safety factor B to provide extra confidence that there will be no flashover across an electrical-clearance distance between the circuit and ground. The safety factor used tends to be around 1.25 times the electrical-clearance distance. A nearby surge arrester should shunt a transient overvoltage to ground to provide an additional assurance that there will not be a flashover across a normal electrical-clearance distance.

The Human Factor (A) in a Minimum Approach Distance

6.9.4 The human-factor minimum approach distance is a space needed to prevent a person from encroaching into the electrical-factor ($F \times B$) distance. A human-factor distance is subjective and has not necessarily been studied scientifically. It could be said that it is unlikely that a powerline maintainer would inadvertently encroach more than 3 feet (one meter) toward a live conductor. If 3 feet, representing A, is added to the electrical factor ($F \times B$), in the formula, a minimum-approach distance can be established for each voltage.

At transmission line voltages, 3 feet added to the electrical-factor distance would establish a realistic minimum approach distance table and allow transmission lines work procedures to be carried out. At transmission line voltages, there is no cover-up available, and maintaining a safe approach distance is a necessary barrier for work on or near live transmission lines.

At distribution voltages, the human-factor distance is larger than the electrical-factor distance. It is often necessary to go closer than 3 feet to an exposed conductor. A human-factor distance of less than 3 feet requires extra barriers—such as the use of cover-up, an insulated platform, or a dedicated observer—to reduce the risk of inadvertent contact.

6.10 Fire in an Electrical Environment

Line-Crew Involvement with Fires

6.10.1 It is not unusual for a line crew to be called out to a pole fire or transformer fire. They are also called by fire departments to cut off the power to buildings that are on fire. *Isolating a circuit in or near a fire creates the safest environment for line crews and firefighters.* There are safe methods to fight a fire in an electrical environment, but they should only be carried out by trained people in an emergency.

An Electrical Fire

6.10.2 The only real electrical fire would be an electric arc, but that is not the type of fire put out with fire-fighting equipment. An electrical fire refers to a fire in an electrical environment. Any extinguisher used in an electrical environment must have a C rating in combination with another rating. Figure 6–13 shows the three ratings important to fighting fires a powerline worker may be involved with.

For example, a transformer fire is actually a flammable-liquid fire with the added hazard of it being in an electrical environment. A fire extinguisher rated as BC is needed for a flammable-liquid fire in an electrical environment. However, if a transformer is on fire, the transformer is not salvageable. Workers should stay upwind because if the transformer oil is contaminated with polychlorinated biphenyls (PCBs), then the smoke has some very dangerous chemicals in it.

A pole fire is a wood fire in an electrical environment and would require an AC extinguisher, or the more common ABC extinguisher. Be aware, however, that any spent dry powder lying on a cross-arm or on other hardware can become contaminated and conductive, especially in damp conditions. Wet powder could become a path to ground and cause an explosive arc near any worker trying to put out a fire.

Stay back when using a fire extinguisher. Even a small extinguisher can have pressure equal to a large extinguisher, it only has less powder.

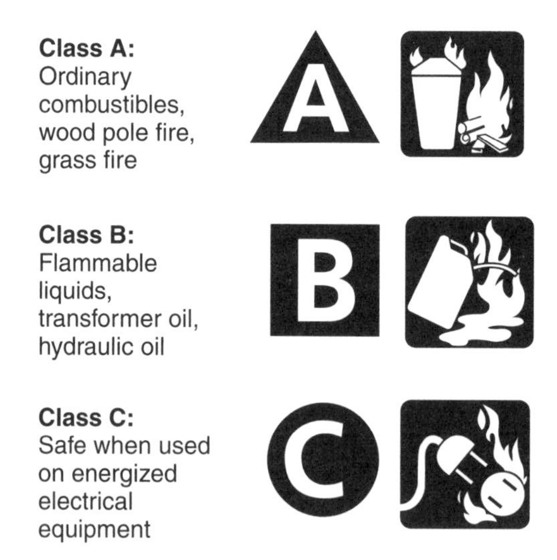

Figure 6–13 Types of fire extinguishers.

Water and Electricity

6.10.3 Everyone has been made aware that water and electricity do not mix. However, exposed electrical apparatus and water do mix every time it rains. When insulators are clean, the rain does not cause the circuits to trip out. It is the combination of water and some kind of contaminant that causes electrical current to flow. Water becomes a conductor when it dissolves salts from contaminants.

When a line crew washes insulators on live circuits, the water does not normally short out the circuit. A high-pressure pump (600 to 850 psi) is used to blast the dirt off the insulators. Instead of a solid stream of water, a special nozzle is used to break up the stream into fine particles. This allows a closer limit of approach for the powerline worker. The nozzle must be grounded because:

1. A grounded nozzle drains the static charge that can be generated by the water flowing through the nozzle.
2. A grounded nozzle drains any current flowing from the live circuit through the stream of water to ground.

Firefighters can be trained to use water for fighting fires near powerlines. Some fire crews have been trained to use water with a spray or fog stream on 230 kilovolts from 15 feet away or up to 50 kilovolts from 10 feet away. A solid stream using a 5/8-inch nozzle can be used on circuits up to 230 kilovolts from 30 feet. Before firefighters are trained, the water in their jurisdiction is tested for unusually high salt or mineral content.

Live conductors lying on the ground are very dangerous when water is applied to them. Water flowing on the ground is conductive and is a hazard to anyone touching the wet soil. There will, however, be less voltage between the ground gradients when spread over a longer distance.

6.11 Electrical Influence on Corrosion

Corrosion Hazards

6.11.1 Corrosion reduces the life of metal hardware used in electrical utilities. Corrosion due to road salt, fertilizers, seawater, and damaged paint surfaces is a normal occurrence. There is, however, an electrical component to understanding the cause of premature corrosion of metal hardware.

Line materials such as tower footings, anchor rods, electrical connections, and underground/submarine cable sheath normally last a long time, but occasionally these items corrode prematurely and create a severe hazard. The most vulnerable locations for premature corrosion are near pipelines and in marshy areas. The corrosion is usually at its worst just below the ground line, so it is reasonably easy to inspect vulnerable locations.

When a tower footing or anchor rod corrodes, there is an unseen hazard that could cause a structure failure. When the metallic sheath of an underground/submarine cable is also the neutral for the circuit, an electrical hazard exists when the sheath has corroded away and there is a voltage difference across the open point.

Dissimilar Metals in a Circuit

6.11.2 Two dissimilar metals can generate DC when placed in contact and connected together into a circuit. There is a small, self-generated electric current without any external source. For example, a lead-acid battery consists of two different

TABLE 6–8 Galvanic Series of Some Metals

Anodic End (Most Active) Positive Polarity
Magnesium
Zinc (Galvanized Metal)
Aluminum
Cadmium
Steel or Iron
Lead
Brass
Copper
Graphite
Silver
Cathodic End (Least Active) Negative Polarity

metals (lead and lead oxide) in an electrolyte (a conductive solution such as diluted sulfuric acid). The battery becomes part of a circuit when it is connected to a load. One post of the battery gives off ions, eventually deteriorates (corrodes), and disappears.

Similarly, two dissimilar metals connected together with some kind of electrical bond and placed in a conductive solution are like a battery. A circuit is created when a self-generated current leaves one metal (an anode), flows through the electrolyte, and enters another metal (the cathode). The self-generated current flows because of a natural difference between the potentials of the metals in an electrolyte. As current leaves the metallic anode and flows into the electrolyte, metal ions separate from the anode and cause the anode to be consumed (corrode).

Galvanic Series of Metals Used in Line Hardware

6.11.3 Metals have a tendency to dissolve into ionic form and lose electrons when they become part of a circuit capable of conducting electricity. Some metals (such as aluminum) have a strong tendency to lose electrons, while other metals (such as copper) have a weaker tendency. An active metal with a strong tendency to lose ions is *anodic* in an electrical circuit and will corrode. A less-active "noble" metal with a weak tendency to lose ions is *cathodic* in an electrical circuit and does not corrode.

Metals used in electrical utilities can be listed showing the relative placement of their activity in relation to other metals (Table 6–8). The wider apart the metals are on the list, the greater the rate of corrosion.

Galvanic Corrosion

6.11.4 Premature corrosion of metal hardware occurs when the metal becomes part of a corrosion cell. A typical corrosion cell involving line hardware must have certain conditions present:

1. There must be an anode and a cathode. The anode and the cathode must be connected to each other by a wire or some other kind of metallic contact.
2. An electric circuit must be completed between the anode and the cathode through an electrolyte. An electrolyte is a conductive solution such as the earth in a wet marshy area.
3. There must be a DC potential causing current to flow in the circuit. Dissimilar metals connected in a circuit generate a DC potential. When current leaves the anode, it takes minute particles of metal with it and the anode corrodes.

A flow of DC through the ground can cause corrosion to vital components of an electrical system.

Copper-to-Aluminum Connections

6.11.5 It is well known in the lines trade that when copper is connected directly to aluminum, the connection corrodes prematurely and fails. Aluminum is an active metal, while copper is less active. To prevent galvanic corrosion of the electrical connection, there is normally a spacer with another metal, such as cadmium, between the copper and aluminum. A conductive grease can also be used to slow down any corrosion. With an aluminum-to-copper or aluminum-to-brass connection, the aluminum should always be placed on top to prevent copper salts from leaching and attacking the aluminum.

Cathodic Protection

6.11.6 A tower leg or anchor in wet soil becomes vulnerable to corrosion. Cathodic protection is installed at vulnerable locations. Cathodic protection consists of electrically connecting a sacrificial anode, such as magnesium, to a tower leg and burying it nearby. A circuit is set up between the sacrificial anode and the tower leg through the surrounding soil (electrolyte). The magnesium is a more active metal than the galvanized tower leg. Thus, magnesium anode corrodes instead of the tower leg. The anodes need to be replaced on a regular basis as they corrode away.

Stray-Current Corrosion

6.11.7 Stray-current corrosion is due to DC of external origin flowing from a metal into an electrolyte. When a DC leaves a metal and enters an electrolyte, some metal leaves and the metal corrodes. Alternating current is not significant in producing corrosion because the reverse flow will build up the anode again.

One source of stray DC can be a natural current due to the magnetic fields in the earth. Long pipelines can be vulnerable to its current. Man-made DC that can stray into the earth can come from a nearby DC transmission line, a mining operation, or a welding operation.

Pipeline Cathodic-Protection Hazard

6.11.8 Buried electrical utility line materials are subject to corrosion when a pipeline is nearby. The cathodic protection used to protect a pipeline from corrosion accelerates the corrosion of nearby utility anchors, cable sheaths, or ground rods. To protect a pipeline from corrosion, a negative potential is applied to the pipeline. This prevents a current from flowing into the surrounding earth (electrolyte) from the pipeline and prevents any metal ions from leaving the metal pipe. Any stray current flows toward the pipeline.

Review Questions

1. Why does a full-line voltage (recovery voltage) appear across the two ends of a cut conductor?
2. When making a cut to an isolated and grounded conductor, is it necessary to install a jumper across the cut?
3. Any time a person is in parallel with any load in a circuit, that person is exposed to some voltage and some current regardless of the amount of resistance being offered to the current. True or False?
4. Name the two forms of induction that make up electromagnetic induction.
5. When working in a high-voltage environment, what form of induction causes a spark from one's body every time a grounded object is touched? What type of personal protective equipment will reduce this hazard?
6. Why is induction from a magnetic field a hazard on distribution lines as well as a hazard on high-voltage transmission lines?
7. What is the hazard when working at a location where a ground fault results in electrical energy flowing into the earth?
8. Which ground gradient would spread out farthest, one in good, moist earth or one in rocky ground?
9. Why is there a voltage across a break in a neutral?
10. Is a person safe when in contact with a grounded vehicle that makes contact with a live circuit?
11. What is the purpose for grounding a lines truck?
12. When a truck boom or crane is used on transmission line work where a circuit is isolated and grounded, should the truck be bonded to the portable line grounds?
13. What two factors are used to make up minimum approach distance tables?
14. What type of extinguisher must be used for an electrical fire?

CHAPTER 7

Installing and Removing Protective Grounds

Topics to Be Covered	Section
Introduction	7.1
The Grounding Principle	7.2
The Bonding Principle	7.3
Grounding to Provide Protection from Induction	7.4
Specific Grounding Hazards and Procedural Controls	7.5
Protective Grounding of Underground Cable	7.6
A Case Study of a Line Grounding Incident	7.7

7.1 Introduction

Grounding Procedure

7.1.1 The installation and removal of portable grounds has been a key procedure in the lines trade since the earliest days. Grounding is also called *earthing* in some jurisdictions. A good grounding and bonding procedure has to marry electrical theory with practical requirements. The grounding procedure can vary greatly with different utilities; therefore, this chapter concentrates on the electrical theory involved with grounding and bonding.

Reasons for Grounding

7.1.2 There are three reasons to install protective grounds and bonds on an isolated circuit before work is started:

1. *Install grounds to prove isolation.* After isolation and testing, the installation of grounds is the final proof that you are about to work on the correct circuit.
2. *Install grounds to have protection from accidental reenergization.* When working on a grounded circuit, it is necessary to have protection from accidental reenergization, which can come from operator error, contact with neighboring circuits, lightning, or backfeed.

3. *Install grounds to have protection from induction.* Grounds are needed to provide protection from two kinds of induction hazards. Electrostatic induction will induce a voltage on an isolated circuit. Electromagnetic induction can induce a current flow in the conductor and grounds of an isolated and grounded circuit.

Control of Current and Voltage When Grounding

7.1.3 On an isolated circuit that has been grounded, the voltage and current on the circuit are not zero. There can be a voltage relative to a remote ground, and there can be a current flow in the conductors.

To be assured of not being exposed to any hazardous current or voltage:

1. Adhere to the *grounding principle,* which controls the current around a worker.
2. Adhere to the *bonding principle,* which limits the voltage exposure to a worker.

7.2 The Grounding Principle

The Grounding Principle Defined

7.2.1 When grounds are to be installed on a circuit, check your plan to see if the grounds will control the *current* as described in the grounding principle.

The Grounding Principle: Grounds are installed to reduce any current flow through a worker to an acceptable level by providing a low-resistance *parallel shunt* around the worker. At the same time, if the circuit is or becomes energized, the grounds must be big enough to withstand any *fault current* in the circuit.

Grounding for Fault Current

7.2.2 During accidental reenergization, a set of grounds on a circuit provides a major short circuit. The ground wires and clamps are subject to all the current that the circuit can deliver to that point.

Size of Grounds

When the sizes of portable grounds are specified for a circuit, they are sized to carry a fault current for a time long enough to trip out the circuit. It is important that the grounds are maintained regularly because even the correct-size grounds will burn off at bad connections.

Circuit Protection

Circuit breakers, fuses, and reclosers open when a current feeding a fault goes through these devices for a designated period of time. Reclosers and fuses on a distribution system are devices that will trip out a circuit for an overcurrent fault. It is important to connect the portable grounds to a good ground electrode to provide a high-fault current through the protective device.

Circuit breakers and electronic reclosers are more likely to trip quickly because they can be set to detect other variables, such as an increase in ground current.

These protective devices will trip out a circuit quickly even if conditions at the work site are not ideal for grounding.

Grounding to Provide a Parallel Shunt

7.2.3 When a person is working on a grounded circuit and the circuit becomes alive, there are two paths the current can take to ground (Figure 7–1). By far, most of the current will go through the low-resistance grounds; however, because a person working from a structure is a parallel path to earth, some current will go through the worker.

Does All the Current Take the Easiest Path to Ground?

7.2.4 In a circuit, the *voltage* across any components connected in parallel will be equal and the *current flow* will divide in inverse proportion to the resistance of each parallel path. Even when a live-line tool makes contact with a circuit, it is a parallel path to ground and will have some leakage current through it.

The common belief has been that electricity takes the easiest path to earth. This is basically true, but the easiest path does not take all the electrical current to earth. Some current flow will go through each connected parallel load.

Example: A person is working on a grounded circuit from a wooden pole (Figure 7–2). The resistance of the worker's body is an average 1000 ohms, and the resistance of the wooden pole to ground is about 2000 ohms per foot for a total of 100,000 ohms. The resistance of the working grounds to the neutral and

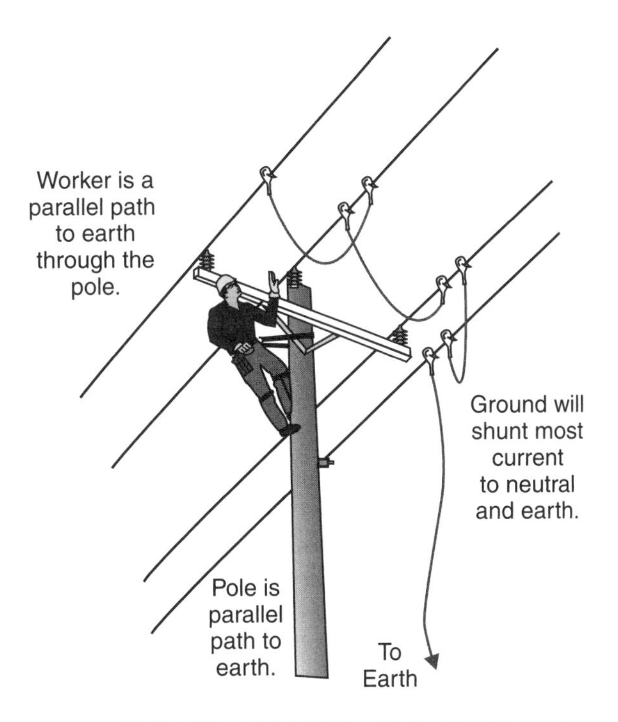

Figure 7–1 Grounds as a parallel shunt around worker.

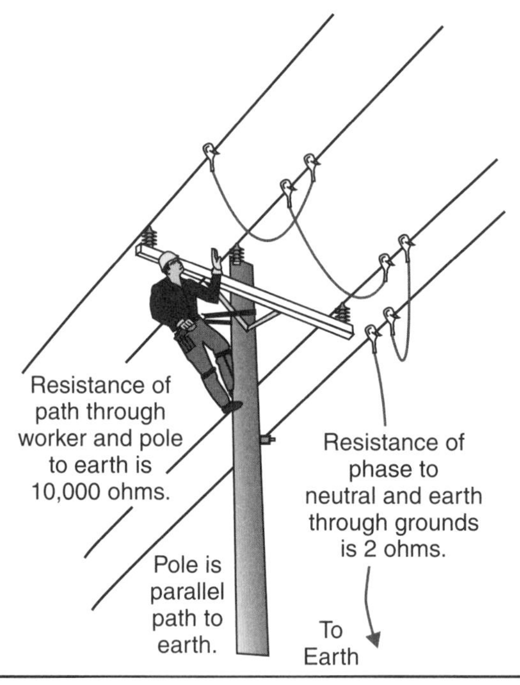

Figure 7–2 Example calculation of current flow in parallel paths.

earth is about 2 ohms. If the circuit was reenergized at 10,000 volts, what is the current through the person?

If the circuit had the capacity to deliver the full-fault current, the grounds at 2 ohms resistance would draw:

$$I = \frac{E}{R} = \frac{10,000}{2} = 5000 \ amperes$$

The worker's body and the wooden pole at 100,000 ohms of resistance would draw:

$$I = \frac{E}{R} = \frac{10,000}{100,000} = 0.1 \ amperes$$

A current flow of 0.1 ampere or 100 milliamperes through a body is the level at which ventricular fibrillation of the heart can occur. When working on a pole with a downground, the hazard of a parallel path to ground is reduced. The downground would be connected to the neutral, and the grounded conductors would be connected to the neutral. The pole and the conductors would be at the same voltage, which keeps the powerline worker at the same voltage. This concept is explained in the *bonding principle.*

Resistance of a Human Body

7.2.5 In most calculations, 1000 ohms is used to represent the body's resistance and probably assumes contact with bare, perspiring skin. A worker with winter clothes and gloves could have a resistance 30 times higher.

While it takes a certain amount of voltage to push an electrical current through a human body, it is the current, exposure time, and the path the current takes through the body that does the damage. The let-go current threshold is about 10 milliamperes and will damage a human body if the victim is not rescued. Considering that a typical household circuit is fused at 15 amperes or 15,000 milliamperes, the potential for a lethal electrical shock is available on most electrical circuits.

7.3 The Bonding Principle

The Bonding Principle Defined

7.3.1 When grounds are to be installed on a circuit, check your plan to see if the grounds will control the *voltage* as described in the bonding principle.

The Bonding Principle: Bonds must be installed so that a worker is kept in an *equipotential zone.* A worker must not be able to bridge between a grounded circuit and any unbonded structure, vehicle, boom, wire, or any other object not tied into the bonded network.

Voltage Rise Due to Induction or Reenergization

7.3.2 The rule says that "it is not dead unless it is ground" is not totally accurate because a grounded line can have potential on it and a current flowing in it. *If* a grounded circuit is exposed to induction or accidental reenergization, *then* the voltage will rise in the complete circuit and everything electrically attached to it, the grounds, the neutral, and the earth where the ground rod is installed. If a vehicle is connected to the grounding setup, the voltage will rise on it as well.

Any component not connected electrically to the bonded network will be at a different potential until the circuit trips out or the source of induction is removed. A person working on a pole or structure that is not bonded to the grounds (as in Figure 7–1) will be a parallel path to ground through that pole or structure.

Note: The momentary rise in potential when a grounded circuit is accidentally reenergized is, in reality, quite a bit less than the rated voltage of the line. For example, if a grounded 7200 volts primary is energized, the phase, neutral, and anything else attached to the grounds will probably see about 3400 to 4500 volts. The low resistance of the neutral and grounds will draw down the voltage. However, this reduced voltage is still at a lethal level and shows that grounds alone do not provide the required protection unless the grounds also bond together everything that can be reached.

Equipotential Bonding

7.3.3 The purpose of bonding is to keep everything a person is likely to touch, from a working position, at the same potential. If there is little or no potential difference across a person's body, there can be no current flow.

For example, a person can put a hand on each post of a 12-volt car battery and not feel anything. Even though a car battery can generate about 400 amperes, the potential difference between the two posts is not great enough to break down skin resistance and cause a current flow.

When working on a pole or steel structure, a powerline worker is a parallel path to ground. To reduce the amount of current flowing through a person, the structure should be bonded to the grounded conductors. When a structure is bonded to the grounded conductors, there is little or no voltage difference for a worker to bridge across. Attaching the grounds to the pole, as shown in Figure 7–3, keeps the worker in an equipotential zone.

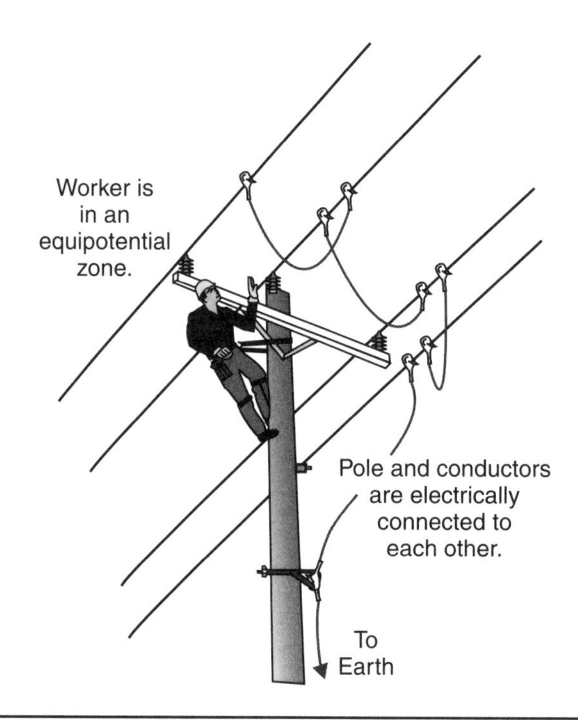

Figure 7–3 Worker in an equipotential zone.

Physical Makeup of Bonds

7.3.4 Portable grounds also act as the bonds, keeping all grounded conductors at the same potential. Theoretically, bonds can be very small and still keep different objects at the same potential. From a practical point of view, they need to be rugged enough to be handled in the field. Standard connectors, such as live-line clamps, would be more rugged and reliable than spring-loaded battery clamps. When a lighter cable is used as a bond to keep different objects at the same potential, it must not be used in a position where it would carry fault current.

7.4 Grounding to Provide Protection from Induction

Grounding to Control Induction

7.4.1 The electromagnetic field from an AC powerline causes two kinds of induction: electric-field (capacitive) induction and magnetic-field (inductive) induction. Section 6.4, *Electrical and Magnetic Induction,* describes the source and effects of induction on linework. Proper grounding and bonding will control, but not eliminate, voltage and current induced on isolated circuits from nearby powerlines.

Grounding for Electric-Field Induction

7.4.2 A powerline will induce a *voltage* on a nearby isolated conductor (Figure 7–4). There is no current flowing in the isolated conductor if it is not part of a circuit.

One set of grounds on an isolated circuit will collapse the voltage due to electric-field induction. One set of grounds on an isolated conductor still does not create a circuit, so there is no circuit for current to flow in. A relatively tiny amount of current will flow through the one set of grounds to earth depending on how long the isolated circuit parallels the nearby live circuit. In other words, if a worker

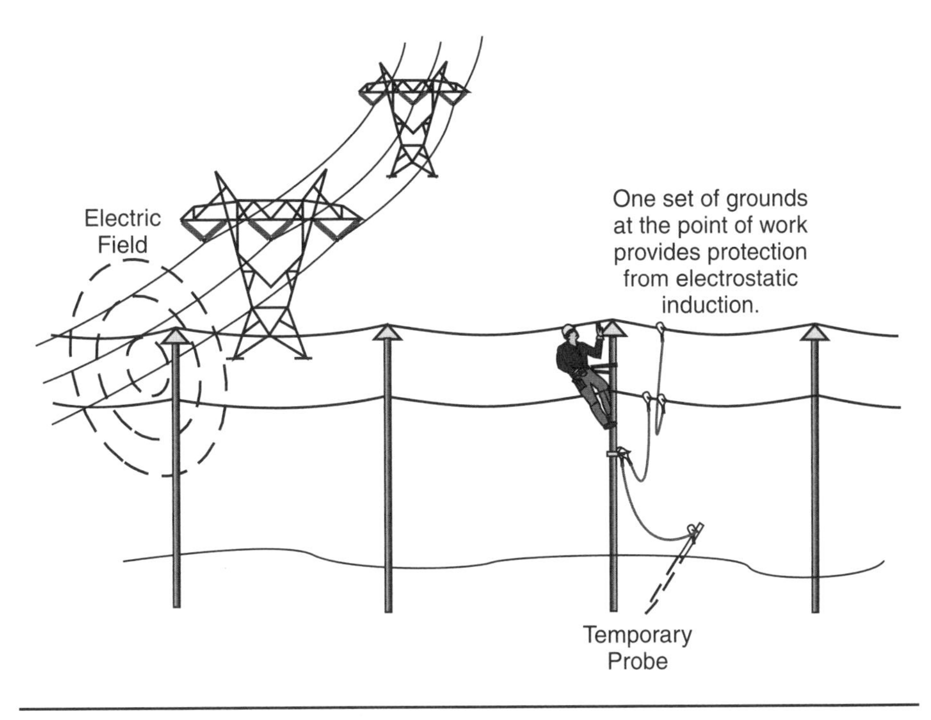

Figure 7–4 Protection from electric-field induction.

was to get between an isolated conductor and the earth, the exposure to the high-induced voltage combined with the relatively small amount of current available could be fatal.

Grounding for Magnetic-Field Induction

7.4.3 A powerline induces *current* on a nearby conductor when that conductor forms part of a circuit. Grounding a conductor on each side of a work zone (*bracket grounding*) creates a circuit through the conductor, down one set of grounds, through the earth, and back up the other set of grounds (Figure 7–5). The first set of grounds installed on an isolated conductor collapses the induced voltage from the electric field; a second set of grounds creates a circuit, which causes an induced current to flow from the magnetic field.

When bracket grounds are removed, the first set of grounds removed opens the circuit and interrupts the current flow. An immediate voltage appears (recovery voltage) across the gap between the ground clamp and the conductor. In high-induction areas, a long arc could be drawn. When there are more sets of grounds installed along the line, it is at the location of the second-to-last ground removed that a long arc could be drawn.

Working and Grounding Between Bracket Grounds: Once bracket grounds are installed, the conductor between the two sets of grounds will have very little induced voltage but will have a potentially high electrical current flowing through it. When grounds are installed at a work site (between the bracket grounds), the work site set of grounds will have (relatively) very little current flowing in it.

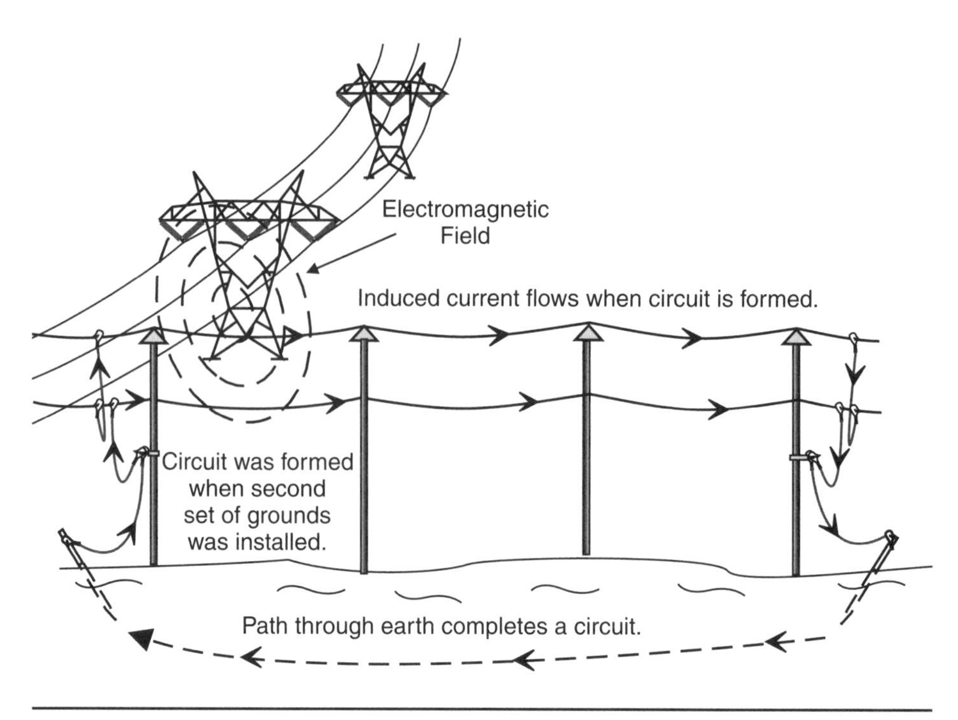

Figure 7–5 Protection from magnetic-field induction.

The work site set of grounds creates two circuits (Figure 7–6) for the ground current to flow where previously there was one. The current from one circuit goes up the work site set of grounds, and the current from the other circuit goes down the work site set of grounds. The two currents tend to cancel one another out, and, therefore, very little current flows in the work site set of grounds.

7.5 Specific Grounding Hazards and Procedural Controls

Cutting or Joining a Grounded Conductor

7.5.1 Working on a grounded conductor is similar to working on a neutral conductor; the voltage may be low but there could be a substantial current flow. If the current flow is interrupted by a break in the circuit, then there could be an immediate high voltage across the break. Anyone bridging across the open circuit will be exposed to this recovery voltage and will be establishing a current flow through the body.

When a live-line tool is used to install a jumper across an open point in a circuit, the hazard is eliminated. Even if a set of grounds is installed on each side of a break in the conductor, there is still likely to be a potential difference across the break unless the two grounds are connected together electrically. Any switchgear or fuse in the work zone that could operate during reenergization should be treated as a potential break or open point when planning the location of the working grounds.

Working on a Circuit One Span from a Single Set of Grounds

7.5.2 In cases of reenergization or induction, there would be a rise of potential on the grounded conductors and on the bonded network. There would be a difference of potential between anything bonded to the grounded circuit and any grounded objects outside of the bonded work zone.

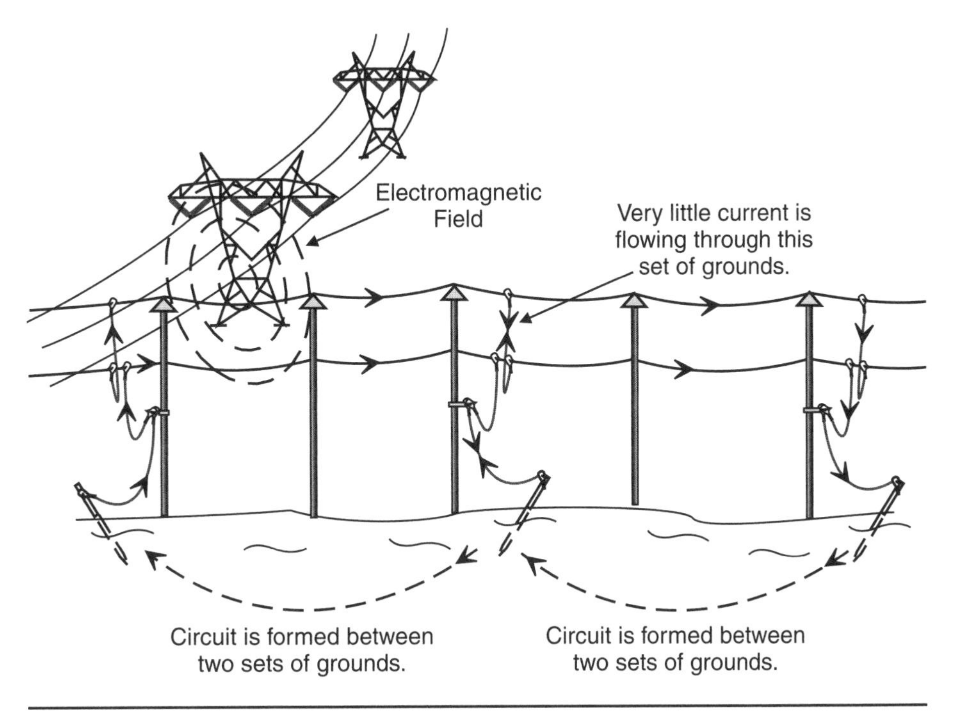

Figure 7–6 Working between bracket grounds.

At the site where a set of grounds is installed, the phase and the neutral or the phase and the steel structure are tied together and will remain at an equal potential during reenergization of the circuit. When working farther away from the installed set of grounds, the potential on the phase will remain high while the potential on a multigrounded neutral or a remote grounded structure will drop quickly, as illustrated in Figure 7–7. The difference of potential between the phase and a neutral or structure can become dangerous when working more than 100 meters (300 feet) from the installed set of grounds.

Electrical Hazards to Workers on the Ground

7.5.3 Accidental reenergization or induction will create *touch and step potentials* around any ground probes, vehicles, wire, and so on, that are bonded to the portable grounds. Most work procedures emphasize protection for the powerline workers aloft, but in rare cases of reenergization or high induction, workers on the ground are very vulnerable unless they are in the bonded zone.

A worker on the ground needs to continuously apply the *bonding principle* for protection from electrical shock. Examples of applying the bonding principle while working on the ground are:

- Stay clear of everything attached to the bonded network.
- When operating a boom, stay on the truck platform.
- If tools are required from a bonded vehicle, use rubber gloves.
- If a ground wire needs to be sent aloft, use a ground-gradient mat.

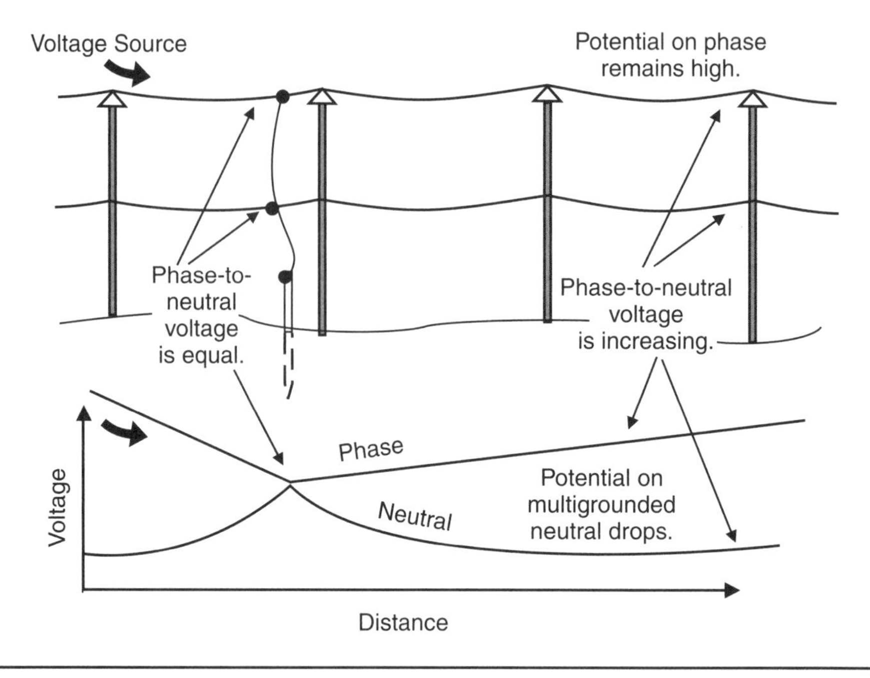

Figure 7–7 Point-of-work grounds.

In a station yard, the ground network and the graveled yard provide some protection from ground gradients.

Wooden Pole Work in Poor Soil or in High-Induction Areas

7.5.4 When working on a three-wire system, especially on transmission lines, a set of grounds will not necessarily reduce the voltage of the grounded conductors to acceptable levels. The voltage on the grounded power conductors and everything bonded to them can be quite high when working in a high-induction area with poor earth for a ground electrode.

A crew could be working on a well-grounded and bonded installation and not be aware that the voltage on their bonded network is high compared to a remote ground. A truck boom or unbonded down guy being sent aloft could become a source of a remote ground if they are not bonded to the grounding setup.

Lowering the voltage on the bonded network provides more protection in case of accidental contact with an unbonded object. When available, grounding to the shield wire (static wire, ground wire, or counterpoise) will lower the voltage on the bonded system. The shield wire can be the best ground electrode available at the work site because it is multigrounded and a direct path to the station ground.

Installing bracket grounds (grounds somewhere on each side of the work site), in addition to installing grounds at the work site, is a very effective method to lower the voltage at the work site. Once the bracket grounds are installed, the electrostatic induction on the circuit becomes negligible. The bracket grounds can be installed in any convenient location that has a good ground electrode.

Grounding Only One Phase

7.5.5 If an accidental reenergization occurs from a source such as a three-phase circuit breaker, then a circuit with all three phases grounded will trip out quickly. There would be no need to rely on the resistance of the earth, because the return path for the fault current would be the phase conductors. Just as there is very little current flow in the shield wire (or neutral) on a balanced three-phase system, there is also very little current flow through the shield wire or ground when the breaker "sees" three grounded phases as a fault. Grounding all three phases results in the desired fast clearing in case of reenergization and reduces the dangerous ground currents for the workers on the ground.

On steel transmission lines, some utilities allow grounding only of the phase being worked on. The steel provides a good ground electrode, and grounding only one phase will trip the three-phase circuit breaker during a reenergization. The workers on the steel structure will be working in a bonded area, but workers on the ground should stay clear of the tower.

Grounds Whipping during a Fault

7.5.6 During a high-fault-current condition, the portable ground cable can whip violently. The ground cable can also burn off if it is wrapped around a steel structure or is left coiled. The huge magnetic field around the ground cable when it is carrying a high fault current results in major mechanical forces being exerted on the cable. Use a rope to tie off grounds where workers could be exposed to the whipping action. Keep grounds as short as practical and lay out the ground so it is not coiled or wrapped around steel.

Grounding for Backfeed from a Portable Generator

7.5.7 Backfeed from a generator improperly connected at the customer can feed back through the secondary of the transformer and create a high potential on the primary side. When working on an isolated circuit, grounding the circuit will reduce the voltage due to backfeed from portable generators to acceptable levels. However, grounding the primary conductors will probably not cause the generator to fail. The transformer impedance and the resistance of the conductor between the grounds and the generator will not cause a large enough fault current to cause the generator to overload.

Backfeed from larger industrial or commercial generators would cause higher voltages to appear on grounded conductors at the work site, but because these generators are more likely to be installed and inspected properly, backfeed is extremely unlikely. The hazard from backfeed is from the unknown generators.

During a backfeed situation, the voltage on a grounded conductor may be acceptable but the generator also produces a current flow in the conductor. Working on a grounded conductor is like working on a neutral. The conductor should not be cut or joined without first installing a jumper.

7.6 Protective Grounding of Underground Cable

Reasons to Ground Underground Cable

7.6.1 The reasons to ground underground cable are:

1. Install protective grounds to prove isolation. Underground cable is difficult to trace. Therefore, it is critical to test the cable at a riser, test at the

capacitive test point at an elbow, and where applicable, spike the cable before placing grounds.

2. Install protective grounds to have protection from accidental reenergization. When working on a grounded cable, a set of protective grounds must always be in place to have protection from accidental reenergization.
3. Install protective grounds to provide protection from induction. Underground cable is a capacitor that can maintain a charge for a long time. A ground must be installed and maintained to drain these charges.

Applying the Grounding Principle

7.6.2 A good ground is needed to trip out the protective switchgear in case of accidentally grounding a live circuit or in a case of accidental reenergization.

Apply grounds to cable wherever it is possible to get access. Install grounds at risers or at any terminations where the conductor is exposed. Install grounds at switching cabinets where there is air-insulated switchgear and an exposed location to place grounds. Use a clamp on the bottom of the switchgear similar to the device shown in Figure 7–8.

Take advantage of using grounding elbows (Figure 7–9) where applicable because they are equipped with arc-quenching tips similar to standard load-break elbows.

When working on a cable between grounded terminations, spike the cable. Before making any cut or exposing the cable, install a jumper across the intended cut on the sheath and apply the bonding principle before handling the conductor.

Applying the Bonding Principle

7.6.3 Applying the bonding principle means that a powerline worker establishes and works in an equipotential zone. Bond the phase conductor to the sheath (con-

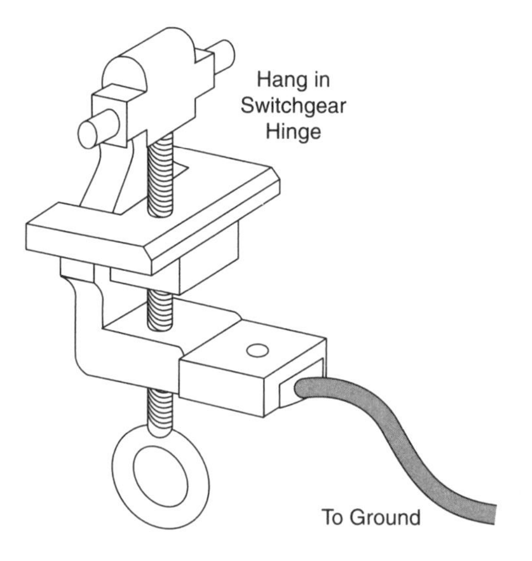

Figure 7–8 Ground clamp attached to live front switchgear.

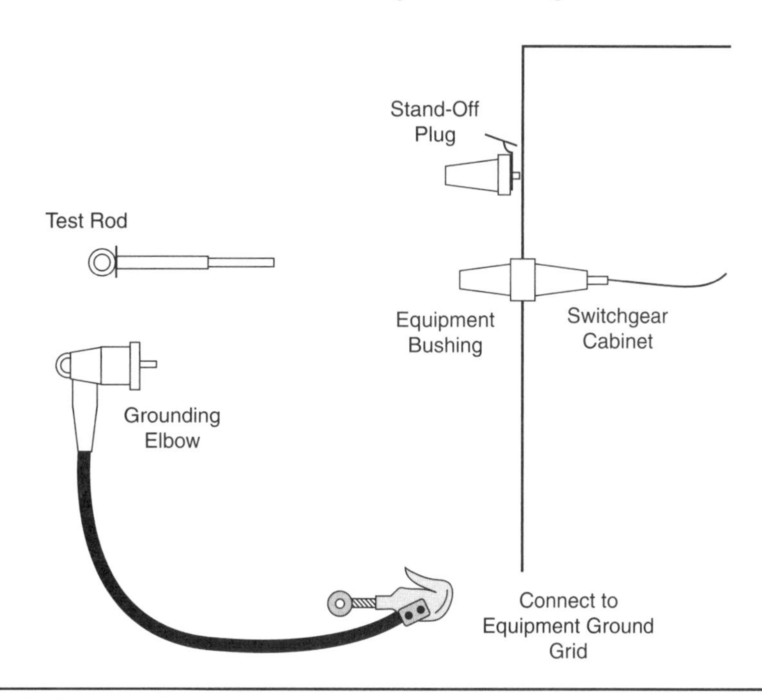

Figure 7–9 Dead front grounding jumper and elbow.

centric neutral) and to the cabinet and grounding network of the cabinet. If the sheath has to be disconnected from ground at the termination, install a portable protective ground on it.

When working on a cable, after the terminations are grounded and after the cable is spiked, install and stay on a ground gradient mat bonded to the concentric neutral and the spike. The grounds at the terminations will help keep the cable and sheath bonded together when the spike is removed and the bond transferred to the cable when practicable. With the bonds in place, a powerline worker on the mat should not experience a difference in potential when working on the cable from the mat. Maintain the equipotential bonded zone by placing a jumper across a cut or splice.

7.7 A Case Study of a Line Grounding Incident

The Incident

7.7.1 A powerline worker received a severe electrical shock when he reached out to grab the winch cable of a crane truck while his other hand was holding on to an isolated and grounded conductor. The analysis of this accident illustrates some of the complexities involved in line grounding.

The job was to change timbers on a 115kV twin pole line during a clearance. The 115kV line ran parallel to a live 230kV line for about 35 miles. A crane truck was being used to lower the conductors to the ground. When the lineman grabbed the winch in one hand and the conductor in the other hand he shouted and seemed to be frozen onto the winch cable and conductor (Figure 7–8). The supervisor on the job sensed what happened and had the crane operator take up the winch to release the victim's grip. The victim was very shaken but said he was all right and planned to climb down. The crew insisted on putting a rope around him, as practiced in pole-top rescue, as a backup while the victim climbed down. He was taken to a medical center, examined, and released.

How the Structure Was Grounded

7.7.2 The crew had been trained in equipotential grounding and was aware that there could be a lot of induction from the neighboring 230kV line. The three 115kV power conductors were grounded to each other, connected to a snubbing band on each pole, and then connected to a temporary driven ground probe in the earth. The three grounds on the conductor and the grounds tied back to the pole meant that everything the linemen were expected to touch was bonded together. The shield wire, a rusty 3/8-inch steel, was grounded as was every structure on the line, with a #4 copper downground to a ground rod buried at the base of the pole. The crew did not tie their working grounds to the shield wire. The grounds on the conductors and the bands around the pole kept the area where the linemen were working aloft all bonded together and in an equipotential zone. The temporary

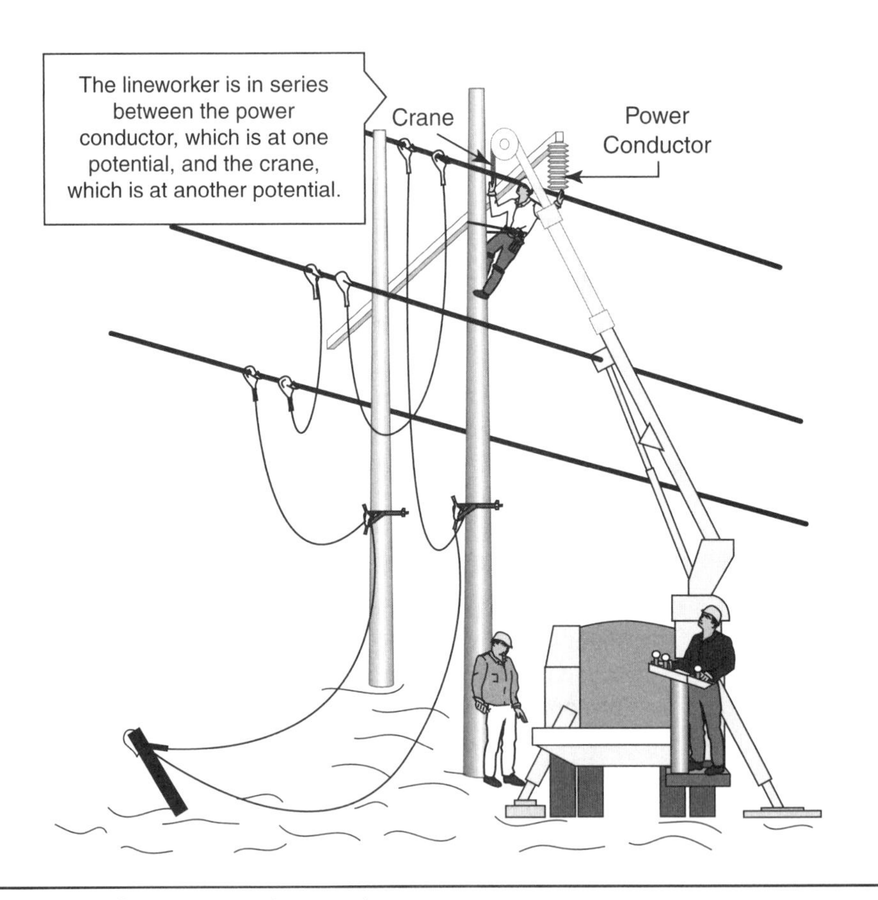

Figure 7–10 Line grounding incident.

ground probe was driven in relatively dry sandy and rocky soil. The truck was not grounded because that was only done when working in the vicinity of a live circuit.

The Source of Potential Difference

7.7.3 The induction from the 230kV line was quite intense, but from the crew's point of view the line they were working on was grounded and everything on the pole was bonded together giving them an equipotential zone. The soil where the temporary ground rod was driven was not a good ground electrode, and while it may have lowered the potential on the power conductors, it did not bring it to zero in relation to a remote ground. The equipotential grounding, with everything bonded together, meant that the linemen working aloft did not feel any discomfort.

When the boom from the crane truck was brought into the equipotential zone, it represented a remote ground and was at a different potential than the power conductors. It was not bonded into the equipotential zone.

The linemen working on the poles did not feel anything when working on the conductors. Everything on the pole, however, was not at the same potential because the shield wire was not bonded into the equipotential zone. The working grounds tied into the snubbing band around the pole, and around the downground, probably prevented the linemen from receiving shocks from the remote ground represented by the shield wire and downground. The shield wire was probably the best ground electrode available at this site.

The conductors were safe to work on because everything was bonded together and there was nothing that introduced another potential until the ungrounded truck boom was brought into the equipotential zone.

Another Source of Potential Difference Was Created

7.7.4 After the accident, the crew grounded (bonded) the truck to the same ground probe as the conductors. This eliminated the difference in potential between the truck boom and the conductors.

Later, when a worker walked up to the truck to get material from a bin he received a relatively uncomfortable shock from the truck. The truck was now at the same potential as the grounded conductors.

To protect workers touching the truck, a ground rod was driven near the truck and the touch potential was reduced to a level that was not noticeable.

Conclusions

7.7.5 One of the conclusions from this accident investigation is that the shield wire should also be grounded and be part of the equipotential zone. In fact, even though a rusty 3/8-steel wire does not look like a good ground electrode, it was demonstrated that the voltage on the conductors was lower after grounding to the shield wire.

It is interesting to note that for this accident scenario, had the truck been grounded to a driven ground probe but not bonded to the conductor grounds the victim would probably have had a more severe shock because there would have been a higher potential difference between the power conductor and a truck boom that was grounded than the actual situation where the truck was not grounded.

It is not common to ground a truck when working on an isolated and grounded line. This accident illustrates that under some circumstances a truck still needs to be grounded and bonded to the grounds on the conductor. While this accident happened on a transmission line, this situation can easily apply to lower-voltage, double-circuit lines when working on one circuit and the other is alive. The proximity and the load on the live circuit will determine the amount of current that will be induced on the other circuit.

The old expression "it's not dead unless it's grounded" is not quite accurate. A grounded circuit can still have a voltage on it in relation to a remote ground. Grounds can also carry a current that would not be noticeable until the grounded circuit is opened or broken. Maybe a better expression would be, "it's not safe unless grounded and bonded."

Review Questions

1. What are three main reasons to install grounds and bonds on an isolated circuit before work is started?
2. When using protective grounds, does the application of the *grounding principle* control current or voltage?
3. Does all the current in a grounded conductor take the easiest path to earth?
4. What is the *bonding principle*?
5. When removing protective grounds on an isolated circuit, which set of grounds is more likely to create an arc, the first set or the second set?

CHAPTER 8

Supplying Quality Power

8.1 Introduction

More Sensitivity to Power Quality

8.1.1 Electrical disturbances in powerlines and on customer premises appear to be increasing. This is due to a greater amount of disturbance-producing electrical equipment and to an increased sensitivity of certain customer loads. Other than power outages, a variation in voltage is the most noticeable and obvious power-quality problem. There are also power-quality problems that are not picked up by a voltmeter. It is worthwhile for powerline workers to recognize the signs of poor power quality when checking customer complaints of erratic power problems.

8.2 What Is Power Quality?

Definition of Power Quality

8.2.1 A quality power supply is one where the AC and voltage rise and fall at a rate that can be represented by a sine wave. Any deviation in the magnitude or frequency of the 60-hertz sine wave is considered a power-quality disturbance. Poor power quality affects the performance of electrical equipment in an adverse manner.

Power Supply Disturbances

8.2.2 There have always been electrical disturbances in the supply of power. The increased sensitivity of certain loads causes these disturbances to be unacceptable to the customer. Momentary disturbances that affect the quality of power can be due to:

- Switching surges, fault clearing, and capacitor switching
- Voltage flicker from starting large motors or arc welders
- Transient faults and operation of surge arresters

Continuous disturbances that affect the quality of power can be due to:

- An unacceptable range of voltage rise and fall
- Intentional voltage reductions (brownouts) during periods of peak load
- Unbalanced voltages between phases
- Harmonic distortion
- Tingle voltage
- Electrical noise (radio and television interference)

Modes Where Disturbances Are Measured

8.2.3 There are two modes or means where the voltage can be erratic. These modes refer to the points where the unwanted voltage is measured. The *differential mode* (also called the *normal mode*) refers to disturbances between phases or between the phase and the neutral. Most voltage problems on the supply system would be differential mode and the source of trouble calls looked into by the lines trade.

The *common mode* refers to disturbances between the neutral and the ground. This is also called *noisy ground.* Most common-mode problems would be on a customer's premises. For example, tingle voltage involves a voltage between the neutral and the ground. Trouble in this mode would normally involve engineering staff.

Corrective Measures Available

8.2.4 Customers understand that a utility cannot completely eliminate power outages, but some types of customers are becoming less tolerant to momentary outages and other disturbances. Often a power-quality problem suffered by a customer comes from equipment on the customer's own service, such as large motors or arc-welding equipment.

A utility can improve power quality on a feeder by installing voltage regulators, capacitors, and surge arresters. It can even build a dedicated substation or dedicated feeder to a customer who would be willing to pay more for a quality power supply with fewer disturbances.

Some customers own an uninterruptible power supply (UPS), which is a back-up power supply to maintain service from batteries or some kind of generator during a power failure. A UPS will not protect the customer from voltage surges. Surge arresters can be installed on customer equipment for extra protection from voltage surges. Filters are available for installation on customer equipment to limit harmonic interference.

A Frequent Fix for Power-Quality Problems

8.2.5 Poor grounding is often a cause of power-quality problems. In most cases, poor or improper grounding is within the customer's premises. Good grounding is important to the customer because grounding balances the electrical system by bleeding off overvoltage or overcurrent.

There is a difference between a neutral and a safety ground. The neutral is intended to carry current. A safety ground is usually a small wire intended to drain away voltage that occurs during a failure of equipment such as appliances and tools.

The neutral and the safety ground are normally tied together both at the transformer supplying the service and at the service panel. The neutral and the ground should not be tied together anywhere else on the load side of the main service disconnect because a circuit will be formed through the ground and the neutral. The safety ground wire will then share with the neutral the job of carrying current back to the source. The resulting ground currents can cause tingle voltage as well as disturbances to the normal operation of electronic equipment in the customer's premises.

Supplying Custom Power

8.2.6 *Custom power* is the ultimate solution to power-quality problems. Instantaneous voltage regulation, voltage flicker control, reduced harmonics, and no momentary outages are possible with specialized equipment installed on the utility system. The equipment uses the technology developed for high-voltage DC (HVDC) transmission and flexible AC transmission systems (FACTS). An electronic controller can convert DC from a back-up source to AC with any wave shape needed.

Modern electronic rectifiers and inverters convert AC to DC and vice versa. High-speed switching (less than 10 milliseconds) is possible with solid-state breakers (SSB), which are practically instantaneous. A back-up DC source provides power to an electronic controller, which can counter voltage dips, momentary outages, voltage flicker, and changes in reactive power.

For example, a dynamic voltage restorer (DVR) is an injection-transformer device that is installed in series with a circuit. Capacitors in the DVR maintain an internal bus, which is a DC power source that an electronic controller can draw on to supply AC power. During a disturbance, the electronic controller reshapes the power wave back to a proper sine wave by injecting real or reactive power as needed.

Similar technology is used in a distribution static compensator (DSTATCOM), which is connected into the system like a shunt capacitor. An electronic controller will input and take out real power and reactive power as needed from a rechargeable energy-storage system.

Other devices are static VAR compensators (SVC) and adaptive VAR compensators (AVC), which are electronic devices that instantaneously input reactive power to counter changes in a supply system.

8.3 Factors Affecting Voltage in a Circuit

Voltage as a Measure of Power Quality

8.3.1 There is a voltage drop along every circuit and through every transformer. The extent of the voltage drop depends on how much the current flow is impeded by the device the current is flowing through. For example, more current flow is impeded in a long length of a conductor than in a short length of a conductor and more current flow is impeded by a small-diameter conductor than by a large-diameter conductor.

Voltage fluctuates throughout the day and throughout the seasons in proportion to fluctuation of the electrical load (current flow). Almost all circuits need some kind of additional voltage regulation and control. Some loads are very sensitive to voltage variations. It is important to keep the voltage to a customer within an acceptable range. High or low voltage is noticed by a customer, and both high and low voltage can damage customer equipment.

Voltage must be maintained within a set standard. For example, the American National Standards Institute (ANSI) is +6 percent or −13 percent for 120, 208, 277, 480, or 575-volt services.

Voltage Control and Voltage Regulation

8.3.2 The terms *voltage regulation* and *voltage control* tend to be used interchangeably. *Voltage control* refers to the direct method of voltage change, such as changing a transformer output with transformer taps or changing the feeder voltage with line-voltage regulators.

Voltage regulation refers to the indirect method of keeping voltage at a proper level. Voltage is regulated by ensuring that the conductor size and distance and the power factor of a circuit are adequate. Improving the power factor in a circuit reduces the amount of apparent power or current needed to supply the load. Less current in the circuit reduces the voltage drop.

Voltage Drop

8.3.3 When the load in an electrical system increases, the voltage at the load decreases. The amount of current flow affects the voltage drop. The following equations show that line loss and voltage drop are related to current flow.

$$Power\ loss = I^2R \qquad Voltage\ drop = IR$$

Current flow varies according to the amount of customer load and the impedance offered by the powerlines and transformers feeding the load. Resistance is the largest component to the total impedance of current flow in a circuit. The design of the electrical-power system includes keeping losses as low as practical because line loss in a powerline or transformer is wasted energy.

Voltage Regulation

8.3.4 *Voltage regulation* is defined as the difference between no-load voltage and full-load voltage, expressed as the percentage of full-load voltage.

$$\%V\ regulation = \frac{(no\text{-}load\ V) - (full\text{-}load\ V)}{no\text{-}load\ V} \times 100$$

For example, if a station transformer delivers 4900 volts at no load and 4800 volts at full load, the voltage regulation on the transformer is:

$$\frac{4900 - 4800}{4900} \times 100 = 2\%$$

If there is no load on a transformer, it would have a near-perfect voltage regulation of 0 percent. However, because load changes constantly, an electrical system should be designed so that the voltage regulation does not exceed a range of 2 to 3 percent.

Conductor Size and Length

8.3.5 Conductor size and circuit length affect the magnitude of the voltage drop in a circuit. A large-diameter conductor offers less resistance to current flow than a small-diameter conductor. Less resistance reduces line loss and therefore reduces voltage drop. The distance to a customer affects the voltage, because a longer conductor imposes more resistance to current flow than a shorter conductor.

Daily Changes

8.3.6 Because a load increase can result in increased voltage drop, voltage-control equipment must adjust the voltage to reflect the load changes during the day. A load profile of a customer at the end of a residential feeder is displayed on the graph in Figure 8–1.

Seasonal Changes

8.3.7 The heat in summer and the cold in winter produce the peak-load periods for an electrical system. Voltage-regulation studies by a planning engineer consider these peak loads as the benchmarks for worst-case, low-voltage conditions.

Peak-load statistics are used to determine where an upgrade is needed to substation transformers, the conductor size of feeders, or the number of feeders, voltage regulators, and capacitors.

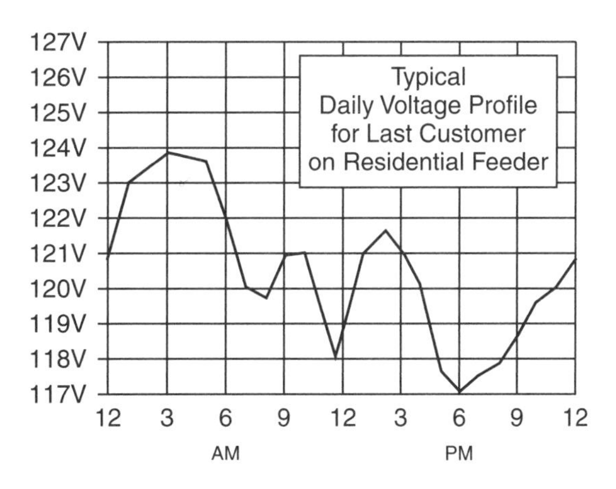

Figure 8–1 Daily voltage profile.

Reactance in a Circuit

8.3.8 The amount of reactance in a circuit adds to the total impedance of the circuit. If the current was kept in phase with the voltage, there would be very little voltage drop due to reactance. Transformers, motors, and fluorescent lighting have an inductive component in their energy demand, and, therefore, an inductive reactance is set up in the distribution feeder. Additional impedance caused by reactance in a circuit results in the need for additional current to feed the load. A higher current will cause a greater voltage drop.

8.4 Voltage on the Transmission Lines System

Voltage Control of Transmission Lines

8.4.1 Transmission line voltage is regulated and controlled by equipment at substations. The voltage at the source of a transmission line can be boosted so that the voltage at the load end of the line is at the required level. This method would not work on a distribution line because customers are usually spread out along the whole circuit. Voltage can be boosted by changing the transformer voltage taps to alter the transformer ratio. Most of these taps are designed to be changed under load. Transformer taps need to change position regularly to keep the voltage constant as the load changes.

At the end of the line, voltage can be regulated by installing capacitors. Capacitors improve the power factor on a circuit. An improved power factor results in a lower current and, therefore, a boost in voltage. Relays control the amount of capacitance needed to maintain a relatively constant voltage.

Parallel conductors on very long transmission lines increase capacitance on a circuit to a level where the capacitive reactance causes a high impedance to current flow. At transmission substations, reactors are installed to input an inductive reactance, balancing the capacitive reactance of the long line.

8.5 Distribution Substation Voltage

Voltage at the Distribution Substation

8.5.1 A distribution substation is the source for distribution feeders. The voltage at the source of the feeder must be high enough to provide an adequate input voltage to transformers and voltage regulators feeding customers downstream. The voltage at a substation can be corrected by:

- Adjusting the subtransmission line voltage
- The automatic operation of a load tap changer (LTC) at the substation transformer, if equipped
- The changing of the no-load tap changer at the substation transformer, if equipped
- The automatic operation of a feeder voltage regulator in the distribution substation, if equipped

Subtransmission Line Voltage

8.5.2 The voltage of a subtransmission line can be controlled at the source substation. If a subtransmission line is short and does not feed many distribution substations, the voltage stays fairly constant along the full length of the line. The voltage on a long subtransmission line that feeds multiple distribution substations cannot be controlled to suit the needs of each substation. Each distribution substation will need its own voltage control to supply the distribution feeders. Voltage regulators can be installed along a subtransmission line to ensure that the distribution substation receives an acceptable voltage.

Voltage Control at a Distribution Substation

8.5.3 Tap changers at a distribution substation transformer can adjust the voltage by changing the ratio between the primary and the secondary coils of the transformer. Similar to a line voltage regulator, a transformer LTC can boost or buck voltage as needed while the transformer is in service.

A no-load tap changer requires all the load to be dropped from the transformer while the tap change is made. A transformer with a no-load tap changer cannot make regular adjustments during the day and would depend on the subtransmission line source or line step voltage regulators to regulate the voltage for daily adjustments.

Feeder Voltage Regulator in a Substation

8.5.4 A feeder voltage regulator is sometimes used in small or lightly loaded substations where the substation transformer is not equipped with an LTC. A feeder voltage regulator is more commonly installed downstream from the substation on long, individual feeders. The voltage is boosted or bucked as needed to ensure that the customers receive a voltage within the standard range.

8.6 Distribution Feeder Voltage

Voltage Profile of a Feeder

8.6.1 The design of a distribution feeder facilitates keeping the voltage drop along every element of the circuit to a minimum. To provide an acceptable voltage to the customer, each part of the distribution system (Figure 8–2) must also have an acceptable voltage.

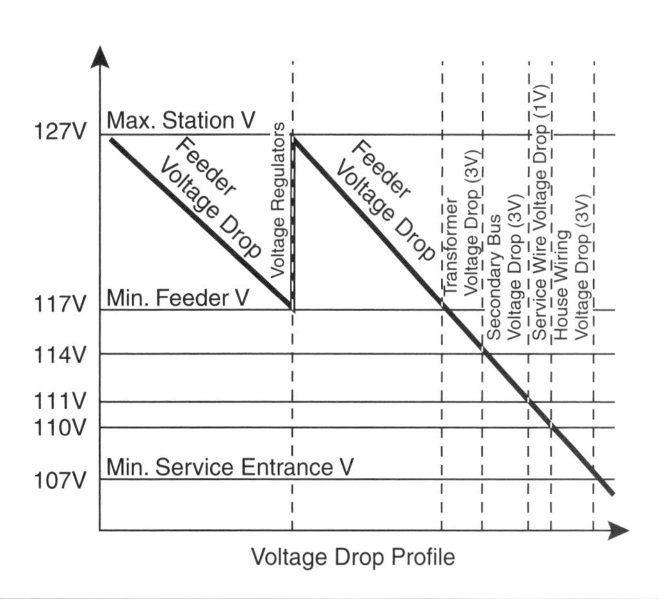

Figure 8–2 Feeder voltage profile.

Conductor Size

8.6.2 The amount of current in a circuit is a prominent factor affecting voltage drop. The less resistance there is to the current flow, the less line loss or voltage drop there will be. A large conductor over a short length offers the least resistance.

An ampacity chart to determine conductor size for a utility circuit is not very applicable. For example, the ampacity chart for conductors may show that a 3/0 aluminum conductor has a capacity to carry 255 amperes. However, at 255 amperes, the voltage would drop about 1 percent every 10 meters (39 feet), which means it can carry a 120-volt service at that current for about 60 meters before the voltage is below standard.

Distribution circuits have larger conductors than are needed for ampacity. A larger conductor is used to reduce the voltage drop and line loss on a circuit. Less line loss also allows circuit breakers or fuses to "see" a short circuit downstream. For example, a short lead in and out of a set of voltage regulators would need to be sized to carry the current but does not need to be the same size as the main line conductors. The use of a smaller conductor for the short distance involved in and out of the voltage regulator would not reduce the voltage significantly.

The cost of conductors affects the size of a conductor chosen for distribution feeders. Usually, a utility decides on a few standard sizes for their feeders and selects conductors large enough to keep the voltage drop within practical and economical limits.

Locating Feeder Voltage Regulators

8.6.3 To improve the voltage on a feeder, it is usually more economical to install voltage regulators than to carry out a major betterment—such as stringing larger conductors, converting single-phase to three-phase, or installing additional feeders. Feeder voltage regulators should be installed in a location where the voltage is still high enough that the voltage is boosted from an adequate base. Voltage regulators step up voltage in percent; for example, 2 percent of 117 volts boosts the voltage to 119 volts, which stays at a higher level for a longer distance than 2 percent of 110 volts (112 volts).

As a point of interest, a rule of thumb states that it requires 1000 volts to feed one mile of a normally loaded circuit before the voltage needs boosting. A 4800-volt circuit should be able to feed about 5 miles before needing a boost.

Locating Capacitors

8.6.4 Installing a capacitor is the most economical way to improve the voltage on a feeder and also benefit the whole electrical system. When the power factor of the circuit improves, there is less apparent power needed to feed the load and, therefore, less current and less voltage drop. Capacitors can be in a substation or downstream on a distribution circuit.

A capacitor is a capacitive load on the system used to offset the inductive load put on the system by motors, transformers, and fluorescent lights. An improved power factor of the circuit affects the amount of current and the power factor of the circuit upstream. On a distribution feeder, capacitors are located near load centers to reduce the need for apparent power on the circuit. The capacitor reduces the current flow and reduces the voltage drop in the circuit.

Transformers

8.6.5 Transformer losses can add to the total line loss in a circuit. The loss presented by a transformer can be unnecessarily excessive when larger-than-needed transformers are installed. The current required to excite the iron core is constant regardless of load. The greater the load on the circuit, the greater the voltage drop.

A transformer with a tap changer can be a quick solution to a voltage problem for an individual customer. Individual distribution transformers with tap changers change the output voltage by changing the number of turns on the primary coil. Depending on the manufacturer, each tap change raises or lowers the secondary voltage by 4.5 percent or 2.5 percent. Tap changers on distribution transformers are *no-load* tap changers. The transformer must be deenergized before turning the tap-changer handle.

8.7 Feeder Voltage Regulators

Operating Principle

8.7.1 The most common feeder voltage regulator used on a distribution feeder is a step voltage regulator. A step voltage regulator corrects excessive voltage variation and raises or lowers the voltage during low- or high-voltage conditions. Step voltage regulators are normally used on long rural feeders or at small, older distribution substations where the transformer is not equipped with an automatic under-load tap changer.

A step voltage regulator works on the same principle as an autotransformer. On an ordinary distribution transformer, the primary and secondary coils are coupled magnetically; on an autotransformer, the primary and secondary coils are connected magnetically and electrically, as shown in Figure 8–3. The secondary coil is connected in series with the primary. If the transformer ratio was 10 to 1, then the voltage on the load side would be 10 percent higher than the source voltage.

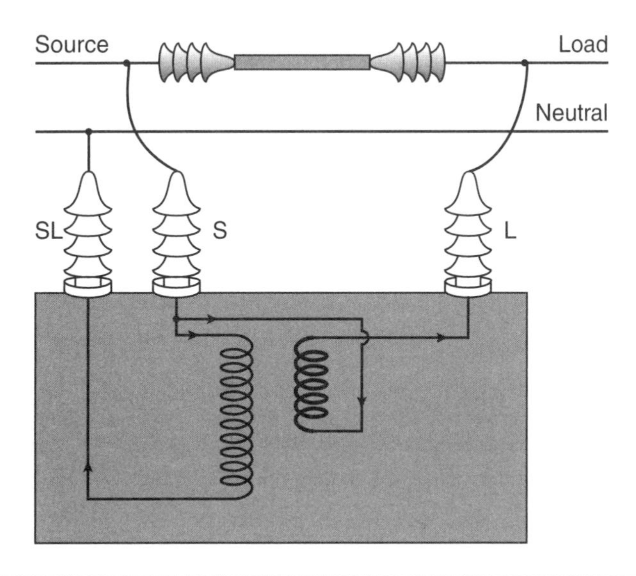

Figure 8–3 An autotransformer.

Step-Voltage Regulator Defined

8.7.2 A voltage regulator is an autotransformer with an under-load tap changer on the secondary coil. It has the ability to step the voltage up or down in small, incremental steps by moving a contact along the series-connected secondary coil sections. If the ratio of the primary to the secondary is 10 to 1, then the secondary voltage at the highest tap would be 110 percent. In Figure 8–4, the 10-to-1 ratio regulator has the secondary coil divided into eight equal sections. Each tap change would change the voltage by 10 ÷ 8 = 1.25 percent. A reversing switch allows the regulator to either boost or buck the voltage eight steps each way to make this regulator a 16-step regulator.

Typical Step-Voltage-Regulator Nameplate

8.7.3 A typical step-voltage-regulator nameplate is shown in Figure 8–5. It shows the voltage settings of the regulator and the types of connections available.

Regulator Controls

8.7.4 A potential transformer (PT) is installed on the output of the regulator. It sends a representative voltage to the control box. For example, on a 7200-volt primary system, a 60:1 PT is installed. If the voltmeter at the control box reads 120 volts, then it is known that the primary voltage is 7200 volts. This meter, therefore, tells what the *actual* voltage level is on the regulator output. The voltage-level control knob is set to the *desired* output voltage, and the regulator raises or lowers the actual output to the desired level.

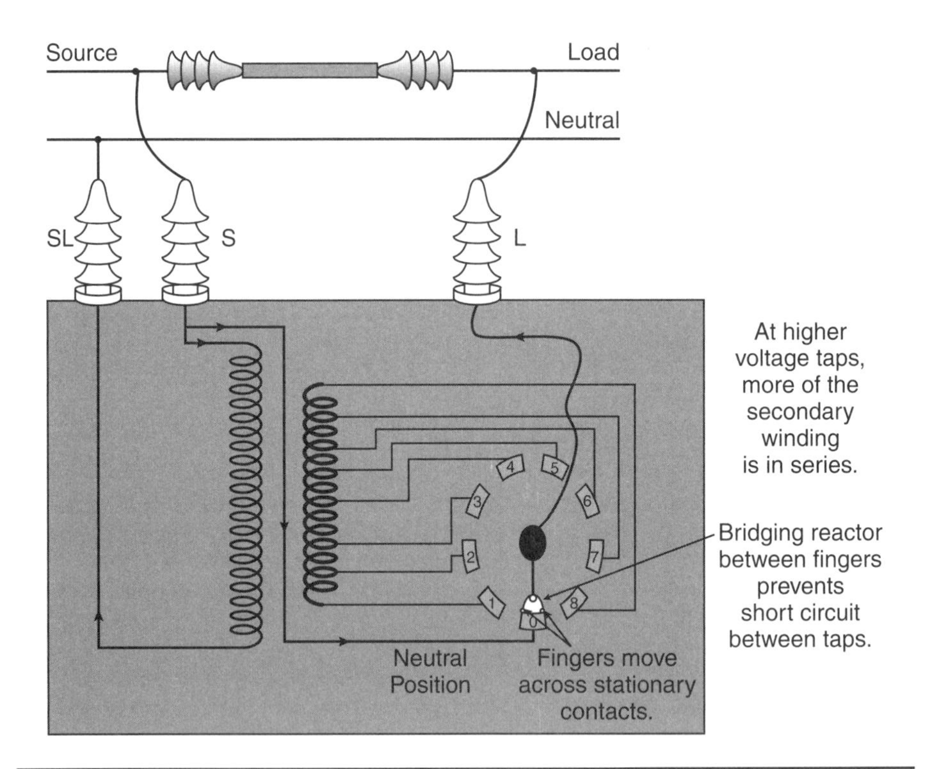

Figure 8–4 A step-voltage regulator.

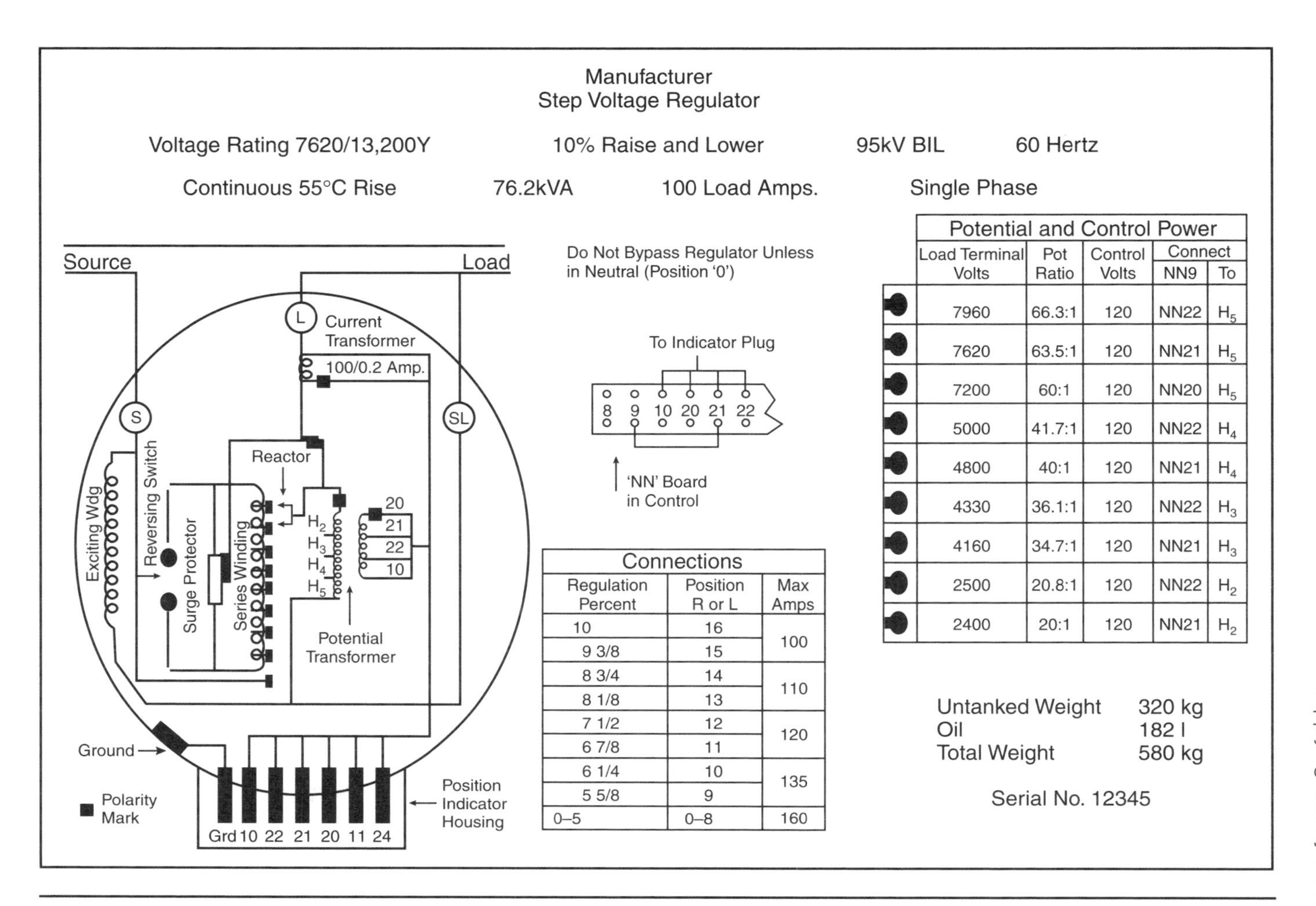

Connections		
Regulation Percent	Position R or L	Max Amps
10	16	100
9 3/8	15	
8 3/4	14	110
8 1/8	13	
7 1/2	12	120
6 7/8	11	
6 1/4	10	135
5 5/8	9	
0–5	0–8	160

Potential and Control Power				
Load Terminal Volts	Pot Ratio	Control Volts	Connect NN9	To
7960	66.3:1	120	NN22	H_5
7620	63.5:1	120	NN21	H_5
7200	60:1	120	NN20	H_5
5000	41.7:1	120	NN22	H_4
4800	40:1	120	NN21	H_4
4330	36.1:1	120	NN22	H_3
4160	34.7:1	120	NN21	H_3
2500	20.8:1	120	NN22	H_2
2400	20:1	120	NN21	H_2

Figure 8–5 Regulator nameplate.

The settings in a control panel, shown in Figure 8–6, indicate the desired operation of the regulator. The *voltage-level* setting is the desired voltage output of the regulator. If the voltage level drops below the voltage shown on the voltage level setting, the regulator automatically moves up one tap to boost the voltage.

Bandwidth Setting

8.7.5 A voltage regulator changes taps in steps that do not result in the exact desired voltage. For example, on a typical voltage regulator, each step changes the voltage 5/8 percent, which brings the voltage close to the desired voltage but is probably not exact. The regulator would keep on boosting and bucking (hunting), trying to reach the desired voltage if some tolerance for an inaccuracy was not available.

The *bandwidth* setting allows some variation from the actual desired voltage setting. A voltage setting of 125 volts and a bandwidth setting of 2 volts will cause the regulator to maintain a voltage between 124 and 126 volts. The difference between the minimum and maximum voltages allowed is the bandwidth. The bandwidth indicator lights in the control panel indicate when the output voltage is outside of the bandwidth.

Time-Delay Setting

8.7.6 The voltage on a circuit can dip temporarily, such as when a customer's large motor is started. To save the regulator from reacting to every voltage change and to reduce unnecessary operations, a time-delay switch, found in the control box, is set to delay the operation of the regulator. The *time-delay* setting delays the operation of the regulator long enough to avoid needless operations. A time delay of 30 seconds is common and prevents the regulator from starting to adjust the voltage each time the output voltage is outside of the bandwidth setting.

When a downstream voltage regulator makes a voltage adjustment, it does not affect the upstream regulator. However, each time an upstream regulator makes a voltage change, the downstream regulator also senses a need for a voltage change and starts an unnecessary operation. Therefore, the time delay on a regulator downstream from another regulator should be set at least 10 seconds longer so that it does not react immediately to the upstream regulator.

Compensation Settings

8.7.7 The line-drop compensation settings are an option used to supply a constant voltage at a point downstream, remote from the regulator. A voltage regulator *without compensation* keeps the voltage constant at the output terminal. As the load current changes, the voltage at the output terminal stays constant, while the voltage drops at the end of the line. As shown in Figure 8–7, the voltage starts to drop and continues to drop as the distance from the source regulator increases. To supply the end of the line with an adequate voltage, the voltage setting at the regulator would need to be increased. Customers close to the regulator would then have a constant high voltage.

A regulator with *line-drop compensation* keeps the voltage swings on the circuit to a minimum when the load current changes. Instead of keeping a constant voltage at the output terminal, the compensation settings give the regulator the ability

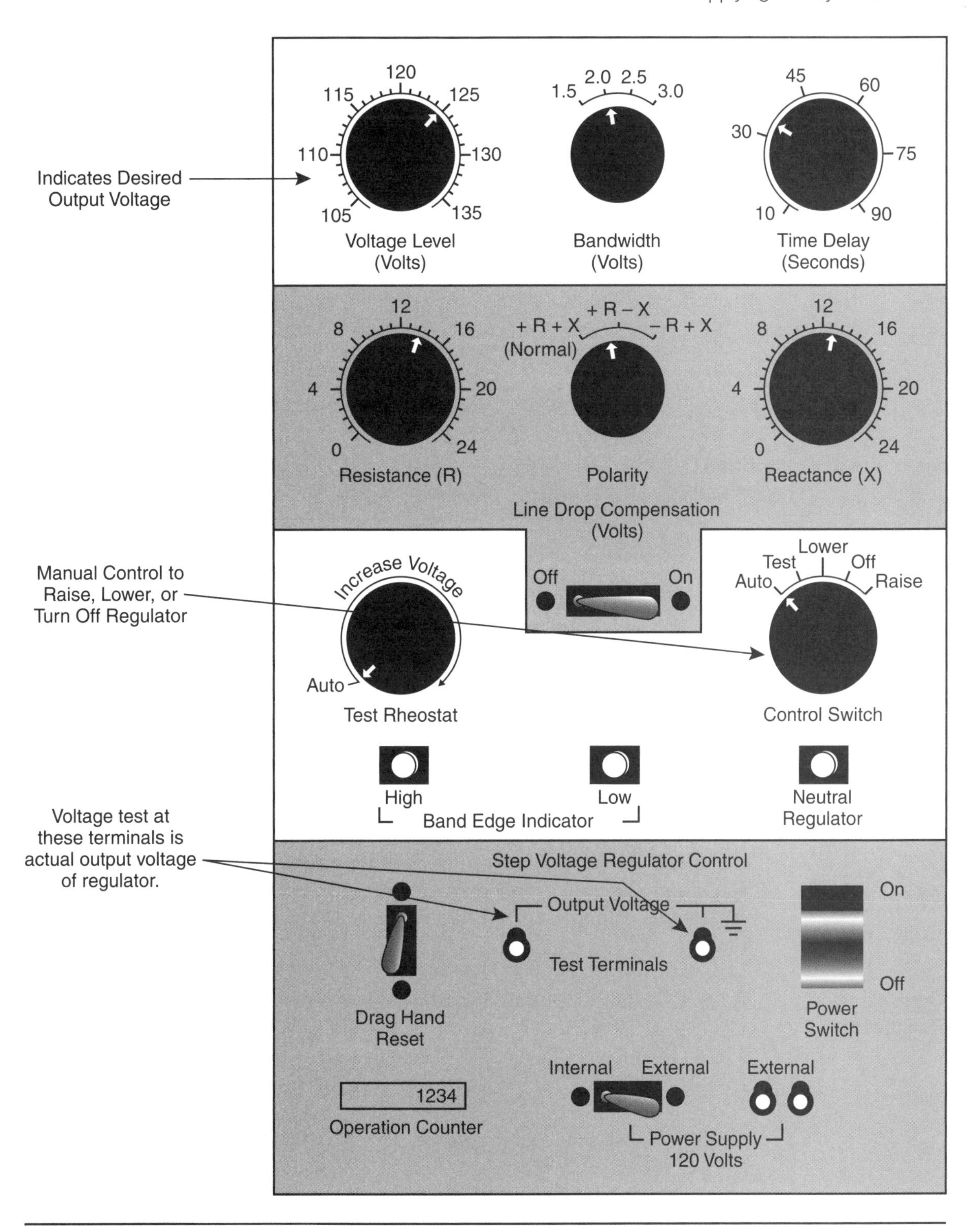

Figure 8–6 Regulator control.

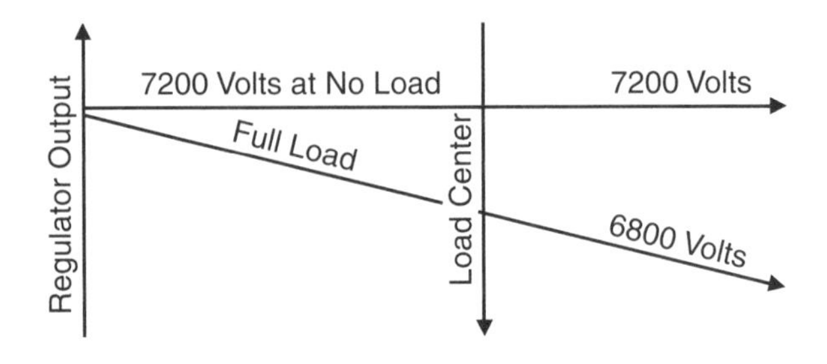

Figure 8–7 Voltage profile without line-drop compensation.

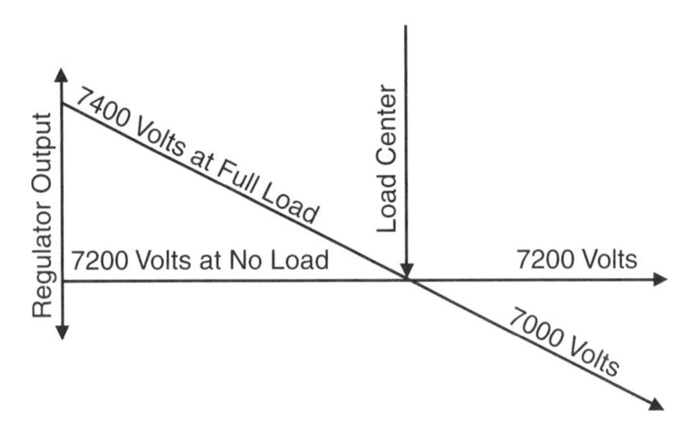

Figure 8–8 Voltage profile with line-drop compensation.

to keep a constant voltage at some point downstream, as shown in Figure 8–8. The resistance and reactance between the regulator and a point downstream are calculated. These values become the compensation settings in the control panel. The projected voltage drop for that distance is automatically added to the voltage output by the compensator circuit in the regulator. When the load current increases, the voltage at the regulator terminal increases so that the load center downstream continues to have a relatively constant voltage. The planning engineer must choose settings that will not cause the customer near the regulator to get a voltage that is too high.

Operating a Regulator

8.7.8 Operating a regulator normally involves putting the regulator into service or taking it out of service. The operation involves a bypass, a source (S) switch, and a load (L) switch (Figure 8–9). The most critical operation involving a regulator is ensuring that the source and load voltages are equal when the bypass is about to be closed. If the source and load voltages are not equal when the bypass is closed, the regulator will be subjected to a short circuit.

Note: A bypassed regulator is damaged more quickly when it is one step up or down from the neutral tap (shorting 45 volts) than when it is at full boost or buck (shorting 720 volts). When the series winding of the regulator is shorted out, there is more impedance to the current flow when it travels through all of the windings.

Figure 8–9 Single-phase voltage regulator.

With the regulator at the number one tap, there is not enough impedance to reduce the current and prevent the series winding from burning out.

The *control switch* is used to manually raise or lower the regulator to the neutral position. The *neutral indicator* light should come on when the regulator is in neutral position. To equalize the voltage, the auto-manual switch in the control box is turned to manual operation. The voltage then can be raised or lowered until it reaches the zero or neutral tap. When the neutral tap is reached, the switch is turned to the off position. The bypass switch (as shown in Figure 8–10) can then be closed and the input and output cutouts opened to isolate the regulator. Some utilities require that a test be made to prove that the electric neutral of the regulator coincides with the neutral-position indicator before carrying out any switching.

Troubleshooting a Regulator

8.7.9 The troubleshooting guide shown in Table 8–1 assumes that the trouble crew does not actually maintain the regulator or calculate the required setting for bandwidth, time delay, or compensation. A temporary fix means that the regulator problem will be reported to have the unit fixed as soon as practical.

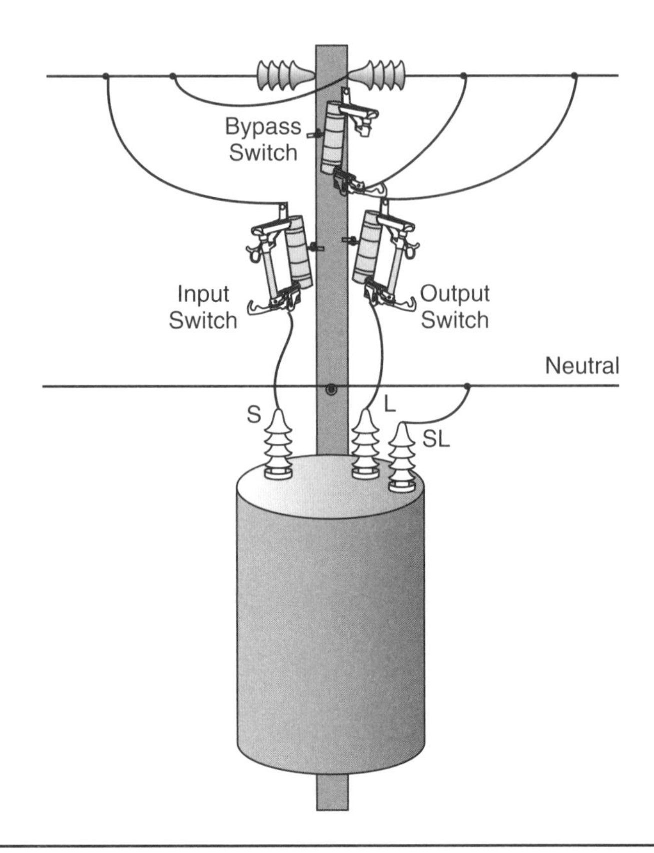

Figure 8–10 Typical regulator switching arrangement.

Reverse Feed through a Voltage Regulator

8.7.10 When a circuit is temporarily fed in reverse through a regulator, the regulator continues to try to adjust the voltage on its load side, which is now the source. The voltage sensor measures the input instead of the output voltage. The regulator tries to change the input voltage, though it is not able to do so.

Typically, the regulator goes to the maximum boost or to the maximum buck position. To avoid these problems, the regulator should be bypassed and removed from service before the circuit is fed in reverse.

The *S* terminal must always be connected to the source and the *L* terminal always connected to the load. It is possible to set up a *reverse-power-flow-switching* arrangement, which swings the input and output around so that the new source goes into the *S* terminal. This fairly complex switching arrangement is useful in cases in which reverse feeding is a common requirement.

8.8 Capacitors

The Purpose of Capacitors

8.8.1 Most capacitors are installed to provide power-factor correction on an electrical system, which in turn boosts the voltage. An electrical-power system must supply the apparent power needed to meet the customer's needs. Customers tend to have motors that cause an inductive reactance in the circuit, which increases the overall impedance of the circuit.

TABLE 8–1 Guide for Troubleshooting a Regulator

If	Then
There are excessive regulator operations.	The regulator could be overloaded. Phase balancing may unload the affected regulator.
There is a voltage complaint on a circuit where there is a regulator.	Often the tap-changing mechanism gets stuck on a tap. Raise and lower the voltage using the auto-manual switch, and then return the switch to automatic to see if the tap changer will move of its own accord.
The voltage problem was not due to a sticking tap changer.	The compensation settings may be out-of-date due to changes to the circuit, such as the installation of a capacitor, an upstream regulator, new conductors, or an increase in load. As a *temporary* measure, adjust the voltage using the auto-manual switch and then turn the switch off. Test the voltage at the customer and at the voltage test studs in the regulator control box. Now customers close to the regulator and/or at the end of the line will be exposed to more extreme voltages when the load current changes.
The regulator is chattering or hunting. The regulator is changing taps continuously.	The bandwidth or time-delay settings are incorrect. As a *temporary* measure, adjust the voltage using the auto-manual switch as before. This fix is temporary because customers may be exposed to extreme voltages when the load current changes.
The regulator does not respond to the previous solutions.	The regulator may be damaged. Take the regulator out of service. If the tap changer cannot be put in the zero or neutral position, arrange to take the circuit out of service by opening the source switch with a load-break tool, then open load switches and close the bypass.
The regulator is at the maximum boost position but is not able to supply an acceptable voltage.	1. An increased load in the circuit may have reduced the input voltage to the regulator. 2. Phase balancing could unload the affected phase. 3. An upstream or downstream regulator may be needed.
The actual position of the zero or neutral tap is not certain.	The pointer on the tap position indicator can be broken and/or the neutral indicating light is not working. Sometimes the specification for the control settings does not have the zero tap at the center. If in doubt, take the regulator out of service by opening the source switch with a load-break tool as before.

When capacitors are installed, a capacitive reactance is introduced into the circuit, which neutralizes the inductive reactance. Therefore, the overall impedance of the circuit is reduced. Figure 8–11 shows that with less impedance, there is less current needed to supply the load. With less current, there is less voltage drop. To improve the voltage on an electrical system, installing capacitors is more economical than installing voltage regulators, stringing larger conductors, or adding more feeders.

Power Factor Improvement or Voltage Booster?

8.8.2 Sometimes capacitors are specified to be installed on a distribution system to help improve the power factor on the transmission system right back to the generator. There is little need for system capacitors today because most utilities have capacitors on their high-voltage systems within the transmission substation yards. There is more control over station capacitors because operators can input a variable amount of capacitance as required by the system. There is very little control over capacitors on distribution circuits from a system point of view. Most capacitors on distribution circuits are installed to boost the voltage and to reduce line loss.

There was very little control over capacitors on distribution circuits from a system point of view. Most capacitors on distribution circuits are installed to boost the voltage and to reduce line loss. The location of the capacitors on a feeder is important because the voltage boost is upstream towards the source. While the best location can be calculated, a typical "location rule" for distribution lines is that the capacitor bank be located about 2/3 of the distance from where the voltage has dropped by 2/3 (2/3–2/3, location rule). The rule is intended to prevent the customers closest to the source from exposure to over-voltage and improving the voltage at the end of the line.

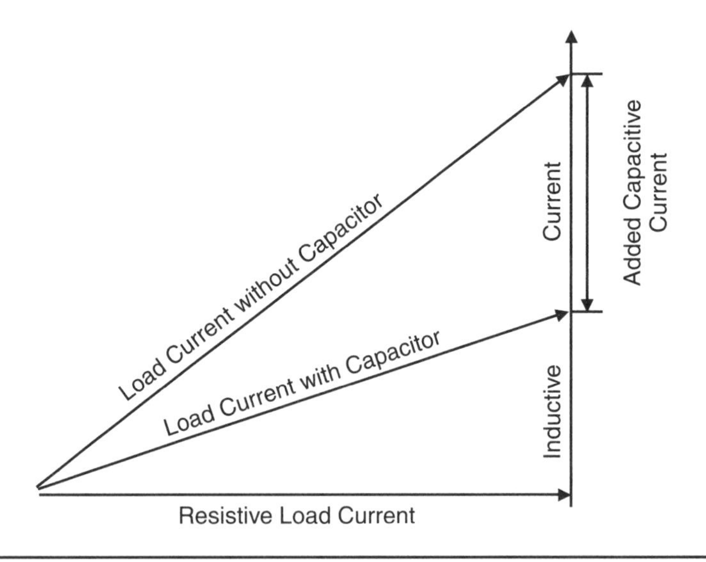

Figure 8–11 Capacitor effect on load current.

Construction of Capacitors

8.8.3 Capacitors consist of two plates with insulation between them. The larger the plates, the more capacitance there is. The closer the two plates are to each other, the more capacitance there is.

A typical distribution capacitor consists of two plates (Figure 8–12) made up of two long sheets of aluminum. The long sheets of aluminum are rolled up with insulation such as oiled paper or polyethylene film between them. Each aluminum sheet or plate is connected electrically to a terminal. The rolls are made up flat so that they are more compact and can be stacked with other rolls in the capacitor tank. Multiple rolls within a unit can be interconnected in series or parallel depending on the capacitor kilovolt-ampere reactive rating and voltage rating. The capacitor case is filled with an insulating oil.

Capacitors store a charge. Inside the capacitor unit, between the two terminals, there are discharge resistors that are designed to drain the electric charge from the capacitor after a capacitor is isolated. It is normal to wait about 5 minutes after isolating a capacitor to let the resistors drain the charge before a powerline worker applies grounds to the capacitor. The terminals of a capacitor should always be left shorted out when the unit is exposed to contact by people.

How a Capacitor Works

8.8.4 The plates of a capacitor are charged and discharged 60 times a second in a 60-hertz circuit. During the first half cycle, one plate is positively charged, which causes a negative charge of equal voltage to be electrostatically attracted to the other plate.

Current flows into the capacitor only while the voltage is rising. When the voltage approaches peak value, the counter-electromotive force is also approaching peak value, which causes the current flow to decrease. There is no current flow when the voltage is at its peak (90 degrees). In a capacitor, the current reaches its

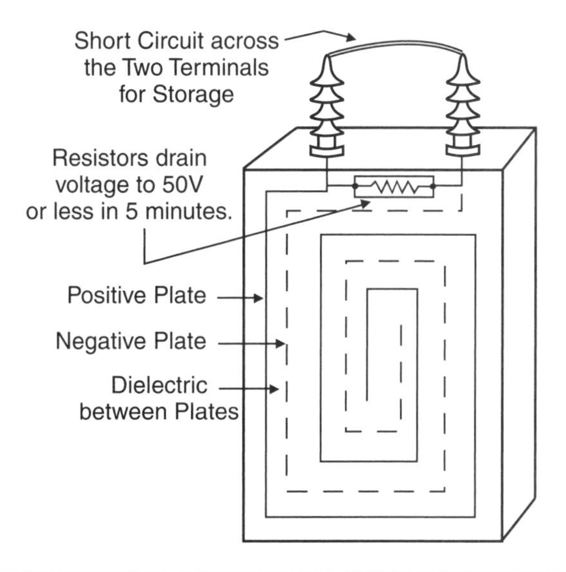

Figure 8–12 Illustration of capacitor construction.

peak before the voltage reaches its peak. A capacitor opposes the *change* in voltage, which in an AC circuit is constantly changing.

Because change to the voltage across the capacitor plates is delayed, this capacitive reaction causes the voltage wave to lag the current wave. In a circuit where the current lags the voltage, the capacitor effect of the current leading the voltage helps cancel the two effects and brings the circuit closer to a unity power factor.

Shunt Capacitors versus Series Capacitors

8.8.5 In an electrical-power system, most capacitors are connected in parallel (shunt), which has the full line voltage applied between the two capacitor plates. On a distribution system, shunt capacitors are connected near the load center to help reduce voltage drop. The voltage drop is reduced because the capacitors reduce the line loss in the complete circuit back to the source.

Series-connected capacitors are used to reduce a severe voltage flicker on radial circuits where frequent motor starting, electric welders, or electrical arc furnaces affect other customers on the circuit. Constructionwise, shunt and series capacitors are the same. Series capacitors are connected in series and therefore conduct the full-line current through them. The voltage drop across a series capacitor changes instantly when the load changes and, therefore, reduces the effect of a voltage flicker.

Switched Capacitors

8.8.6 On distribution lines, shunt capacitors are installed downstream closer to the load center. During peak-load periods, the capacitors are needed to reduce line loss and keep the voltage at the proper level. During light-load periods, the capacitors can improve the power factor to the point where the voltage will be too high. Capacitors need to be switched off before damaging high voltage occurs. Capacitors can be switched on or off with electrically operated oil switches. Control of the oil switches can be by a time control, a radio control, a voltage control, a power-factor control, or a reactive-power (VAR) control.

Daily switching of capacitors in and out of service, especially large capacitor banks in substations, causes a transient overvoltage in the circuit. Sensitive customers, such as those with variable-speed-drive motors, may find these fairly large voltage fluctuations unacceptable.

Operating Capacitors

8.8.7 Unlike transformers, shunt capacitors (Figure 8–13) draw a constant current regardless of the customer load. The current drawn by the capacitor is due to the energization of the relatively large amount of "metal" of the capacitor plates. Large capacitors have oil switches to allow a safe energization or deenergization. Large capacitors should not be energized or deenergized with a cutout unless a load-break tool can be used.

The current a capacitor bank draws can be measured or calculated. Some typical current readings for a 25-kilovolt system are:

600kVAR bank 13.9A
450kVAR bank 10.4A
300kVAR bank 7.0A
225kVAR bank 5.2A

Figure 8–13 Small, three-phase capacitor bank.

To calculate the expected current for other voltages on wye systems:

$$I = \frac{kVAR\ of\ capacitor\ bank}{3 \times phase\text{-}to\text{-}ground\ kV}$$

where I = current in one phase
kVAR = rating of the capacitor bank
kV = phase-to-ground voltage

For example, how much current can be expected in the leads feeding a 600kVAR capacitor bank on a 12.5-kilovolt system?

$$I = \frac{600}{3 \times 7.2} = 27.8\ amperes$$

Operation of the Discharge Resistors

8.8.8 The discharge resistors in a capacitor drain the voltage of an isolated capacitor to a level below 50 volts in about 5 minutes. Draining the voltage allows for a safer installation of portable grounds.

When a capacitor is isolated with the intent to reenergize it, it is important to wait 5 minutes to allow the voltage on the capacitor to drain before reenergizing.

If a *charged* capacitor is returned to service, the line voltage builds up well above normal.

There is a danger when energizing capacitors with a fuse cutout, because, if proper contact is not made the first time, any immediate second strike to retry closing the cutout results in double the line voltage across the cutout. Wait 5 minutes before trying to reclose a capacitor.

8.9 Troubleshooting High- or Low-Voltage Problems

Trouble-shooting High-Voltage Problems

8.9.1 A high-voltage problem becomes noticeable to a customer when the lights seem brighter and there is a premature burnout of lightbulbs, motors, and electronic equipment. A customer close to a distribution station or downstream close to a regulator can experience high voltage. The voltage is often relatively high at these locations in order to provide a proper voltage farther downstream.

Capacitors that are not switched out of service when the circuit is lightly loaded can raise the voltage during off-peak times. A transformer can occasionally suffer some shorted-out turns in the coil and cause a high voltage to customers.

Trouble-shooting Low-Voltage Problems

8.9.2 Low-voltage problems usually show up during peak-demand times, such as during cold and hot weather and during suppertime. A customer may notice low voltage when an electric motor is running hotter than it should. During low voltage, the motor draws more amperes and the torque that the motor produces is reduced. The torque is inversely proportional to the square of the voltage. For example, even though most motors are designed to run at 90 percent of their voltage ratings, a motor running at 90 percent voltage produces 81 percent torque and runs hotter.

There can be many causes of low voltage. Finding the cause of low voltage is a step-by-step process that starts at the customer and works toward upstream equipment (see Table 8–2).

8.10 Harmonic Interference

What Are Harmonics?

8.10.1 With AC power, the magnitude and rate of current and voltage rise and fall are represented graphically by a sine wave. Harmonics are AC and voltage disturbances that can also be represented by sine waves but are at different frequencies than the fundamental 60 hertz.

The harmonic frequency is a multiple of the fundamental 60-hertz frequency. Harmonic frequencies can be 30 hertz, 120 hertz, 180 hertz, and so on. A 120-hertz harmonic is twice the fundamental frequency and is called the second harmonic. Similarly, 180 hertz is the third harmonic.

Many various harmonic waveforms exist in the power supply at one time. These waveforms are out of phase with each other. The many differing waveforms tend to sum and cancel each other to form a distorted, but a predominantly 60-hertz sine wave.

The illustration in Figure 8–14 shows only one harmonic. There are generally many more. The 60-hertz power wave stays predominant in a normal distribution

TABLE 8–2 Finding the Cause of Low Voltage

Step	Action	Details
1	Check the voltage at the customer meter base.	If the voltage is low, check for any recent increase in the load a customer is drawing. An increase in load could make the length or size of secondary conductors inadequate for the additional load. The planning engineer or engineering standards books have voltage-regulation charts for secondary bus and services.
2	Remove all load from the transformer and take a voltage reading at the transformer.	If the voltage reading is normal at the unloaded transformer, the low-voltage problem must be due to the length and/or size of the secondary or possibly an overloaded transformer. A recording voltmeter may need to be installed if the problem is intermittent. If the voltage reading is low at the unloaded transformer, the primary voltage is low and the cause of the problem is upstream. Other customers should be affected by upstream problems. A quick fix to an individual customer is to make a change to the transformer tap if equipped.
3	Check out any upstream voltage regulator or transformer with an LTC.	A voltage regulator could be stuck in a low position. Put the regulator in "manual" position and raise the voltage. Similarly, the LTC (load tap changer) at the substation transformer could be malfunctioning.
4	Check out any upstream capacitors.	Switched capacitors should be in service during daily peak-load times and switched out of service during lightly loaded times. Check to ensure that the time clock or other control is working. Check that the motor-operated oil switches are closed during peak-load periods.
5	On a heavily loaded circuit, check the phase balance.	The planning engineer may show that the circuit is able to carry the load, but if one phase is carrying more than its share, then it could be overloaded and cause an excessive voltage drop on that phase.

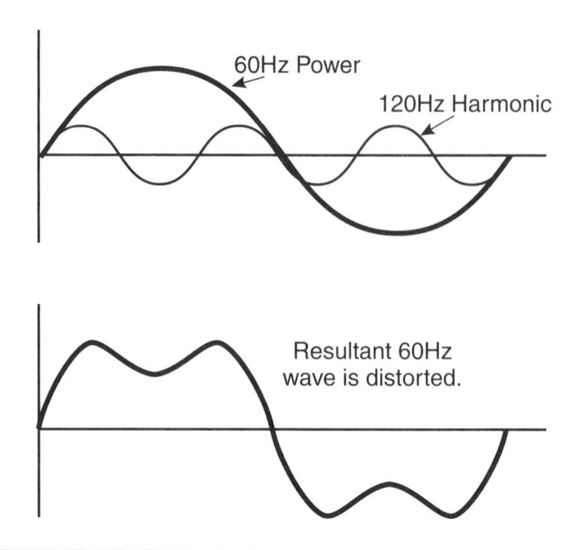

Figure 8–14 Harmonic illustrated as waves.

system. The resultant 60-hertz wave becomes distorted, and, when it becomes severely distorted, it affects sensitive electronic equipment.

Sources of Harmonics

8.10.2 Electric loads that are nonlinear propagate harmonics back into the electrical-supply system. Nonlinear loads are loads that are not purely resistive but have a capacitive or inductive component in them.

There are many kinds of nonlinear loads, and all do not distort the "pure" 60-hertz waveform to the same degree. Examples of nonlinear loads are transformers, arc furnaces, arc welders, adjustable-speed and variable-frequency motor drives, electronic lighting ballasts, converters, rectifiers, and large computer systems.

Harmonics are a steady-state disturbance that exists in the system as long as the equipment generating the harmonics is in operation.

Conductive Harmonic Interference

8.10.3 Harmonic interference can be conducted back into the power supply from the direct connection to the harmonic-producing customer load. The harmonic currents conducted back into the power source add additional current to conductors, transformers, and switchgear. Harmonic currents can also interfere with protective relays, meters, and induction motors.

Capacitors are exposed to overloading because they act as a sink for the higher-frequency harmonics. When capacitors are on the system, there is also a possibility for resonance to occur at one of the other frequencies.

Inductive Harmonic Interference

8.10.4 Just as voltage and current can be induced on adjacent circuits, harmonic voltage and current can be induced on a neighboring power supply or telephone circuit. Unshielded telephone cable running parallel to power circuits is especially vulnerable to induced harmonics from the powerline.

Signs of Harmonic Problems

8.10.5 One of the first signs of a problem due to harmonics is from customers with sensitive electronic equipment. Harmonics distort the needed voltage and current magnitude, which can show up when a computer acts erratically or loses data.

Conductors, transformers, and motors are subject to heating because the high-frequency component of a harmonic disturbance causes an increase to the skin effect of conductors. A larger proportion of current is carried on the outer edge of a conductor, which limits its current-carrying capacity. Harmonics reduce the reliability of electric signals that activate meters, relays, powerline carriers, and equipment such as robots.

Testing for Harmonics

8.10.6 Before remedial action is taken for harmonic interference, tests should be carried out to confirm its presence. A specialist uses a power harmonic analyzer to make tests, interpret the waveform, and predict the most likely cause of the disturbance.

Harmonic Interference Solutions

8.10.7 Once a problem has been diagnosed as a harmonic problem, filters can be installed at the customer to limit the effect on the sensitive loads. The filter consists of an inductor, a capacitor, and a resistor, which are tuned to provide a low impedance to ground for specific harmonic frequencies. For large customers with big power-quality problems, arrangements can be made to supply "custom" power.

8.11 Voltage Flicker

The Voltage-Flicker Problem

8.11.1 Erratic fluctuation in voltage shows up as blinking lights or intermittent shrinking of a television screen. For a customer with sensitive electronic equipment, such as a computer or an electronic cash register, the problem becomes unacceptable. Normally, calculations for potential voltage-flicker problems are carried out by the planning engineer when a customer with large motors or welders applies for service.

Sources of Voltage Flicker

8.11.2 A loose neutral or other bad connections are possible sources of flickering lights, especially during windy or heavy-load conditions. When a customer has a very noticeable dip in voltage on a regular basis, other sources need to be investigated. Large motors, arc welders, X-ray machines, or electrical arc furnaces have loads that have a varying demand, are mostly unbalanced, and have a poor power factor. Under starting conditions, these loads draw considerably more current than when operating. A voltage flicker is noticed by other customers served by the same feeder as the offending motors, welders, and so on. The extent of the voltage flicker depends on the capacity of the feeder supplying the load to the customer.

When Does a Voltage Flicker Become Objectionable?

8.11.3 There are standards for an acceptable or unacceptable voltage dip. These standards may change as more sensitive electronic devices come on the market. The standards combine the percent of voltage dip with the frequency of its occurrence. A typical standard looks like the chart in Figure 8–15. For example, from the chart it can be seen that a 3 percent voltage dip five times per minute would be objectionable.

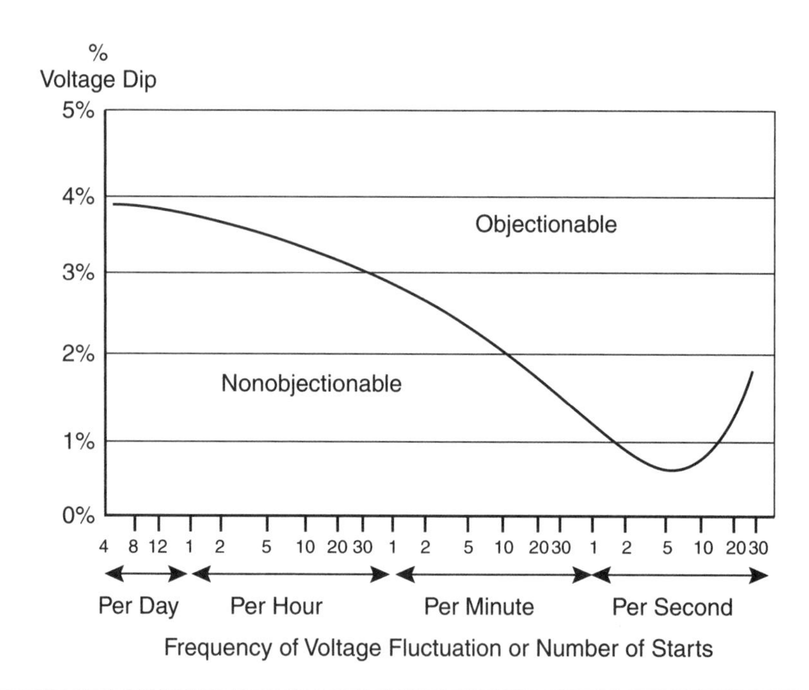

Figure 8–15 Threshold of objectionable voltage.

Factors Affecting Voltage Flicker

8.11.4 The size of voltage flicker is dependent on the type of load and the capacity of the feeder supplying the load. The load itself can have equipment that reduces the start-up inrush current. Inrush current information for motors, arc welders, and so on is normally found on the equipment nameplate. An electric motor has an inrush current of about six times its load current. This can be reduced by using motors with soft-start capability.

The greater the capacity of the feeder supplying the load, the less the feeder is affected by voltage flicker. Voltage flicker is reduced when:

- The source transformer is large and is able to carry a high peak load for a short period of time
- The distribution feeder conductor is large
- The distribution feeder distance to the offending load is short

Calculating Predicted Voltage Dip

8.11.5 To predict a voltage dip, detailed calculations can be made. However, the planning engineer normally inputs the needed information into a computer program or uses an estimating chart. The information needed is the inrush current for the load to be fed, the capacity of the power supply system (which is the phase-to-phase fault current availability of the feeder at the source substation transformer), and the length and size of the conductor from the station to the load.

8.12 Ferroresonance

What Is Resonance?

8.12.1 Objects have their own natural frequencies. For example, a rattle in the dashboard of a car can vibrate annoyingly when a car travels at a certain speed. The frequency of the engine matches the frequency of the dashboard and sets up a sympathetic vibration in the dashboard. Soldiers are told to march out of step when they cross a bridge because when they march in step they sometimes match the natural frequency of the bridge. The sympathetic vibration set up in the bridge can cause severe damage. When the vibration in one object matches the natural frequency of another object, the two objects vibrate together in resonance.

Resonance in an Electrical Circuit

8.12.2 In any series-RLC circuit (circuits that have resistance, inductive reactance, and capacitive reactance), resonance occurs at some frequency. When the frequency rises, the inductive reactance increases and capacitive reactance decreases. As the frequency rises, the increasing inductive reactance will at some point be equal to the decreasing capacitive reactance and vice versa. When this happens the inductive reactance cancels out the capacitive reactance, which means there is no reactive load impeding the current flow in the circuit.

If the resistance in the circuit happens to be low, there is very little to impede current flow and the current increases. As the current increases, the voltage also increases and rises above the source voltage.

The tuner on a radio varies the capacitance in a circuit to match the inductance at a desired frequency. The signal is amplified when it resonates at the desired frequency. Of course, the frequency on an electrical-power system is constant. Resonance only occurs by coincidence when the capacitive reactance is in series with and happens to match the inductive reactance at the standard 60 or 50 hertz. When resonance occurs, the voltage can build up from two to nine times the normal phase-to-ground voltage. The increase in voltage damages equipment, which is evidenced by rumbling and whining noises at transformers, arcing at insulators, and sparkover at arresters.

Resonance with Harmonic Frequencies

8.12.3 The frequency of an electrical-power system may be a constant 50 or 60 hertz; however, there are frequencies superimposed into the system from certain loads. The power at these "harmonic" frequency waves sometimes resonates.

Causes of Ferroresonance

8.12.4 Resonance is a rare occurrence in a circuit because *all* of these factors must be in place:

- The inductive reactance X_L is equal to the capacitive reactance X_C.
- The inductive load and the capacitive load must be in series with each other.
- There must be virtually no resistive load on the circuit.

When these factors are in place there is practically no impedance to current flow in the circuit. Ferroresonance usually occurs when one or two phases are disconnected from the source by a fault; or, by switching a single-pole device, the transformer windings connected to the open phases are excited through the capacitance to ground in the cables and between phases.

The Source of Inductive Reactance in Series

8.12.5 Most electrical-system circuits have loads that are a source of inductive reactance, but these loads are normally connected in parallel. The most common and possibly the only way to get an inductive load in series with a circuit is to open one or two phases feeding a three-phase transformer bank.

As seen in Figure 8–16, the windings in the transformer with a delta primary are in a series configuration when one phase is open. The windings in a transformer with a wye primary do not become a series load when one switch is open. This only applies when the wye point is grounded. However, the wye point on a wye-delta transformer is left ungrounded or floating.

The inductive reactance of a transformer changes as the magnetic field from the coil magnetizes and eventually saturates the iron core. This gives a range of inductive-reactance values, which increases the possibility of matching the capacitive reactance in the circuit. The process of saturating the iron (ferrous) core is where the name ferroresonance comes from.

The Source of Capacitive Reactance in Series

8.12.6 There is some capacitance on any circuit because a live conductor acts as one plate; the air or cable insulation is the dielectric between the plates, and any conductive material near the conductor acts as the second plate. Paralleling overhead conductors causes a capacitive reactance in a circuit. The capacitive reactance from long transmission lines is often countered by the installation of reactors at stations.

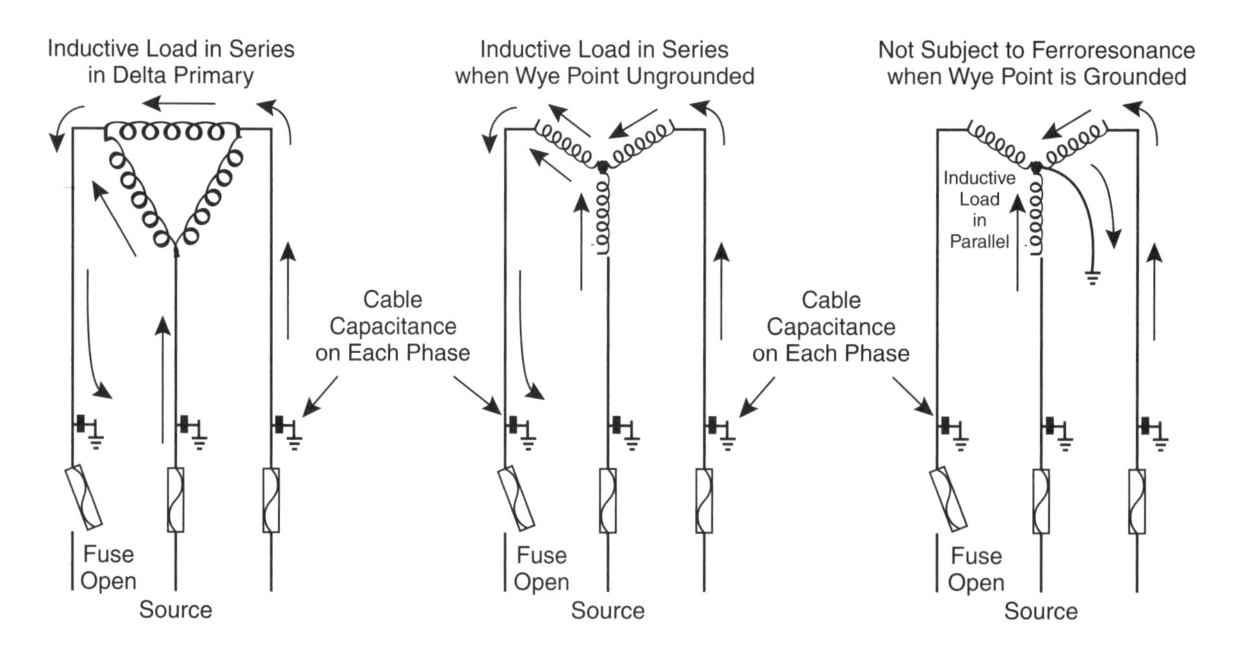

Figure 8–16 Sources of inductive reactance in series.

Underground cable is a natural capacitor with the live conductor separated from another conductor, the grounded sheath, by relatively thin insulation. Long lengths of underground cable need reactors to cancel some of the capacitive reactance.

Field Examples of Ferroresonance

8.12.7 The most common occurrence of ferroresonance involves a three-phase transformer bank fed with a length of underground cable as illustrated in Figure 8–17. A certain critical length of underground cable provides the critical amount of capacitative reactance in the circuit. A transformer bank can provide the inductive reactance matching the capacitive reactance of the cable. The inductive load and the capacitive load must be in series with each other, and this occurs when only one or two phases are energized. Resonance is most likely to occur when switching single pole devices on a high-voltage distribution underground cable feeding a transformer with a delta primary.

One example of ferroresonance on an overhead 115-kilovolt line occurred when one of the blades of a gang-operated air-break switch failed to open during the isolation of a substation transformer. With one phase closed, the unloaded transformer became an inductive load connected in series with the 115-kilovolt line. The paralleling 115-kilovolt conductors supplied the capacitive load. A corona discharge was seen on the 115-kilovolt conductors, and surge arresters operated as the voltage climbed higher.

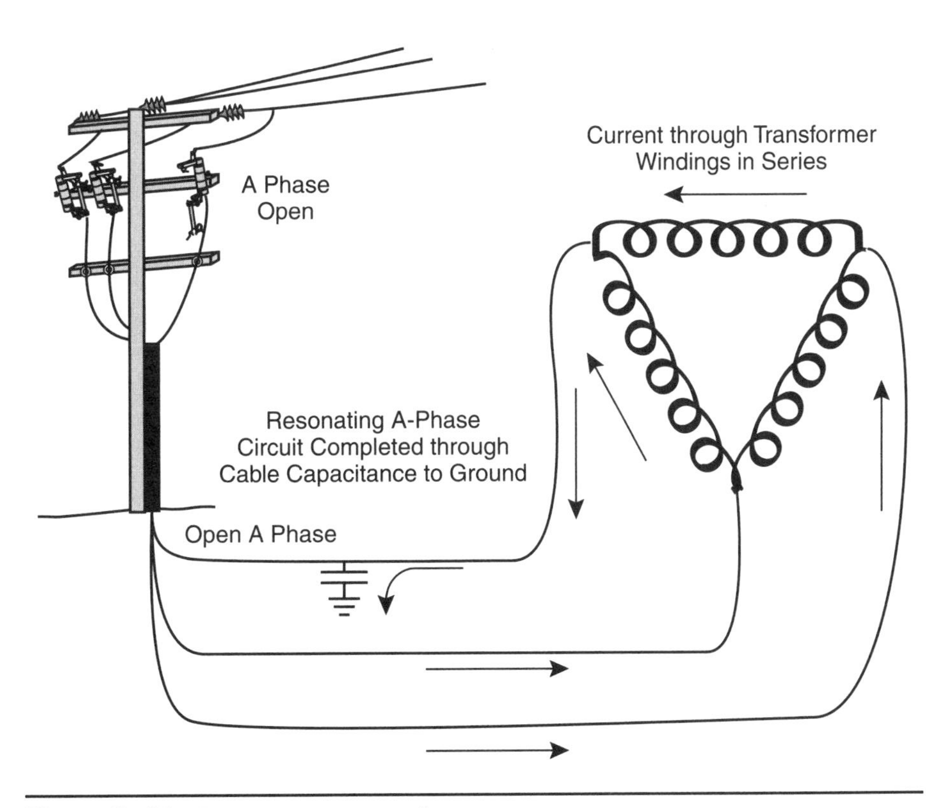

Figure 8–17 Ferroresonance example.

Prevention of Ferroresonance

8.12.8 Ferroresonance can be avoided by removing one of the causes of the phenomenon (see Table 8–3). By changing the design of the installation or changing the switching procedure, the capacitance, inductance, series connection, or loading can be changed.

8.13 Tingle Voltage

What Is Tingle Voltage?

8.13.1 The term *tingle voltage* refers to a small voltage that is noticed by people or animals while contacting certain equipment or hardware at their premises. This occurs when there is an unacceptable voltage between the neutral and earth.

TABLE 8–3 Prevention of Ferroresonance

Causal Factor	Action
Operating single-pole switches to energize or deenergize an underground cable feeding a three-phase transformer allows the transformer coils to be an inductive load in series with the circuit.	Install a three-phase gang-operated switch.
If there is no load on the transformers when they are energized, the low resistance in the circuit causes a higher voltage when resonance occurs.	Keep a load on the transformer when it is being energized or deenergized. The increased resistance in the circuit will lower the effects of resonance.
A three-phase bank with a grounded wye primary shorts out the two windings that are part of the series circuit causing induction.	At the planning stage, where possible, use a transformer with a wye primary and the wye point connected to the neutral.
Changing the length of the underground cable feeding the three-phase transformer will change the amount of capacitive reactance in the circuit.	Changing the length of the cable as a retrofit is probably an expensive option. Calculating the length needed to avoid ferroresonance is complex. The changing inductive reactance that occurs before the transformer core becomes saturated means that there is also a range of capacitive reactance that will at some point be equal to the inductive reactance.
A wye primary with a floating (ungrounded) neutral is susceptible to ferroresonance.	Temporarily ground the floating neutral of the wye primary, which will short out the series circuit. A three-phase bank with a grounded wye primary shorts out the two windings that are part of the series circuit causing induction.

A system neutral on a utility distribution circuit will have some voltage on it in relation to a remote ground. This voltage is kept very low, normally below 10 volts, by grounding at each transformer and at other locations. At a transformer, the primary neutral is bonded to the secondary neutral. This connection takes advantage of the customer-service grounds to lower the voltage on the neutral even more and to prevent a potentially dangerous open circuit in the grounding network. However, depending on the quality of grounds and other factors, there will still be some voltage left on the neutral.

With the ground bonded to the neutral at the service entrance, the safety ground will have some voltage on it because it is connected to appliances, equipment, and fixtures by the many electrical circuits coming out of the customer panel. Tingle voltage occurs between a grounded object and a remote ground. There will always be some current flow back to the circuit's source through earth. Where there is current flow, there has to be some voltage. Figure 8–18 shows how a circuit can be established through a person between the shower control and the main drain.

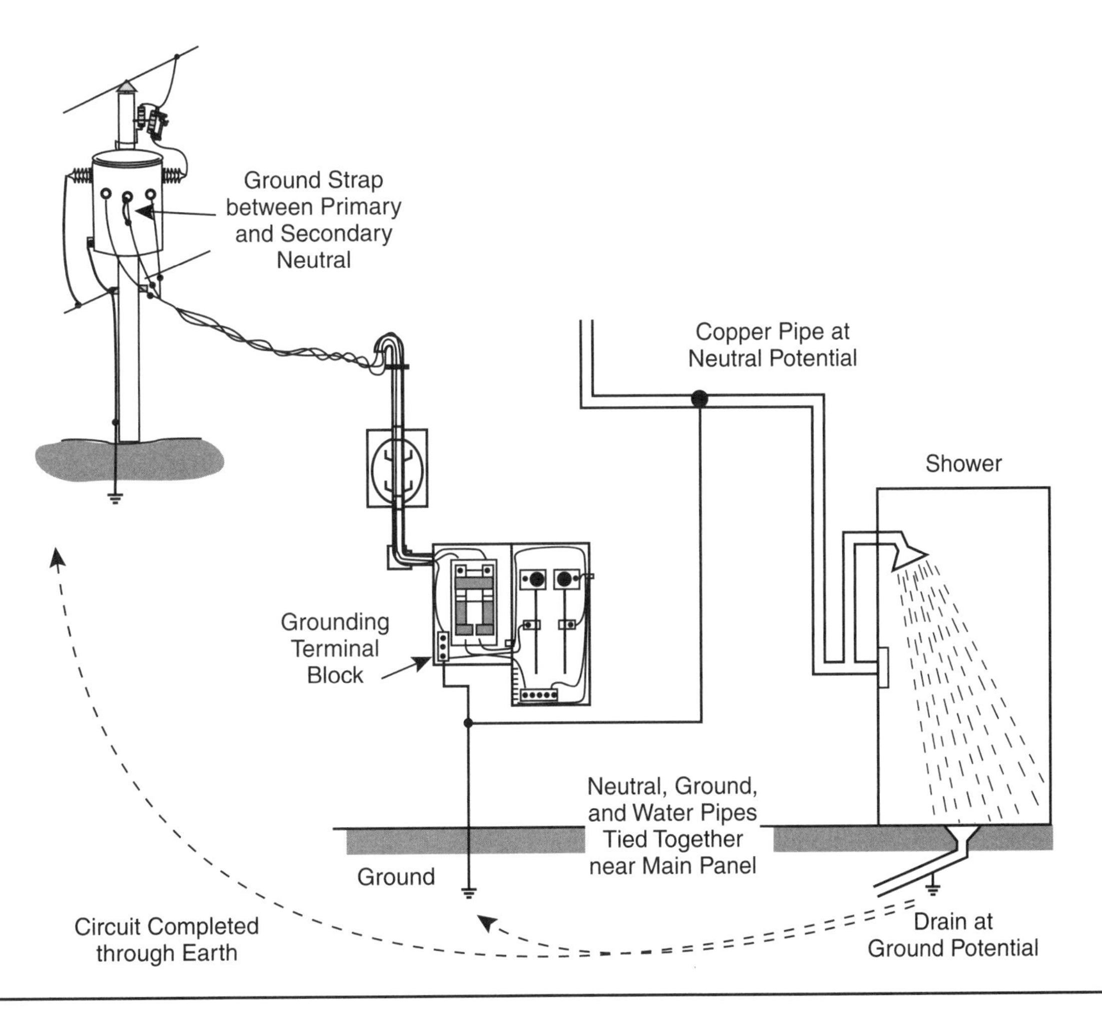

Figure 8–18 Tingle voltage in shower.

Examples of Tingle Voltage Problems

8.13.2 The voltage between the neutral and earth is normally well below the threshold of sensation for the vast majority of customers. Very vulnerable customers, such as dairy farms, require additional efforts to reduce tingle voltage. Milk cows have been the most vulnerable to tingle voltage. A small voltage between the stanchions, water bowl, or milking machine in relation to the floor that the cow is standing on results in a current flow that turns the animal into a "dancing cow." In these conditions, a cow can feel a potential difference as low as half a volt.

Normally, people would not feel tingle voltage because 10 volts or less is not enough to overcome a person's skin resistance. However, the skin resistance is reduced when it gets wet. People have noticed tingle voltage in a shower, kitchen, wet basement, or barn. In a shower, a person can feel a potential difference as low as 2 volts.

The Utility as a Source of Tingle Voltage

8.13.3 Single-phase circuits have current flowing back to the source through the neutral, and a portion of this current flows through the parallel path in earth. When there is a current flow, there is some voltage. A heavily loaded single-phase circuit will have an unacceptable voltage buildup on the neutral if the neutral is poorly grounded. Normally, the voltage on the neutral should not exceed 10 volts.

Delta circuits and three-phase circuits have considerably less current flowing back to the source through earth because the phase conductors carry current back to the source. The more evenly the load is balanced on the three phases, the less current there is in the neutral.

Remedial Action by the Utility

8.13.4 A utility will check the neutral-to-earth voltage to ensure that the voltage is lower than their standard. A voltage check between the neutral and a remote ground location about 15 meters (50 feet) away is taken. *If* the voltage on the neutral has to be lowered, *then:*

- Neutral connections are checked
- Additional ground rods are driven to lower the voltage on the neutral
- A larger neutral conductor can be strung to promote more current flow through the neutral and less through earth
- A conversion to a three-phase circuit will promote more current returning to the source in the other phases and less current flow in the neutral

Neutral Separation at the Transformer

8.13.5 In extreme cases, a high neutral voltage on the utility supply system will be isolated from the customer's neutral by removing the connection between the primary neutral and the secondary neutral at the transformer. The separation of the neutral, as shown in Figure 8–19, is a hazard to utility personnel because there could be a potential difference across the open circuit between the primary and secondary neutral. Under fault conditions, the voltage on the transformer tank or on the downground could be lethal.

A gas-tube-type surge protector can be installed between the secondary neutral and the transformer tank. During a voltage surge, the protector sparks over internally

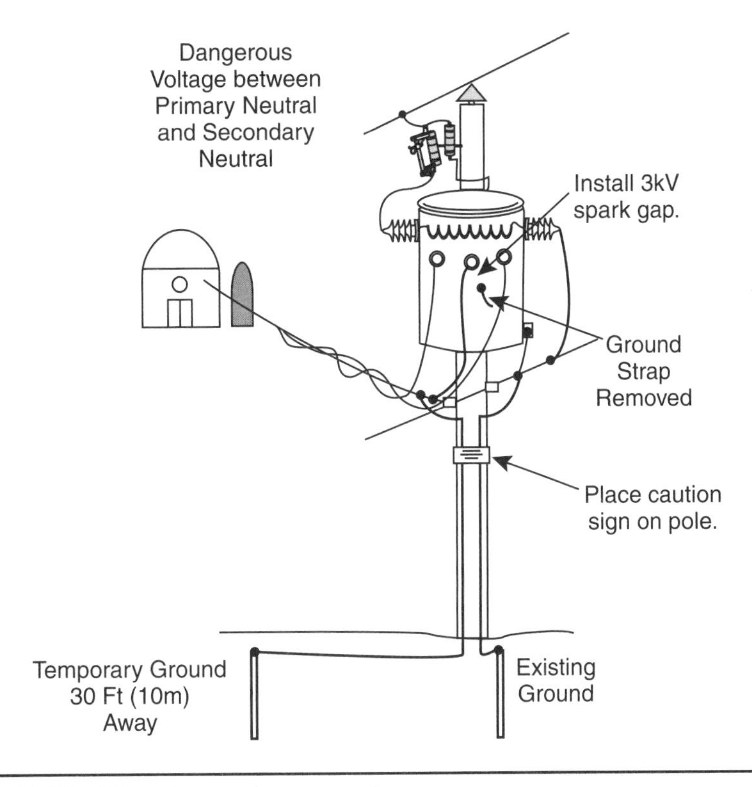

Figure 8–19 Neutral separation at transformer.

and temporarily connects the secondary neutral to the transformer tank. With the transformer tank connected to the downground and primary neutral, the primary and secondary neutrals are temporarily connected together, thereby preventing a secondary insulation failure in the transformer. A caution sign is normally placed on the pole to alert powerline workers to the hazard of the separated neutrals.

The Customer's Facilities as a Source of Neutral Voltage

8.13.6 The cause of excessive ground current can also be found at the customer end of a service:

- There should be only one electrical connection between the neutral and earth: at the service entrance. A connection between the neutral and the safety ground wire at an appliance some distance from the main panel means that the neutral and safety ground are parallel paths. This encourages more current flow through the safety ground wire. More current in the safety ground wire means more voltage in the safety ground network.
- There could be a poor neutral or ground connection at the service entrance or at junctions.
- There could be ineffective grounding due to poor earth, an insufficient number of ground rods, or poor ground-rod connections.

- There could be too much load on one leg of the 120-volt service, which results in more current returning to the source through the neutral and ground.
- There could be a voltage on an electrical appliance because there is no safety ground connection.
- There could be worn or poor insulation in wiring or equipment. Leakage to ground flows back to the source.

Customer Remedial Action

8.13.7 The usual first step to reduce the voltage on the neutral is to improve the grounding of the neutral at the service entrance. An inspection of the insulation of the wiring, neutral connections, and safety ground connections is also carried out.

The neutral-to-earth voltage can be reduced to acceptable levels by installing a tingle-voltage filter, which is designed for this purpose. It is possible to install a grid bonded to the source of the tingle voltage, using the same principle as the grid attached to tension machines when stringing near live circuits. A grid on the stable floor bonded to the stanchion and water bowl would ensure that there is no exposure to any difference of potential. This is not always practical as a retrofit.

8.14 Radio and Television Interference

TVI Trouble Calls

8.14.1 Trouble calls involving finding and fixing the source of radio and television interference (TVI) can be a frustrating experience for the lines trade. Customers that have interference with their reception are quite often on the fringe of being too far from the radio or television transmitter. The source of the interference is not always obvious and is not always due to the electrical-utility facilities. Hitting poles with a sledge hammer to listen for the source of TVI is one way of trying to find the problem. However, having a technician who uses specialized instrumentation is the best way to find difficult TVI sources.

Three Main Sources of TVI on Utility Facilities

8.14.2 The three main sources of TVI on utility-owned electrical circuits are:

- Loose hardware
- Defective insulators
- Corona discharge

Table 8–4 provides some assistance in finding the source of TVI.

Loose Hardware

Loose hardware is the most common and the most intense source of TVI. Sparking occurs between loose pieces of metal that are not electrically bonded to each other. For example, a washer does not make a good electric connection to a bolt unless the bolt is tight.

When the wood on a wood pole or cross arm drys and shrinks, the hardware loosens. Dielectric gaps build up where metal pieces are not forming a tight bond with each other, and sparking occurs across the gap. The sparking occurs in

TABLE 8–4 Finding the Source of TVI

If	Then
The interference is intermittent.	The probable cause of intermittent noise is loose hardware.
The interference is continuous.	The probable cause of continuous noise is from a stable source such as a defective insulator.
The interference appears to be weather dependent.	The noise is most likely from the powerline and not from within the customer's home if the interference occurs during dry weather but disappears during a rain.
The interference is prevalent throughout the neighborhood.	The TVI has a strong source, and the powerline is a probable cause when the noise is prevalent throughout the neighborhood.
The interference affects only one customer.	The noise probably comes from equipment in the building when a weak source only affects one customer.
The interference occurs at similar specific times.	The TVI could be from industrial equipment such as an arc welder when the noise occurs during specific times, such as during working hours. This type of source should also produce a noise throughout the neighborhood.
The interference on the customer's television occurs on both the video and the audio.	The source is strong when both the audio and the video are affected. The audio signal of a television is an FM signal and is not as vulnerable to interference. If only the audio is affected, the problem is probably within the customer's equipment.
The interference is from devices in the customer's premises.	Devices in the home that have been known to cause noise are: an electric motor, a fluorescent light, electronic equipment, a doorbell transformer, flashing decorative lighting, an aquarium pump, heating pad, dimmer switch, or a refrigerator butter conditioner.
The interference is from an outdoor source other than a powerline.	Check for potential sources such as radio and television transmitters or two-way radio transmitters used by police, utilities, and other businesses.
The noise is the same across all television channels.	Depending on the strength of the noise, channels 2 to 6 are affected first, channels 7 to 13 are affected next, and UHF is almost never affected. Suspect the television itself if the noise is constant for all channels.

repetitive bursts, which results in a radiated noise that interferes with radio and television reception.

Defective Insulators

Pin-type insulators on subtransmission lines have been a common source of TVI. The problems typically occurred with older insulators or poor conductor ties. Newer pin-type insulators have a semiconducting glaze (Q glaze) on the top surface of the insulator. The semiconducting glaze provides an equipotential area that prevents sparking between the conductor, the conductor ties, and the insulator.

Corona Discharge

Corona discharge as a source of TVI is most likely from higher-voltage circuits. The discharge emanates from sharp points on live hardware. It is most likely to show up when a new line is first put into service or after maintenance work is carried out on the line. The design of the live hardware on a high-voltage circuit requires everything to be smooth or rounded off. The usual cause of a corona discharge is a connection that is not properly smoothed off. Radio noise because of corona discharge should be a rare occurrence.

Review Questions

1. What does quality power look like?
2. The neutral and the safety ground are normally tied together at the transformer supplying the service and also at the service panel. What happens when the safety ground is tied together downstream on the customer's premises?
3. At a customer's meter base, 127 volts is measured. Is that an acceptable voltage?
4. Why do distribution circuits have larger conductors than what is needed for ampacity?
5. The leads going in and out of a set of voltage regulators are very small compared to the main line conductors. Will the relatively small leads affect the capacity of the feeder to supply the load?
6. When a circuit is temporarily fed in reverse through a regulator, what adjustments should be made at the regulator?
7. A regulator is at the maximum-boost position but is not able to supply an acceptable voltage. What options are there to fix the voltage problem?
8. How does a set of capacitors installed on a distribution feeder benefit the electrical system?

9. How much current will be interrupted when trying to open a 300kVAR capacitor bank on a 13.8-kilovolt feeder?
10. When a capacitor is isolated with the intent to reenergize it, why is it important to wait 5 minutes before reenergizing to allow the voltage on the capacitor to drain?
11. Name two possible sources of a high-voltage problem.
12. What are two possible sources for flickering lights?
13. Name one way a lines crew can reduce the risk of ferroresonance when energizing a three-phase transformer fed from a length of underground cable.
14. Why is separating the primary neutral from the secondary neutral at the transformer a poor way to solve a tingle-voltage problem?
15. When the source of TVI noise is from a distribution line, what is the most common cause?

CHAPTER 9

Conductors and Cable

9.1 Introduction

Electric Current in a Conductor

9.1.1 When an electric current flows in a conductor, there are three basic effects:

1. A magnetic field is set up around the conductor.
2. Heat is generated to some extent.
3. A drop in voltage occurs to some extent.

Some kind of conductor is used in every electrical circuit. A conductor can be wound into coils such as in a transformer coil, it can be a large aluminum pipe such as in a station bus, or it can be in an electronic circuit board of a protective relay switch. This chapter deals with conductors used in overhead and underground transmission and distribution circuits.

9.2 Electrical Properties of a Conductor

Conductor Selection

9.2.1 From an electrical perspective, the selection of a conductor is based on the ampacity and voltage requirements of the circuit. The larger the diameter of the conductor, the greater the capacity to carry large amounts of energy with the least amount of line loss. However, some compromise has to be made between a large-diameter conductor and the mechanical properties of a conductor.

A larger-diameter conductor increases overhead line tension, adds to sag, allows for more ice buildup, and provides more surface for wind. Depending on the length of a span and the expected electrical load on a conductor, the mechanical strength requirements can be more critical than the electrical properties of a conductor. On underground circuits, current-carrying capacity and voltage regulation are the dominant factors involved in specifying a conductor.

Ampacity of a Conductor

9.2.2 *Ampacity* can be defined as *the current in amperes a conductor can carry continuously without exceeding its temperature rating.* When a conductor temperature reaches the annealing point, its strength, brittleness, and elasticity are permanently changed.

There are two factors that affect the amount of heat in a conductor:

1. There is a resistance in the conductor itself. This resistance impedes the current flowing in the conductor and causes heat to be produced based on the formula

$$Watts = I^2 R$$

In other words, a large-diameter conductor has less resistance to electrical current, which means that less heat is being generated within the conductor.

2. Heat generated in a conductor can transfer by convection or conduction to the surrounding environment. With an overhead conductor, heat transfers to the air; therefore, on a cold day, a conductor can carry more current than it can on a hot day. Underground cable has thermal barriers that slow the cooling of a conductor, for example, conductor insulation, the soil surrounding the cable, or poor air circulation in a duct.

Capacity Rating of a Transmission Circuit

9.2.3 Each transmission circuit has a continuous current rating. A normal electrical load causes some heating of the power conductors, and the conductor sag increases as the load increases. Overloaded conductors heat and sag into crossing circuits or trees. Depending on the span length and conductor type, a conductor sag can change about 10 meters (30 feet) with load and weather changes. The circuit is rated to a maximum permissible loading based on conductor size and sag.

As the load on a circuit and the ambient temperature change, a control-room operator uses precalculated tables to maintain a circuit within one of three ratings. There are also computer programs that give a real-time thermal rating for a circuit. These programs continually calculate the maximum permissible loading as the load, temperature, and wind change.

A utility also has transmission capacity ratings for emergency loading when other circuits are lost. An emergency rating allows additional sag as a temporary measure.

Capacity Rating of a Distribution Circuit

9.2.4 It is not normal to have a capacity rating for each distribution circuit. On rare occasions, an overloaded distribution circuit may sag into a neutral or secondary, but normally a protection problem or a voltage problem would indicate a possible overloaded conductor.

Secondary buses and services are not as closely monitored as a distribution feeder, and a trouble crew will be alerted to problems during a low-voltage trouble call. An underground system should be monitored more closely because an overloaded cable will heat up, shorten the life of the cable, and eventually lead to failure.

Voltage Rating of a Conductor

9.2.5 The voltage to be used on a bare overhead conductor is not a factor in specifying conductor size until it is used on high-voltage transmission lines. On high-voltage lines, a minimum diameter is needed to avoid corona loss. Extra-high-voltage circuits use two, three, or four cables bundled together with spacer dampers to form a group, which creates a virtual large conductor.

On the underground cable the voltage rating is based on the property of the insulating material. The cable insulation as well as other design considerations must be suitable for the voltage of the circuit. Voltage induces electrical stresses on the insulating material and an inadequate amount of insulation, a sharp bend in the cable, or external pressure from a rock will eventually cause a rupture in the insulation. Insulation will break down faster because of AC voltage stress than for DC voltage stress. The DC voltage rating of a specific cable insulation is three to four times higher than the AC voltage rating. The stresses induced on the insulation by DC voltage are constant while the stresses induced by AC are multidirectional and fluctuating.

Voltage Drop in a Conductor

9.2.6 When current passes through a conductor, there is resistance opposing the flow and there is a resultant voltage drop. The amount of voltage drop can be calculated using Ohm's Law ($E = IR$). The amount of resistance offered by a conductor depends on the conductor's diameter and length.

Voltage-drop calculations are normally done by computer or by the use of voltage-regulation tables. Table 9–1 illustrates the impact of conductor size and length in a single-phase line with a 50-ampere load. The table also illustrates why voltage regulation is a bigger consideration than ampacity when choosing a conductor.

For example, the voltage at an unloaded distribution transformer is 120 volts. What would be the voltage of an unloaded transformer 3 miles downstream if the primary conductor is 3/0 ACSR?

TABLE 9–1 Voltage-Regulation Table

Percent Voltage Drop per Mile for 4.8kV, 50A Load, at 0.9% Power Factor					
	One Mile	*Two Miles*	*Three Miles*	*Four Miles*	*Five Miles*
#4 ACSR	5.6%	11.2%	16.8%		
#2 ACSR	3.8%	7.6%	11.4%	15.2%	
1/0 ACSR	2.7%	5.4%	8.1%	10.8%	13.5%
3/0 ACSR	1.9%	3.9%	5.8%	7.8%	9.7%

Table 9–1 shows that there would be 5.8-percent voltage drop over a 3-mile length of 3/0 ACSR. Therefore,

$$120V - (0.058 \times 120) = 113.04V$$

Voltage Drop in a Three-Phase System

9.2.7 To use Table 9–1 to calculate the voltage drop for a *balanced* three-phase circuit, the voltage drop for a single-phase circuit is divided by two because the flow back to the source is no longer on the neutral. The result is then multiplied by the square root of three, which is the voltage drop per phase. Because the square root of three divided by two is 0.866, the voltage drop on one phase of a balanced three-phase system can be calculated by multiplying a single-phase voltage drop by 0.866.

For example, the voltage at an unloaded three-phase transformer bank is 120 volts per phase. What would be the voltage of an unloaded transformer bank 3 miles downstream if the primary conductor is 3/0 ACSR?

Table 9–1 shows that there would be a 5.8-percent voltage drop over a 3-mile length of 30 ACSR. The voltage drop for a single-phase line is

$$120V \div 1.058 = 113.4V$$

which is a drop of

$$120 - 113.4 = 6.6V$$

For a three-phase line, the voltage drop would be

$$6.6 \times 0.866 = 5.7V$$

Specifying Conductor Size

9.2.8 The current-carrying capacity of a conductor is not normally the controlling factor when specifying a conductor for an electrical utility circuit. Other factors are:

- A conductor must be large enough to keep the voltage drop to an acceptable limit.
- A conductor must be large enough to limit line loss and keep the availability of fault current at the end of the line high enough for the protective switchgear to see a fault.
- A conductor must be large enough to accept future load growth.
- A conductor on transmission lines must be large enough to limit corona loss.

On a very short length of line, ampacity can be a limiting factor because voltage drop or line loss will not be as noticeable. For example, a main line consisting of 336 thousand circular mils (kcmil) AL has a voltage regulator with #2 copper input and output leads. This is acceptable because the #2 copper has the ampacity to carry the load current and the short length will not affect the voltage regulation appreciably.

Conductor Sizes

9.2.9 The numerical systems used to size conductors can be quite confusing. The numbers used to indicate conductor size have no practical application to the lines trade. Fortunately, most utilities standardize on a relatively small number of conductors and provide tables for their weight, die sizes, and grip sizes.

The American Wire Gauge (AWG) system is formed by defining a forty conductor as 0.46 inches in diameter and a #38 wire as 0.005 inches in diameter. There are thirty-eight sizes of wire, spaced in a geometric progression, between these two sizes. Conductor sizes are expressed by numbers. Common sizes used in the lines trade are #4, #2, #1/0, and #3/0.

Conductors larger than 4/0 are referenced to their cross sectional area (CSA), expressed in circular mils. The size of a conductor is not calculated by using the area of a circle formula, $\pi \times R^2$. One circular mil is defined as a circle with a diameter of 1/1000 or 0.001 inches. Based on this definition, a conductor size can be determined by measuring the diameter of a conductor in mils (1/1000 or 0.001 of an inch) and then squaring that number. The size of a conductor is normally expressed in thousands of circular mils, or kcmils.

However, only the conductive material is used when discussing conductor size, so type of stranding and steel wires in a conductor will change the calculated diameter. For example, a 336.4kcmil ACSR conductor with twenty-six strands of aluminum and seven strands of steel is measured as 0.720 inches in diameter, which when squared would indicate that this conductor should be a 518.4kcmil conductor.

In the metric system, a conductor size is referenced to the area of its cross-sectional area (CSA), expressed in square millimeters. The conductor size can be calculated using the formula for the area of a circle, which is $\pi \times R^2$. The aluminum and steel cores of ACSR conductors are measured separately and shown as 40/20 mm^2 for 40-mm^2 aluminum and 20-mm^2 steel core.

Corona Loss

9.2.10 An electric field around a conductor can be strong enough to break down or *ionize* the molecules in the air and cause sparks or a *corona* around the conductor. Under certain weather conditions, on a dark night, corona can be seen around a conductor as a purple glow or as sparking. Corona can also be heard as a crackling or hissing sound, especially on a foggy or damp morning.

When the air between phases, or between phase and ground, is electrically stressed, the air becomes ionized (the air becomes conductive) near the surface of the conductor. The electric field generates a cloud of tiny electric discharges into the air surrounding the conductor.

Corona generates ozone gas (0_3), which is an ionized form of oxygen and is recognized by a distinct caustic smell. Ozone decomposes organic materials, such as rubber, and affects materials that are subject to oxidization. Older rubber cover-up was subject to ozone damage when the cover-up was left on the line for any length of time.

It takes energy to make corona. Corona loss results in power loss and causes radio and television interference. The design of a circuit, its conductor, and its hardware reduce corona loss to a minimum:

- Voltage stress of the air between phases or between phase and ground is reduced by designing the circuit to have enough spacing between conductors. The higher the voltage, the greater the spacing required.

- Voltage stress is reduced at the surface of the conductor by spreading the stress over a larger area. High-voltage circuits, therefore, need a larger-diameter conductor. Extra-high-voltage circuits use bundled conductors to spread the voltage stress over a greater air space. Corona rings or grading rings used on extra-high-voltage circuits also spread the stress over a larger area.
- Voltage stress is reduced at the surface of conductors by smoothing any rough or sharp points where voltage stress of the air concentrates.

Skin Effect

9.2.11 Alternating current does not travel equally distributed throughout the conductor. Electrical current interacts with its own magnetic field. The electromagnetic field around the conductor also cuts through the conductor itself, and the created inductive reactance causes a counter-electromotive force (cemf). The cemf is greater at the center of the conductor than along the outside. The main current flow becomes concentrated along the outer surface of the conductor.

Large-diameter conductors or tubular conductors are used to overcome the higher impedance to current flow caused by the skin effect. Skin effect increases with an increase in frequency. Although the frequency of a powerline is a standard 50 or 60 hertz, the skin effect is increased when the circuit has high-frequency harmonics superimposed on it. Higher-frequency harmonics increase the amount of induction created by the electromagnetic field and increase the skin effect. For the same reason, harmonic distortion causes an increase in the heating of transformer and motor windings.

9.3 Overhead Conductor

Types of Overhead Conductor

9.3.1 Overhead transmission line conductors are always bare. They are mostly made up of aluminum stranding with some kind of steel reinforcement. An overhead distribution conductor can be bare, covered with weatherproofing, or have an insulting jacket. Weatherproof covering is apparently a carryover from early days when local or municipal electric codes called for conductor insulation. The weatherproof covering provides some protection from low voltages, but today it only provides some additional electrical protection from tree contact. A powerline worker should treat weatherproof-covered conductors as though they are bare.

Secondary buses and services are mostly insulated. The live legs and neutral are often wrapped together to form a spun aerial cable or triplex cable.

Advantage of Aluminum

9.3.2 Almost all overhead power conductors strung today are made from aluminum, aluminum alloy, or aluminum with steel reinforcing. The conductivity of aluminum is about 62 percent that of copper, but the weight of an aluminum conductor is about half of a corresponding copper conductor with an equal resistance and length. In other words, aluminum has a conductivity-to-weight ratio that is twice that of copper. The light weight and cost advantage of aluminum have made aluminum the preferred conductor for overhead applications.

Types of Bare Conductors

9.3.3 The bare conductors available are a compromise between a conductor's tensile strength and its conductivity. Other properties considered in a conductor design are conductor weight per unit length, thermal expansion, elasticity, surface shape-drag, fatigue resistance, and its ability to dampen vibration and resist galloping.

Examples of conductors follow:

1. For transmission and distribution lines, the strength and weight of aluminum conductor, steel-reinforced (ACSR, Figure 9–1), allow longer spans, less sag, and, therefore, shorter structures. The reinforcing wires may be in a central core or distributed throughout the cable. A galvanized or aluminized coating reduces corrosion of the steel wires.

   There is a large variety of stranding arrangements available. For example, a 336.4kcmil ACSR can have the number of aluminum strands to the number of steel strands of 6/1, 18/1, 20/7, 24/7, 26/7, or 30/7. Each conductor would have a slightly different diameter. An increase in diameter increases the resistance to wind and ice and decreases the resistance to electrical current.

2. All-aluminum conductor (AAC) is a conductor with a high corrosion resistance but relatively poor tensile strength. It is used in distributions in which shorter spans do not require steel reinforcement for strength. The excellent corrosion resistance of aluminum has made AAC a conductor of choice in coastal areas.

3. All-aluminum-alloy-conductor (AAAC) is a conductor with higher-strength aluminum-alloy strands. This is a popular distribution conductor because, compared to an ACSR of the same diameter, AAAC has lighter weight, comparable strength, better current-carrying capacity, and better corrosion resistance.

4. Aluminum-alloy conductor, steel-reinforced (AACSR), is an aluminum-alloy conductor with steel strands added for more strength. AACSRs have approximately 40 to 60 percent more strength than comparable ACSRs of

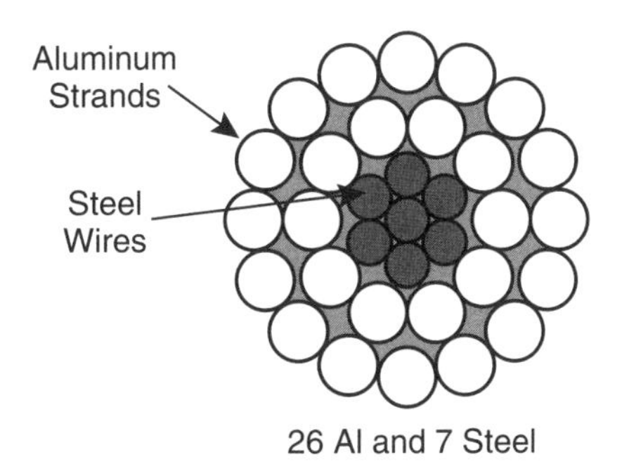

Figure 9–1 ACSR.

equivalent stranding and only an 8 to10 percent decrease in conductivity. It is a very high strength conductor used for extra-long spans or for use as a messenger/neutral cable for spun secondary.

5. Aluminum conductor, aluminum-alloy-reinforced (ACAR), is made up of aluminum stranding and aluminum-alloy stranding. The aluminum-alloy strands provide a conductor with a balance of electrical and mechanical properties. ACAR can have any combination of the two types of strands to provide the best choice between mechanical and electrical characteristics for each application.
6. An ACSR with trapezoid-shaped aluminum wires (ACSR/TW, Figure 9–2) is a compact conductor. With a conductor diameter equivalent to conventional ACSR, there is a 20 to 25 percent increase of aluminum area. This provides a significant decrease in the resistance and an increase in the current-carrying capacity of the conductor.
7. Aluminum-weld (AW) conductor has aluminum cladding bonded on each individual strand of high-strength steel wire. It has the strength of steel with the conductivity and corrosion resistance of aluminum. A wire size of 7 No.5 means there are seven strands of #5 wire. Aluminum-weld conductor is often used as overhead ground wire (shield wire) or as a messenger wire. On transmission lines, AW is used as overhead ground wire (shield wire).
8. Copper-weld (CW) conductor is used as a buried ground wire (counterpoise) at the base of transmission line structures.
9. An ACSR self-dampening (SD, Figure 9–3) is a self-dampening vibration-resistant conductor. It is constructed of steel wires in the core surrounded by layers of trapezoid-shaped aluminum wires. It is designed to keep a small gap between the layers of wire while under tension. The interaction of the different natural vibration frequencies of the steel core and aluminum layers provides an internal dampening effect.

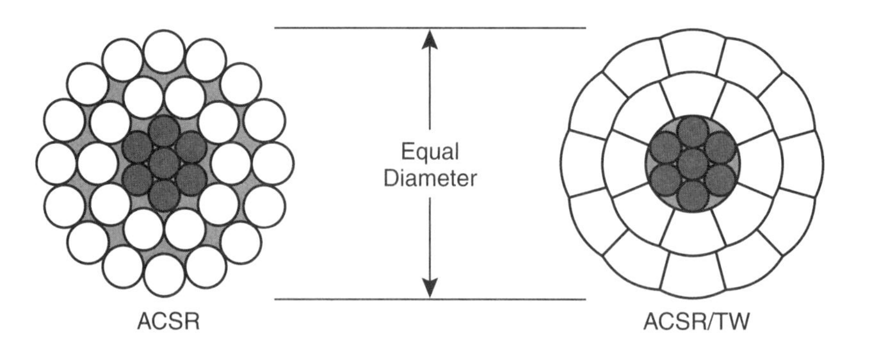

Figure 9–2 Compact trapezoidal conductor.

10. Vibration-resistant (VR, Figure 9–4) conductor is composed of two identical conductors twisted together, giving the conductor a spiraling figure-8 shape. VR conductor is used in areas subject to vibration and galloping due to wind or ice. The spiraled shape presents a continuously changing conductor diameter to the wind, which disrupts the force of the wind on the conductor. This type of conductor can be sagged to full allowable tension without the need for additional vibration protection. A twisted oval conductor provides similar continuously changing diameters to the wind.

Tree Wire or Spacer Cable

9.3.4 Tree wire and spacer cable (Figure 9–5), which are basically the same type of conductor, have an insulated jacket that protects the conductor from tree abrasion. The jacket, however, is not considered insulation but is considered a cover. The cover prevents an immediate short circuit when the conductor makes contact with trees or other grounded objects. If the contact with a grounded object is prolonged, the insulated cover deteriorates and eventually fails. It is estimated that tree cable reduces tree trimming by 60 percent.

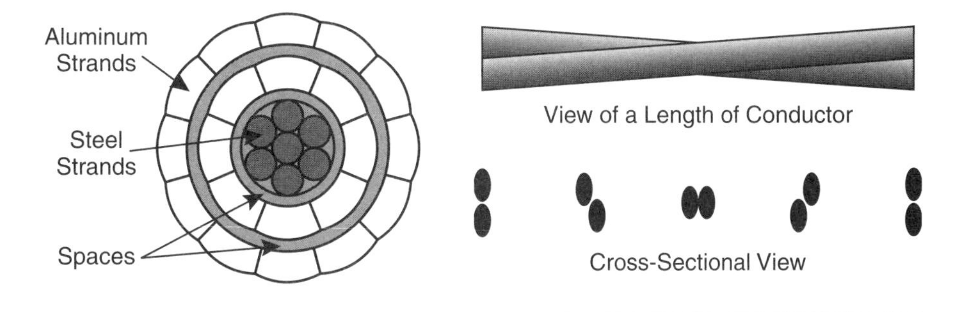

Figure 9–3 Self-damping conductor.

Figure 9–4 Vibration-resistant conductor.

Figure 9–5 Tree wire on an urban street.

The insulation on unshielded cable such as a covered conductor cannot be depended on for personal safety. The cable needs to be treated as a bare conductor for working purposes because the cover may have deteriorated without the deterioration being readily visible.

Bundled Conductors

9.3.5 A bundled conductor (Figure 9–6) is an arrangement of conductors in which each phase has two or more conductors in parallel. The conductors are held a short distance apart by dampers, as shown in Figure 9–7. Bundled conductors are frequently used for high-voltage and extra-high-voltage transmission lines. From an electrical point of view, a bundle of conductors is one very large conductor and has all the advantages of a very large conductor.

A bundled conductor operates at a lower temperature, lower resistance, and lower line loss than a single conductor with the same total amount of material. The inductive reactance in the circuit is reduced. For example, a two-conductor bundle has only about 50 percent of the reactance of a single conductor having the same circular mil area as a bundled pair. The greater the spacing between each conductor in the bundle, the lower the reactance. There is less corona and radio noise from a bundled conductor because corona loss from a conductor is related to the voltage gradient at the conductor surface.

Brass

9.3.6 Brass is an alloy of copper and zinc that can be machined and used for hardware such as live-line connectors. The conductivity of brass depends on the composition of the alloy.

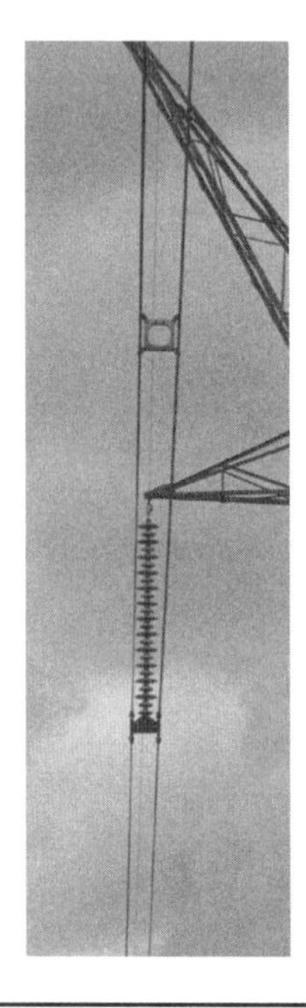

Figure 9–6 Four-conductor bundle.

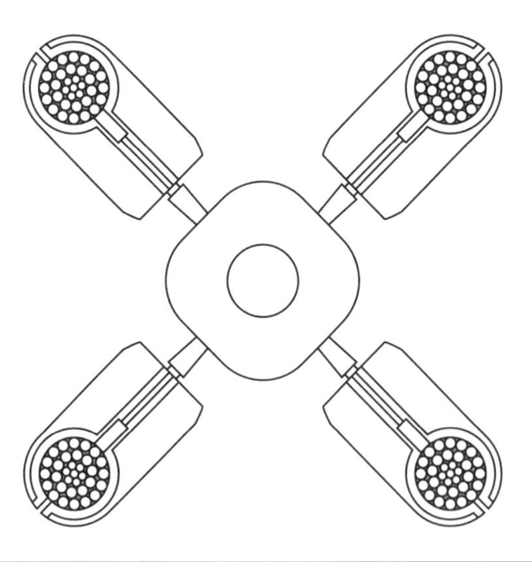

Figure 9–7 Spacer damper supporting a four-conductor bundle.

Working with Conductors

9.3.7 Work with conductors normally involves splicing and making connections. To avoid unnecessary line loss and a possible burn-off, splices and connections must be made on a cleaned conductor to ensure a low-resistance connection.

Aluminum conductor forms black aluminum oxide very quickly on its surface. Aluminum oxide is not a good conductor and needs to be removed before making a splice or connection. Copper oxide, which is the green coating on the wire, is somewhat conductive, and that is why copper connections rarely burn off. When aluminum and copper are connected together, there should be a divider between the two metals. The copper should be on the bottom to prevent the copper oxide from leaching over and corroding the aluminum.

9.4 Underground Cable

Types of Cable

9.4.1 There are high-voltage transmission line cables, distribution-voltage cables, and secondary-voltage underground cables. Cables are generally identified or described according to their type of insulation. For example, there are paper-insulated lead-covered (PILC) cables and cross-linked polyethylene (XLPE) cables.

Transmission Cable

The most common high-voltage transmission cable has been the pipe-type oilpaper insulated cable. The insulated cable is put in a pipe, and the pipe is filled with an insulating oil. Oil-filled cables are referred to as high-pressure fluid-filled (HPFF) cable. Environmental concern about oil leaks has resulted in some of these cables being converted to high-pressure gas-filled (HPGF) cable.

Older cables tended to be low-pressure oil-filled cables. Many high-voltage transmission cables are now XLPE cable.

Working with transmission cable tends to be a specialized cable trade. A powerline worker's involvement may be limited to placing grounds at pot heads (Figure 9–8) or to doing other work where lines equipment is needed.

Distribution Cable

Earlier distribution-voltage cables were paper-insulated lead-covered (PILC) cables. The paper insulation was impregnated with oil, and the lead covering acted as a protective sheath. A lead-covered cable can withstand rainfall runoff or oil from the streets better than the modern polymer cables. There is, however, a concern about the environmental effects of lead and oil. A substitute copper-alloy sheath is available for this type of cable.

Other types of cable insulation are XLPE, tree-retardant cross-linked polyethylene (TR-XLPE), or ethylene-propylene rubber (EPR).

Secondary Cables

Underground secondary cables are usually three insulated conductors bundled together. Each conductor is insulated with a 600-volt polyethylene insulation plus a protective polyvinyl chloride (PVC) jacket.

Figure 9–8 230kV overhead to underground transition.

Cable Design

9.4.2 Underground cables have common design requirements (Figure 9–9) to protect and ensure the continued integrity of the insulation. The electric field within the insulation must be uniform and not become concentrated in any part of the insulation. If the electric field becomes concentrated in one area, the electrical stress in that location will lead to eventual failure.

Cable Shielding

9.4.3 The insulation of a cable has a semiconducting shield at both the inside and the outside of the insulation as seen in Figure 9–10. A semiconducting layer on each side of the insulation spreads the electric field uniformly.

The semiconductor layer next to the conductor is the *conductor shield,* which spreads an irregular electric field uniformly throughout the insulation. Electric stress would otherwise concentrate in certain areas because of the irregular shape of stranded conductor or bends in the cable.

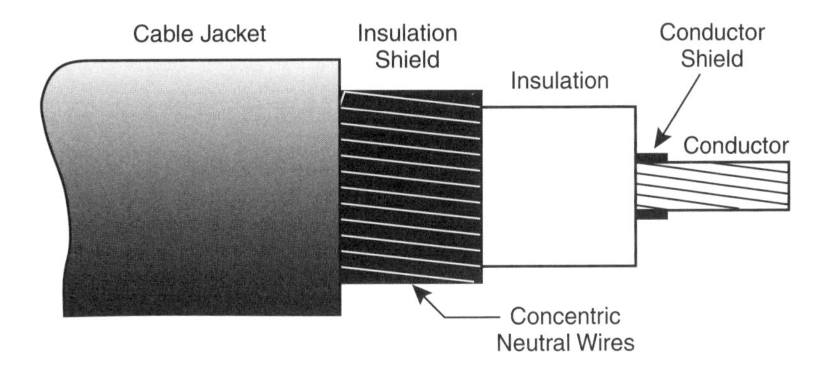

Figure 9–9 Underground cable design.

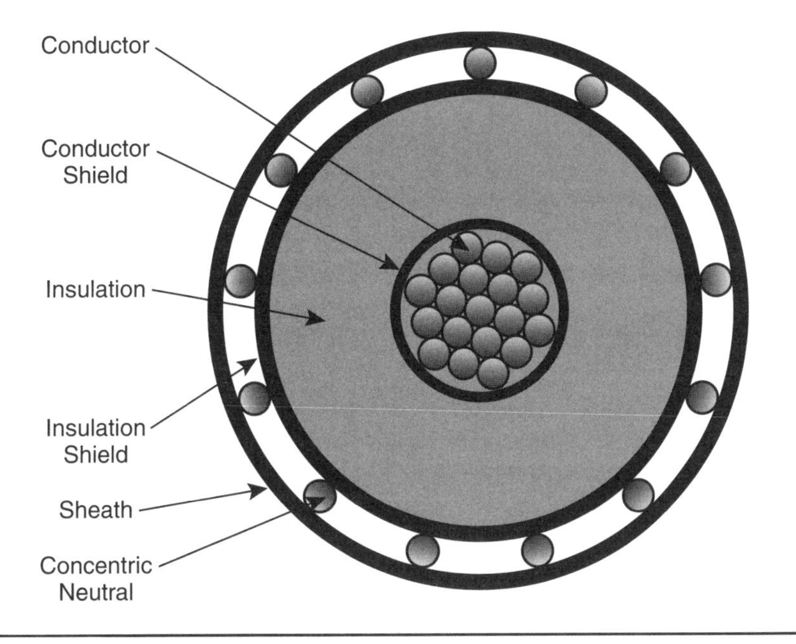

Figure 9–10 Cable shielding.

The semiconducting layer on the outer surface of the insulation is the *insulation shield,* which spreads the influence of the concentric neutral wires or prevents minor cable damage from causing the electric field to concentrate in individual locations.

Cause of Cable Failure

9.4.4 Cable failure occurs when electric stress is allowed to be concentrated in one area. The semiconducting cable shield and insulation shield normally spread out any electrical stress, but any sharp bends, voids, or sharp pressure against the cable concentrate the electric stress at that point.

Water treeing is also a cause of premature failure. Cable can have some water or moisture in it. The moisture enters the ends where the cable is cut during storage or installation. Moisture fills the very fine electric stress lines that occur at voids, sharp bends, and shield irregularities. The stress lines form in the shape or appearance of a tree; the root of the tree is at the cable, and the stress lines spread out from there toward the outer edge of the cable. Distribution submarine cable usually has a solid conductor instead of a stranded conductor to reduce the probability of water migrating between the strands.

Cable Terminations

9.4.5 Underground cable is terminated at a transition to an overhead switch, to a transformer, or to a switching cabinet. The transition to the overhead system occurs at a riser pole (Figure 9–11), also called dip pole, transition pole, or pot head.

The concentric neutral and the semiconducting insulation shield are pulled back from the end of the termination to prevent a short circuit. This has to be done so that the electric stress is spread evenly around the termination. There are many ways of terminating cable, depending on the voltage, the size of cable, and the manufacturer's products. The termination in Figure 9–12 shows a taped termination to demonstrate a generic layout.

Figure 9–11 44kV overhead to underground transition.

Limits to Ampacity

9.4.6 The limiting factor for the ampacity of cable is heat. When the cable is direct buried, the ampacity depends on the *thermal resistivity* of the earth. If the thermal resistivity is low, the heat is carried away from the cable and the ampacity of the cable is higher.

A cable in a duct does not have a direct contact with earth; therefore, the cable will heat up quicker and have a lower ampacity. The center cable in a trench is not able to dissipate the heat as well as the outside cables. The section of cable going up a riser pole, inside a metal cover, is exposed to the sun, which limits the ampacity of the circuit.

Electrical Current in a Cable Sheath

9.4.7 The electromagnetic field in an underground cable can induce unacceptable levels of current and voltage on the metal sheath and concentric neutral wires. The magnetic field from the current-carrying conductor induces a current in the metal sheath or concentric neutral wires when the sheath is grounded at both ends of the cable. Figure 9–13 shows that when the sheath is grounded at both ends, a circuit is created for the induced current. The induced current in the sheath flows back to

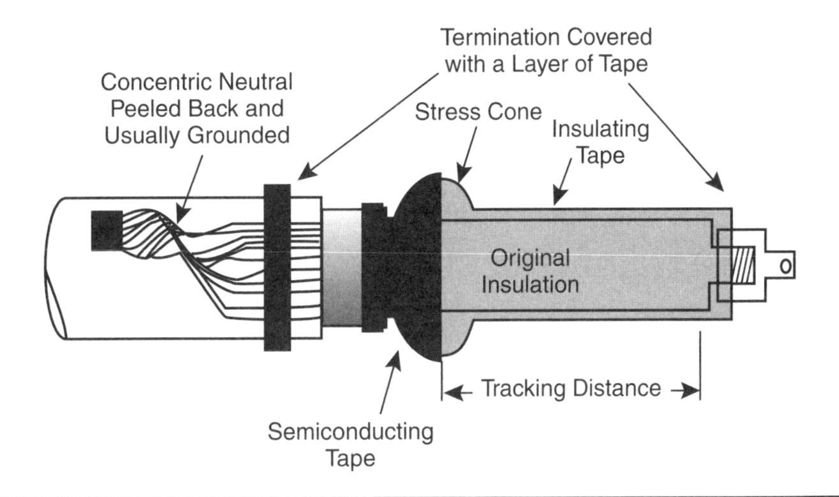

Figure 9–12 Cable termination.

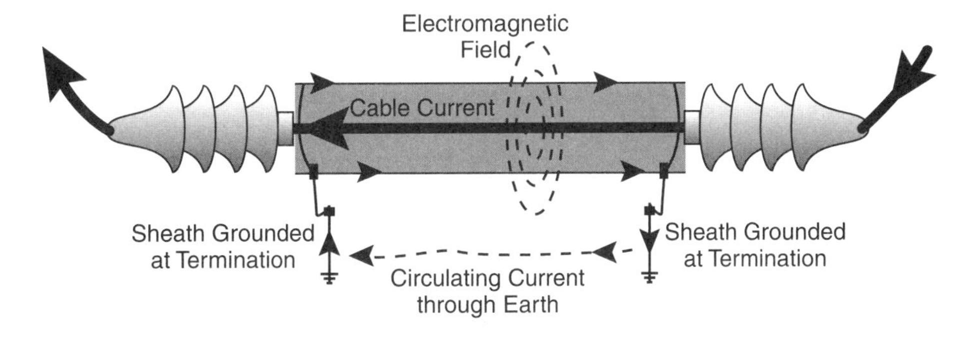

Figure 9–13 Circulating current in sheath.

the source through earth. The current on the sheath can be high enough to cause heating and derate the ampacity of the cable.

Voltage on a Cable Sheath

9.4.8 Grounding the metal sheath at only one end is an option used to eliminate heating caused by the induced current. When only one end of the sheath is grounded, the circuit is open and no current can flow in the sheath. An additional conductor would need to be strung to serve as a neutral for the circuit.

However, at the end where the sheath is not grounded, there is a potentially high voltage between the sheath and the neutral (Figure 9–14). The voltage must be kept at a tolerable level, less than 40 volts under normal operating conditions. Under fault conditions, the voltage across the open point between the sheath and the neutral could be several thousand volts and hazardous to anyone working at the termination. The open circuit should be flagged as a hazard, and a surge arrester could be installed to limit this voltage.

Conductors in SF6

9.4.9 In some station designs where space is limited, bare conductors are suspended inside pipes that are filled with an insulating gas, sulfur-hexafluoride (SF6). The pipe containing the conductor is above ground and is used as a bus in stations. This design has similar advantages to underground cable because less electrical clearance is needed between phases, between phase and ground, and between circuits. The pipes are grounded; therefore, the bus can be laid out along the ground where workers may be in contact with it.

Locating Underground Cable

9.4.10 An underground cable is only visible at the two ends where it comes out of the ground. Record maps and the "water-witching" technique are not reliable enough when someone needs a cable located before starting an excavation. Most cable locators consist of a transmitter, which sends a signal along the cable, and a

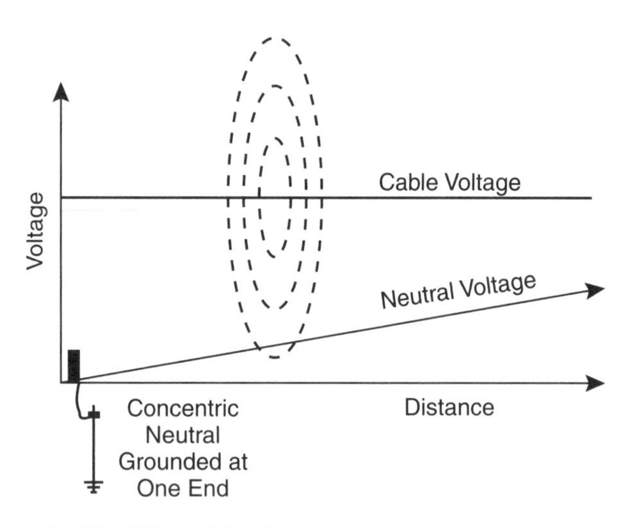

Figure 9–14 Voltage rise in concentric neutral.

receiver, which picks up that signal. Using the receiver, an operator can trace the signal's path directly above the cable.

Transmitters can use a general-induction method, a direct-connect method, or an inductive-coupling method to put a signal on a cable.

1. The *general-induction* method induces a signal into a cable from a transmitter set on the ground near where the cable enters the earth. As an operator with the receiver passes over the cable, it picks up the signal to indicate its location. Unfortunately, the transmitter can also induce a signal on other cables, pipes, and so on, causing the receiver to pick up false signals.
2. The *direct-connect* method requires that the transmitter be directly attached to either the isolated conductor or the sheath. The sheath has to be isolated from ground to prevent the signal from following other ground paths.
3. The *inductive-coupling* method can put a signal right onto a live cable. It uses a donut-shaped coupling device, similar to a clip-on ammeter, that surrounds the cable sheath and emits a signal onto the cable. To use the inductive-coupling method, the cable sheath must be grounded at both ends to form a complete circuit.

The electromagnetic field created by the transmitter can usually be set to a specific frequency. Frequency choices can range from less than one kilohertz to about 480 kilohertz. The lowest frequency is a good starting point. Lower frequencies work better over long distances, and the signal does not transfer to other conducting objects easily. Higher frequencies are more suitable to the general-induction method, but the signal will couple easily with neighboring cables and pipes. The receiver is swept from side to side, and the strength of the signal increases when the receiver is closer to the cable.

Identifying a Cable in a Trench

9.4.11 When underground cables are exposed for work in a trench, it is necessary to distinguish between the cable to be worked on and other cables buried in the same trench. A specialized instrument with a transmitter and a receiver is used. The transmitter sends a pulsing DC voltage down the sheath of the cable. The pulses can be varied with different time delays that help to ensure that the signal received is the signal being sent. The receiver is a DC voltmeter that is connected between the cable sheath and a remote ground probe.

Even with this reasonable assurance that the proper cable has been identified, most utilities require the cable to be "spiked" before working on it. A spike is forced into the cable using a spiking tool fired by a cartridge or turned in with a live-line tool. A spiked cable will cause a short circuit if the cable is alive, and it provides a positive physical identification of a desired cable.

Finding Cable Faults

9.4.12 A cable fault is an insulation breakdown or a break in the conductor. Underground cable faults can vary from a dead short to a high-resistance fault of several megohms. An insulation megger can be used to verify if there is in fact a fault.

Low-voltage, nonshielded cables can be tested using a battery-powered ohmmeter or a megger. Handheld ohmmeters generally have outputs up to 24 volts. They work for detecting a direct short or other low-resistance fault in the kilohm range.

An insulation megger measures resistance in the megohm range using a higher voltage than an ohmmeter. Manual or motor-driven meggers are available for a range of fixed DC voltages. Typical fixed DC voltages are 500, 1000, 2500, and 5000 volts.

There are two main methods for finding the location of a cable fault, the terminal method and the tracer method. The *terminal method* involves sending a low-frequency, high-voltage pulse down a cable and then measuring the time it takes for an echo to come back (radar principle).

The *tracer method* involves putting a voltage impulse on the cable with a transmitter (thumper) and tracing the cable route with a receiver while checking for a change in the form of the signal. The fault creates a ground-gradient potential at the fault location. Difficult variations such as cable faults in a duct, under pavement, or in frozen ground can make this an unsuitable method.

Secondary Cable Faults

9.4.13 A fault on a secondary underground cable often starts out as a complaint of erratic power. A voltage reading at the customer may show a normal voltage.

A break in the insulation of a secondary cable is often a high-resistance fault. When water gets into an aluminum conductor, it slowly corrodes the aluminum and turns it into aluminum hydroxide, which is a high-resistance, white powdery material. The voltage may be normal until a load is put on the conductor because the corroded higher-resistance cable causes an increased voltage drop as the load increases. Using an insulation megger rated at 600 amperes, the cable can be tested. A reading of anything less than one megohm is an indication of a faulty cable.

Primary Cable Faults

9.4.14 A no-power call on a primary underground system probably means that a protective device has opened automatically. The first step is to sectionalize the faulted cable by opening and closing switchgear until the faulted section is known.

The faulted section should be isolated and grounded before any testing is done. A primary cable is like a capacitor and will keep a voltage after the cable is isolated. After grounding, the concentric neutral can be isolated from ground at both ends to prepare for meggering the cable. An insulation megger is connected between the isolated neutral and the grounded phase conductor. A reading of less than one megohm indicates damage between the neutral and the conductor.

Testing Cables

9.4.15 A new or repaired cable installation is usually tested before putting it into service. A common testing method for new cable installations is a high-voltage DC (HVDC) test. An HVDC test is more common than other testing because it causes less damage to the cable in case of a flashover. This is a high-potential test used to detect weak spots in the cable insulation, splices, and terminations to avoid service failure later. The voltage level for an acceptance test is higher than the voltage rating of the cable but lower than the basic-impulse level of the cable. An insulation megger is a common test after repair or maintenance of a cable. An HVDC test can damage a cable that has been in service.

Remember that a cable is a capacitor and has a voltage on it long after a high-voltage test is completed. The cable must be grounded while making the connections for a test and later when removing the test equipment.

Fault Indicators

9.4.16 A distribution underground system does not normally have a protective device for each section of cable between transformers and junctions. Therefore, when one section of cable has a fault, a fuse or breaker somewhere in the system will trip out many sections of cable. To find the faulted section of cable requires opening all the cable sections and then closing each section progressively until the protective device trips out again, that is, unless *fault indicators* have been installed.

A fault indicator is a device that will give a visual "flag" after a current higher than its setting goes through it. The high magnetic field of the fault activates the fault indicator. Fault indicators are put on the cable at various strategic locations, such as shown in Figure 9–15. Instead of opening and closing switching devices, the system can be traced from the source, checking at each fault indicator until one is found in the no-trip position. Depending on the type of fault indicator, some types need to be reset to normal position manually while others will reset after a certain amount of current goes through them or after exposure to a voltage.

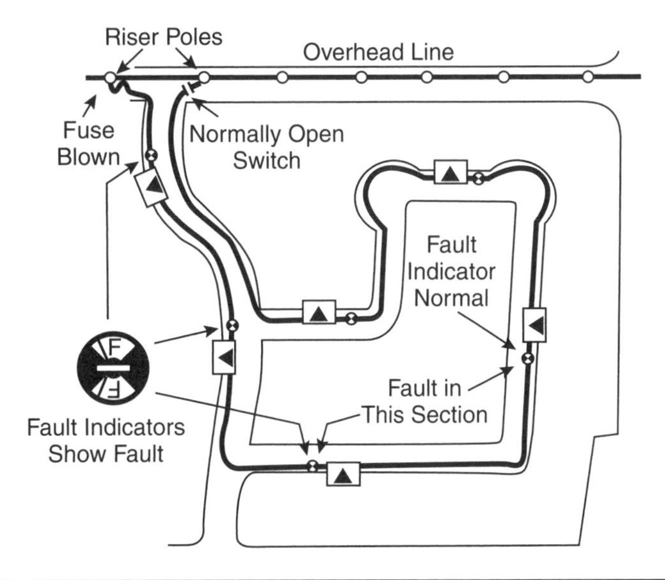

Figure 9–15 Fault indicators in an underground system.

Review Questions

1. What are three basic effects when an electric current flows in a conductor?
2. If a person working on a transmission line leaves any rough or sharp points on a conductor or other live hardware, what effect will that cause?
3. Can the insulation on unshielded cable, such as a covered conductor, be depended on for personal safety?
4. Why is it important to clean aluminum conductor before making a connection or splice?
5. Why may sharp bends, voids, or sharp pressure against a cable cause an eventual failure?
6. Why should the cut end of a cable be sealed when it is stored on a reel in the yard?
7. The magnetic field from the current-carrying conductor induces a current in the metal sheath or concentric neutral wires when the sheath is grounded at both ends of the cable. What can be done to reduce this current flow?
8. When the ground is removed from the sheath at one end to reduce heating in a cable, what hazard exists at the ungrounded end of the cable?
9. Before cutting a cable in a trench, how can a powerline worker be sure the correct cable has been isolated?
10. A customer fed by a secondary underground cable complains about erratic power. A voltage reading at the customer may show a normal voltage. What else should be checked?

CHAPTER 10

Transformers

Topics to Be Covered	Section
Introduction	10.1
Transformer Basics	10.2
Transformation Effect on Current	10.3
Transformer Losses and Impedance	10.4
Transformer Protection	10.5
Single-Phase Transformer Connections	10.6
Three-Phase Transformer Connections	10.7
Three-Phase Secondary-Voltage Arrangements	10.8
Troubleshooting Transformers	10.9

10.1 Introduction

The Purpose of a Transformer

10.1.1 Transformers are used throughout the electrical system, first to raise voltage for efficient transmission of electrical power, then later to reduce voltage to a manageable level for local distribution along roadways and streets and eventually to reduce the voltage to a utilization level.

The power carried in a circuit is equal to volts × amperes. To transmit a large block of energy would require an extremely large conductor if the voltage was not stepped up. Likewise, it would be extremely costly to distribute power to customers along residential streets at high transmission voltages. The transformer is the link between the different voltage systems.

This chapter discusses the distribution transformer, which is used to reduce the voltage on local distribution lines to a utilization voltage. It is the most common piece of lines equipment that the lines trade installs and maintains. Large transformers found in substations work on the same principle as distribution transformers. The substation transformer often has under-load tap changers, additional cooling radiators, and fans. Their reliability affects many customers; therefore, most are monitored for temperature and voltage output by control-room operators.

Other Transformer Applications

10.1.2 In addition to transformers that are used for stepping up or stepping down voltage, there are transformers in an electrical system that are used for other purposes. Some of these are discussed here.

A *neutralizing transformer* is a transformer that is used where telecommunication (telephone) cable enters a substation. The grounded network in a substation is subject to a voltage rise when there is a fault in the electrical system. Any communications cable coming in from outside the station would be seen as a remote ground, and there would be a dangerous potential difference between the communications cable and anything attached to the grounded network in the substation. The neutralizing transformer is a one-to-one-ratio transformer that removes the direct connection between the cables in the substation from the cables leaving the substation.

Fiber-optic-communication cable will eliminate the need for these specialty transformers because fiber-optic cable does not transmit electricity and will not propagate voltage and current out from a station.

A *grounding transformer* is used as an indirect way to ground one phase of a delta circuit. The grounding transformer provides a means for a phase-to-ground fault to get back to the source. A ground-fault relay senses this ground current and trips out the circuit.

Instrument transformers are potential transformers and current transformers that reduce voltage and current to lower, manageable levels. The lower voltage and current represent the primary voltage and current at a given ratio for use in relays and metering.

A *constant-current transformer* is used for applications such as series street lighting where the current must be kept constant and the voltage is allowed to fluctuate.

A *voltage regulator* is a tapped autotransformer that can regulate voltage under loaded conditions.

A *compensator starter* is a tapped autotransformer that is used to soft-start large induction motors. Soft-starting a large motor will prevent a voltage flicker for other customers on the circuit.

10.2 Transformer Basics

Components of a Transformer

10.2.1 A transformer consists of a tank, a steel or iron core, and two or more separate coils or windings of wire (Figure 10–1). The windings are placed in a common magnetic path.

On a single-phase transformer, two separate coils of wire are mounted on a common iron core. When a voltage and current are applied to one coil of wire, a strong electromagnetic field is created in the iron core. The electromagnetic field in the iron core induces a voltage and current into the second wire coil. Therefore, even though the two coils are not electrically connected to one another, voltage and current in one coil induce voltage and current into the second coil.

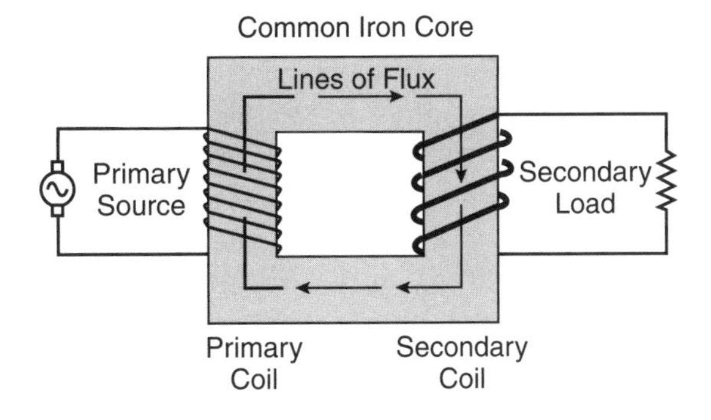

Figure 10–1 A basic transformer.

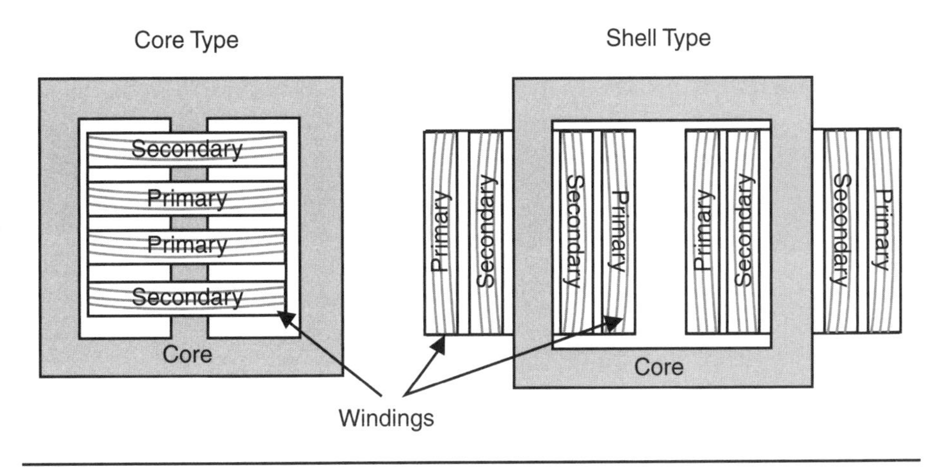

Figure 10–2 Steel core configurations.

A transformer is an electromagnetic device that provides a magnetic linkage between two electrical circuits. Energy is transferred from one circuit to another by the magnetic field.

Two Types of Steel Core Configurations

10.2.2 The actual configuration of the steel core and the windings are of two main types. The steel cores are either shell type or core type. The primary and secondary coils are not necessarily wound around separate legs of the steel core as shown in Figure 10–1 but are wound separately around the same core as shown in Figure 10–2. The shell-type design is more economical for low voltage and a high kilovolt-ampere (kVA) rating, and the core-type design is more economical for high voltage and a low kVA rating.

Turns Ratio

10.2.3 The same electromagnetic field that is in the iron core cuts through both the primary (connected to the source) and the secondary (connected to the load) coils. The same voltage, therefore, is induced in each turn of both coils. The total voltage induced in each coil is proportional to the number of turns in the coil. When the secondary coil has fewer turns than the primary coil, the voltage output is reduced at the same ratio as the ratio of the number of turns of wire on the primary coil to the number of turns of wire on the secondary coil.

For example, a transformer selected to step down 4800 volts primary to 240 volts secondary must have 20 times more turns on the primary or a 20 to 1 ratio, because 4800 divided by 20 is 240.

Input Equals Output in a Transformer

10.2.4 Regardless of the turns ratio, and ignoring some transformer losses, the energy input into the transformer is equal to the energy output.

$$\textit{Input voltage} \times \textit{input current} = \textit{output voltage} \times \textit{output current}$$

When the voltage is stepped down, the secondary current is increased. On a step-down transformer, the secondary coil and leads carry more current and are larger than the primary coil and leads.

Test for Turns Ratio

10.2.5 Every transformer has a nameplate showing the rated primary voltage and the expected secondary voltage. A field test can confirm that the actual turns ratio is equal to the ratio shown on the nameplate. A ratio test will ensure that there are no shorts between turns in the windings.

To conduct a turns-ratio field test on a transformer (Figure 10–3), energize the *high-voltage* coil with a low voltage such as 120 volts. Measure the input voltage and the output voltage with a voltmeter. The ratio is calculated as:

$$\textit{Input} \div \textit{output} = \textit{transformer ratio}$$

Caution! Transformers work both ways! For this test, the low-input voltage must be connected to the *high*-voltage winding. If the input voltage was connected to the low-voltage winding, the transformer would be a step-up transformer and a lethal voltage would appear at the high-voltage terminal.

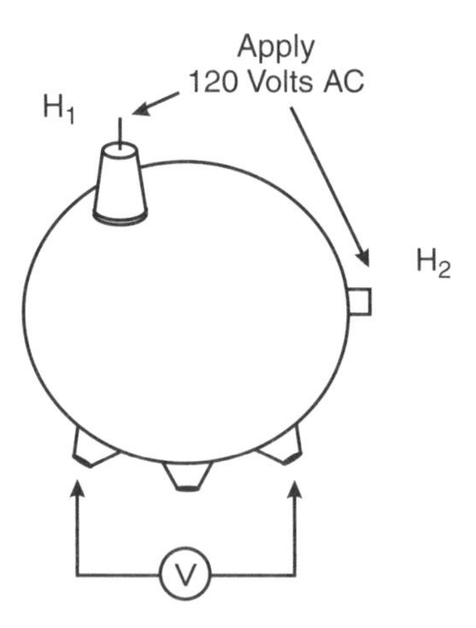

Figure 10–3 Turns ratio field test.

Transformer Backfeed

10.2.6 A powerline worker must always be alert to a transformer that is accidentally fed backward and to the fact that a secondary voltage can be stepped up to a full primary voltage at the transformer primary terminal.

Portable generators available at hardware stores are likely sources of transformer backfeed because unqualified people often connect the generators to the service. When grounds are applied to a primary circuit that is backfed from a portable generator, the generator keeps on running and does not short out. The resistance of the circuit between the generator and the grounds is too high to short out the generator. It can still be safe to work on the grounded conductors because the voltage at the grounded location has been lowered to an acceptable level. However, there will be current flowing in the conductor and the conductor must be jumpered before cutting or opening the circuit.

Large, permanently installed generators do not normally create a hazard because they are more likely to be properly installed with double-throw switches so that the generators cannot feed back into the electrical system.

Sources of secondary backfeed are not always obvious. Removal of the transformer secondary leads, when working on a transformer, will ensure that there is no possibility of backfeed.

Examples

- A recreational vehicle (RV) with a generator is sometimes connected into nearby home wiring.
- An extension cord from a neighboring house can be a source of backfeed.
- When working on a transformer that is networked with other transformers to a common secondary bus, the primary bushing remains alive after the transformer is disconnected from the primary. The secondary bus continues to feed into the secondary of the isolated transformer.

Transformer Taps

10.2.7 Some distribution transformers have off-load tap changers. By changing the position of the tap changer, additional or fewer turns are applied to the primary winding. The nameplate of the transformer will show the various positions and the percentage change for each tap (Figure 10–4). Depending on the manufacturer, each tap raises or lowers the secondary voltage by 4.5 percent or 2.5 percent. The nameplate shown in Figure 10–4 has 2.5-percent taps.

An external switch knob can be turned to the various positions to change the taps. *The transformer must be isolated before turning the knob.* On some older transformers, the cover has to be removed to have access to the tap-changer switch knob.

Feeder voltage changes during the day and at different seasons. Therefore, changing the taps at the transformer could produce extreme voltages when the primary voltage returns to its normal level in off-peak periods.

Taps	
%	
105	A
102.5	B
100	C
97.5	D
95	E

Figure 10–4 Nameplate showing tap settings.

Example

A customer is continuously receiving about 210 volts instead of the rated 240 volts. The transformer, bus, service, or upstream regulators are not the cause of the problem. The 4800/240-volt transformer at this location has a 20-to-1 ratio. The primary voltage at this location must be very low and can be calculated to be:

$$20 \times 210 = 4200V$$

The transformer has taps that will adjust the voltage in 4.5-percent increments. Raising the voltage by three tap settings would change the secondary voltage by:

$$3 \times 4.5\% = 13.5\%$$

Therefore, 210 volts can be boosted to:

$$210 \times 13.5\% = 238.4V$$

Caution

If the primary voltage returns to 4800 volts during off-peak periods, then the secondary voltage will be:

$$240 \times 13.5\% = 272.4V$$

Transformer Polarity

10.2.8 There is no fixed polarity with AC as there is with DC. The polarity of AC changes 120 times a second on a 60-hertz system. However, the coil terminals always have a relative polarity in relation to other terminals. In other words, when the voltage at one end of a coil is positive, the voltage at the opposite end of the coil is negative.

Confirming the actual transformer polarity is important when installing transformers on a secondary network or when banking the transformer together with (an)other transformer(s). By convention, the high-voltage terminals are H_1, H_2, and so on, and the secondary terminals are labeled X_1, X_2 and so on.

Additive and Subtractive Polarity

10.2.9 The polarity of a single-phase transformer is considered to be either *additive* or *subtractive* (Figure 10–5). Whether the transformer is additive or subtractive is

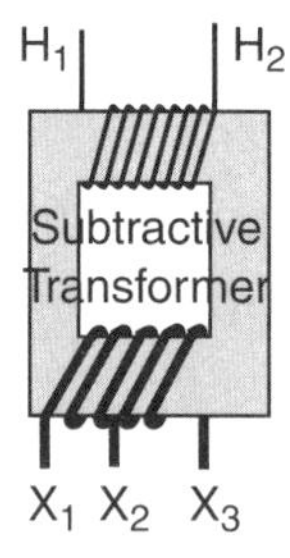

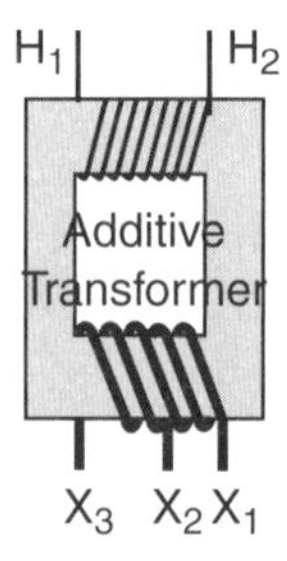

Figure 10–5 Subtractive and additive transformers.

based on the direction the coils are wound. The direction of the coil winding determines the direction of the current flow at the primary terminals with respect to the direction of the current flow at the secondary terminals. From a powerline-worker perspective, it is based on how the secondary leads are brought out of the transformer; or, in simpler terms, the X_1 is either on the left side or the right side of the transformer.

The nameplate on the subtractive transformer shows the H_1 and the X_1 terminals directly opposite one another on the same side of the transformer. Usually, transformers larger than 200kVA or more than 8600 volts are subtractive.

The nameplate on an additive transformer shows the H_1 and the X_1 terminals diagonally opposite one another. Distribution transformers less than 8.6 kilovolts tend to have additive polarity.

Test for Transformer Polarity

10.2.10 A field test can determine which secondary terminal is the same polarity as the H_1 terminal or whether a transformer is additive or subtractive. Figure 10–6 shows a test on a 10-to-1 ratio additive transformer.

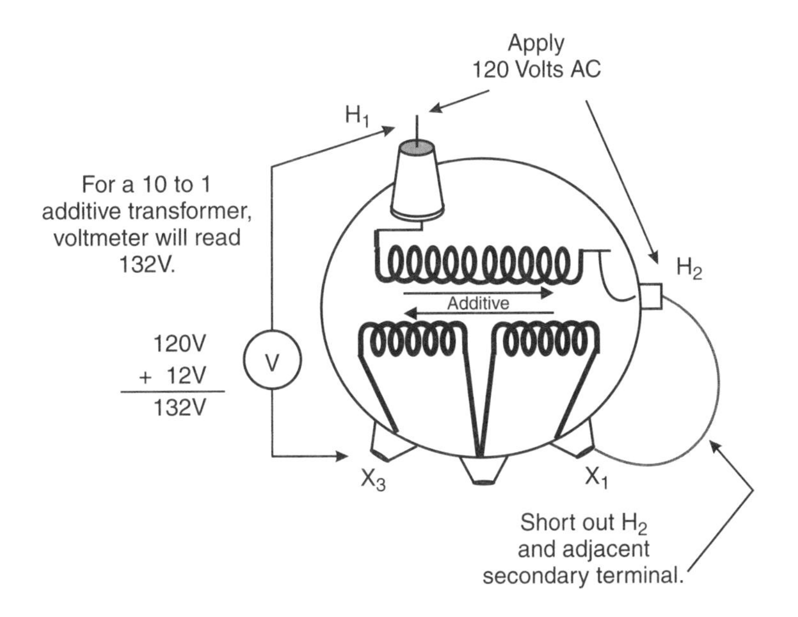

Figure 10–6 Test for transformer polarity.

Step 1

When facing the transformer, install a jumper between the high-voltage neutral (or the right-hand side high-voltage terminal) and the low-voltage terminal on the right-hand side.

Step 2

Apply 120 volts across the primary of the transformer.

Step 3

Measure the voltage between the left-hand side high-voltage terminal and the left-hand side low-voltage terminal.

Connections for Transformers with Different Polarities

10.2.11 When single-phase transformers are to be connected in parallel or connected into a three-phase bank, it is normal to select transformers with the same polarity. It is possible to use transformers with different polarities by ensuring that the X_1 of one transformer is connected to the X_1 of the other transformer (similarly with X_2 and X_3) regardless of the position of the terminals on the transformer tank.

10.3 Transformation Effect on Current

Load Current

10.3.1 When the voltage from one coil is being induced into a second coil, current is also induced into the second coil. The current, however, is transformed in the *inverse* ratio to the voltage transformation. For example, if the turns ratio of the transformer is 20 to 1, the current transformation is 1 to 20. The voltage is stepped down, but the current is stepped up. In a transformer:

$$\text{Input: volts} \times \text{amperes} = \text{output: volts} \times \text{amperes}$$

Example: A 4800-volt transformer with a turns ratio of 20 to 1 has a 200-ampere load on the 240-volt secondary. What current does the transformer draw on the primary side of the transformer?

$$\text{Input: } 4800V \times ?A = \text{output: } 240V \times 200A$$
$$4800V \times 10A = 240V \times 200A$$

The primary coil on a 20-to-1 ratio transformer draws 10 amperes to produce 200 amperes on the secondary coil.

Calculating Load Current

The actual load on a transformer can be measured and calculated in the field (Figure 10–7).

$$\frac{(I_1 \times V_1) + (I_2 \times V_2)}{1000} = \text{total kVA load}$$

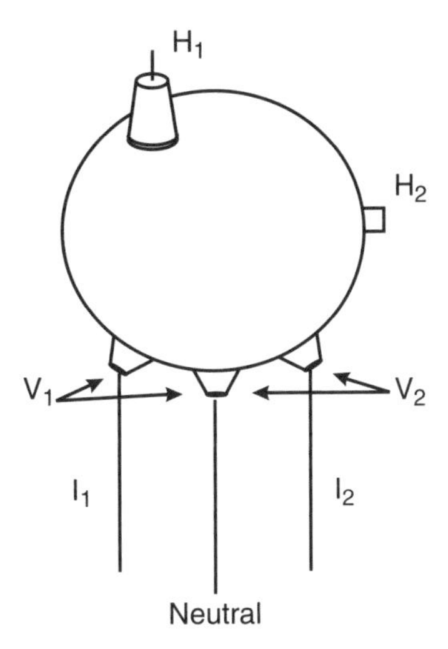

IF: There is a need for a fairly high degree of accuracy.
THEN: The current and voltage on each leg should be taken at the same time to reduce the effects of a load shift from one side to the other.

Figure 10–7 Calculating the acutal transformer load.

Load on Secondary Is Reflected Back into the Primary

When one end of a transformer coil is connected to a live primary source and the other end is connected to a neutral, why is the coil not a dead short? The resistance or impedance of a transformer coil is low and a DC would, in fact, short out between the positive and negative ends of the coil. When AC is applied to the primary coil, there is an induced counter-electromotive force (cemf) from the secondary into the primary to prevent a short circuit.

Lenz's law states that *"an induced electromotive force (emf) always tends to oppose the force that causes the induction."* In a transformer, the magnetic field of the primary coil is inducing a magnetic field on the secondary coil. The secondary coil induces a counter-electromotive force back into the primary coil.

In a transformer, the load on the secondary is reflected back magnetically into the primary. The amount of secondary current that is transferred to the primary is governed by the turns ratio.

Secondary Fault Current

10.3.2 A step-down transformer can generate a very high current on the secondary side if the secondary wires are accidentally shorted. The magnitude of the fault current available on a transformer secondary is mostly dependent on the size of the transformer. The larger the transformer, the greater the capability to generate a large fault current.

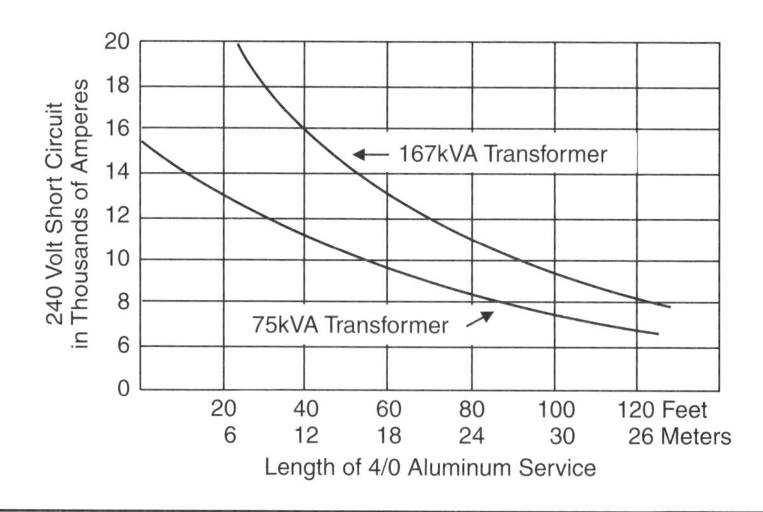

Figure 10–8 Level of fault current at various distances on service.

If the secondary wires are shorted, the transformer can briefly carry a load more than 10 times its rating and supply a high-fault current to the faulted location. Depending on the impedance of the transformer, a short at the secondary terminals of a 100kVA can be as high as 18,000 amperes. The level of fault current at the location of the short depends on the distance from the transformer and the conductor size, as shown in Figure 10–8. The level of fault current available on secondary wires drops quickly with distance away from the transformer.

Secondary services are not always recognized as high-energy circuits by the powerline worker; however, the magnitude of a fault on the secondary is very explosive when a short circuit is close to the transformer. Safety glasses should be worn when working on a live secondary because an eye can be permanently damaged by a large flash.

10.4 Transformer Losses and Impedance

Transformer Losses

10.4.1 From the equation:

$$Input\ V \times A = output\ V \times A$$

it would appear that there are no losses in a transformer. This is essentially true because power transformers are more than 97 percent efficient. However, transformers do account for the largest proportion of energy loss in a distribution system. For example, a typical 25kVA transformer could have losses of about 150 to 350 watts, which would add up, considering the number of transformers in an electrical system. The losses in a transformer are due to core loss and conductor loss.

Core Loss

There is energy consumed when the iron core is magnetized and demagnetized by AC power at 120 times a second. This energy loss is referred to as *hysteresis loss.* To reduce hysteresis loss, the core is made of more permeable iron, which means the core is made with various steel alloys that are easier to magnetize and demagnetize.

When voltage is induced into a transformer coil, voltage is also induced into the iron core. Induced voltage causes current, known as *eddy currents,* to flow in the iron core. Eddy currents cause heating and are a waste of energy. Eddy current flow is reduced by the use of a laminated steel core that has thin strips of steel core material insulated from each other. Eddy currents are, therefore, kept smaller within each separate lamination.

Hysteresis loss and eddy current loss together make up *core loss* or *iron loss.* The core loss of a transformer is constant and is not affected by the load on the transformer. Core loss can be referred to as *no-load loss.*

Relatively excessive core losses occur when a larger-than-necessary transformer is installed for the load to be served. If a 25-kilowatt load was fed from a 100kVA transformer, there would be considerably more core loss than if the load was fed from a 25kVA transformer.

A high-permeability material called amorphous (having no regular form) steel offers 70 percent less power loss than the conventional iron core. It is also called metallic glass because it has a similar atomic structure to glass. It is made when hot liquid iron is forced through supercooled rollers. The steel solidifies so fast that it forms a crystalline structure. The result is an amorphous or irregular structure that is more permeable.

Conductor Loss

There is a resistance to current flow in any wire. A transformer coil is made from a long length of wire, and it offers resistance to current flow. Current flow is also impeded when it flows in a coil because the induced counter-electromotive force in the coil causes a reactance. The resultant resistance and reactance are an impedance to current flow and are called *conductor loss, copper loss,* or I^2R loss. Conductor loss is dependent on the cross-sectional area of the wire in the coils. A lower resistance results in a lower I^2R loss.

Conductor loss can also be called *load loss* because the loss varies with the amount of load on the transformer. Load losses vary by the square of the current (I^2R). That means that a fully loaded transformer has four times the copper loss as one loaded to 50 percent.

Transformer Efficiency

10.4.2 The efficiency of a transformer is equal to:

$$\frac{Output}{Output + conductor\ loss \times pf + core\ loss} = percent$$

For example, a fully loaded 100kVA transformer could have:

- 2000-watt conductor loss at full load and at 90-percent power factor
- 500-watt core loss

The efficiency of this 100kVA transformer is:

$$\frac{100,000}{100,000+2000\times 0.9+500}=97.75\%$$

When the output or load decreases during the day, the iron losses remain constant; therefore, the percent efficiency of the transformer is lower.

Transformer Impedance

10.4.3 The turns ratio of a transformer determines the ratio between the primary voltage and the secondary voltage. When load is applied, the load current is impeded by the resistance and reactance in the windings. When under load, the voltage at the secondary terminals is lower than the voltage indicated by the turns ratio. The impedance of a transformer is expressed as the percentage of the voltage drop at a full load compared to the voltage drop at no load. For example, if a transformer with a 2.2-percent impedance delivers 240 volts at no load, it will deliver 2.2 percent less than 240 volts (234.7 volts) at a full load.

There are practical limits to designing a transformer with lower impedance. There is an advantage for a transformer to have some impedance because the impedance limits current going through the transformer during a fault.

Voltage-Survey Accuracy

10.4.4 When a planning engineer carries out a voltage survey, recording voltmeters are installed at the end of a feeder where a problem low voltage is likely to first show up. This voltage is used by the planning engineers to verify and update their computed feeder calculations.

To get an accurate line voltage, a recording voltmeter is installed on the secondary of an unloaded transformer. An unloaded transformer is used because with a loaded transformer the internal impedance will cause a voltage drop and the actual feeder voltage would still be unknown.

10.5 Transformer Protection

Sources of Transformer Damage

10.5.1 Transformer problems are usually due to an internal insulation breakdown. Insulation in a transformer usually breaks down because of heat or a voltage surge.

Transformer Overheating

Overheating of a transformer is usually due to overload, a short-circuited secondary, or the follow-through current initiated by lightning. The transformer is designed to be fairly tolerant to an overload for short duration. An oil-filled transformer, for example, may withstand an overload of 25 times the rated current for 2 seconds or two times the rated current for 30 seconds.

The kVA rating of a transformer is based on it being at 30°C. When the average temperature is higher or lower than the standard 30°C, the transformer rating can be changed. The kVA rating of a transformer can increase one percent for a decrease of each degree below 30°C and is decreased 1.5 percent for an increase of each degree above 30°C.

Larger transformers found in stations normally have a permissible rating that is calculated based on ambient temperature, load factor, and the existence of external cooling, such as by fans. Figure 10–9 is an example of a transformer rating sheet for a distribution station transformer with no external cooling in a mid-northern climate.

Transformer Loading

The heating effect of current flowing in a transformer coil determines the amount of energy a transformer can supply without causing damage to the insulation. The heat developed in the transformer winding is based on the formula:

$$\textit{Heating effect in watts} = I^2R$$

If a transformer were allowed to carry three times more than its rated load current, the heating effect would be nine times as great as with full-load current.

The full-load current in the windings of a transformer can be calculated for either the primary or secondary coil using the following formula:

$$\textit{Full-load current} = \frac{kVA \times 1000}{\textit{voltage across coil}}$$

Example: The full-load current for a 25kVA, 14400/240-volt transformer is:

$$\textit{Primary full-load current} = \frac{25 \times 1000}{14400} = 1.74A$$

$$\textit{Secondary full-load current} = \frac{25 \times 1000}{240} = 104A$$

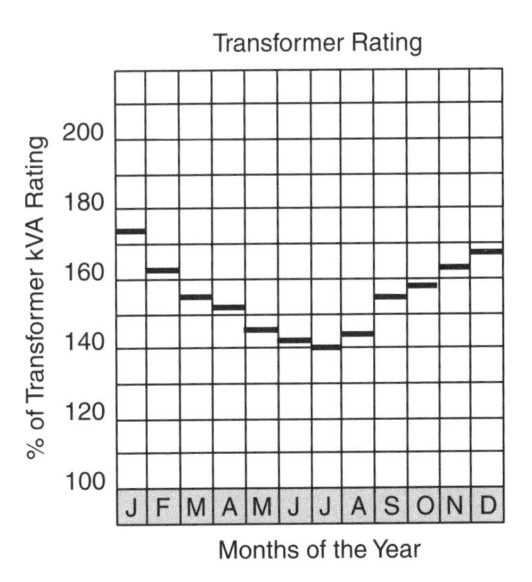

Figure 10–9 Temperature-dependent transformer rating.

Overhead Transformer Fuse Protection

10.5.2 The fuse in a transformer cutout melts when exposed to overcurrent. The fuse melts before the transformer is damaged from the heat generated by an overload or a secondary short circuit. The fuse is also coordinated so that it isolates a faulted transformer before any upstream protection opens the primary circuit.

A link fuse speed and size are specified according to the preferences of the utility. The speed of a transformer fuse is generally specified as a K- (fast) or a T- (slow) link fuse. The slower T-link fuse reduces nuisance fuse blowing due to transients such as lightning.

The kVA rating of the transformer and the primary voltage are two factors that govern fuse size. For example, a transformer on a 2.4-kilovolt system has a fuse size of approximately one ampere per transformer kVA and a transformer on a 14.4-kilovolt system has a fuse size of approximately 1/5 ampere per transformer kVA.

Underground Transformer Fuse Protection

10.5.3 Protection for transformers on underground systems varies depending on whether the transformer is in a vault, is submersible, or is a pad mount. Transformers in a vault often have standard overhead protective switchgear. Live leads, terminals, and switchgear are exposed and need to be in locked enclosures to prevent accidental contact.

Pad-mount transformers are used on underground systems and sit on a concrete pad above the surface of the ground. A pad-mount transformer can be live front or dead front. A live-front transformer has exposed live switchgear when the metal enclosure is opened. The switchgear in a dead-front transformer is insulated. Figure 10–10 shows a dead-front transformer that has a draw-out load-break, bayonet-style, expulsion fuse holder.

A bayonet-style fuse is an under-oil expulsion fuse cutout that has a stab-sheath arrangement to hold the fuse and is field replaceable by a lines crew.

An isolation link or a current-limiting fuse is used in series with a bayonet-style fuse. During a transformer failure, the isolation link or current-limiting fuse opens the primary lead to the faulted transformer. This safety feature prevents a lines crew from reenergizing a faulted transformer. The isolation link or the current-limiting fuse is not replaceable in field conditions.

Current-Limiting Fuse Protection

10.5.4 On overhead transformers, a current-limiting fuse is sometimes installed in series with a fuse cutout to limit the amount of current that can rush into a transformer during a fault. A current-limiting fuse protects a powerline worker from a catastrophic transformer failure that could occur when the worker is trying to energize a defective transformer in a high-fault-current location. A transformer on an underground system often uses a current-limiting fuse to limit arcing, because an arc could easily spill over to a nearby grounded object.

Voltage Surge Protection

10.5.5 A surge arrester is installed to bypass any voltage surge and its associated current away from the transformer. On distribution circuits, the source of a voltage surge is almost always lightning. Occasionally, an accidental contact with a higher-voltage overbuilt circuit results in a surge.

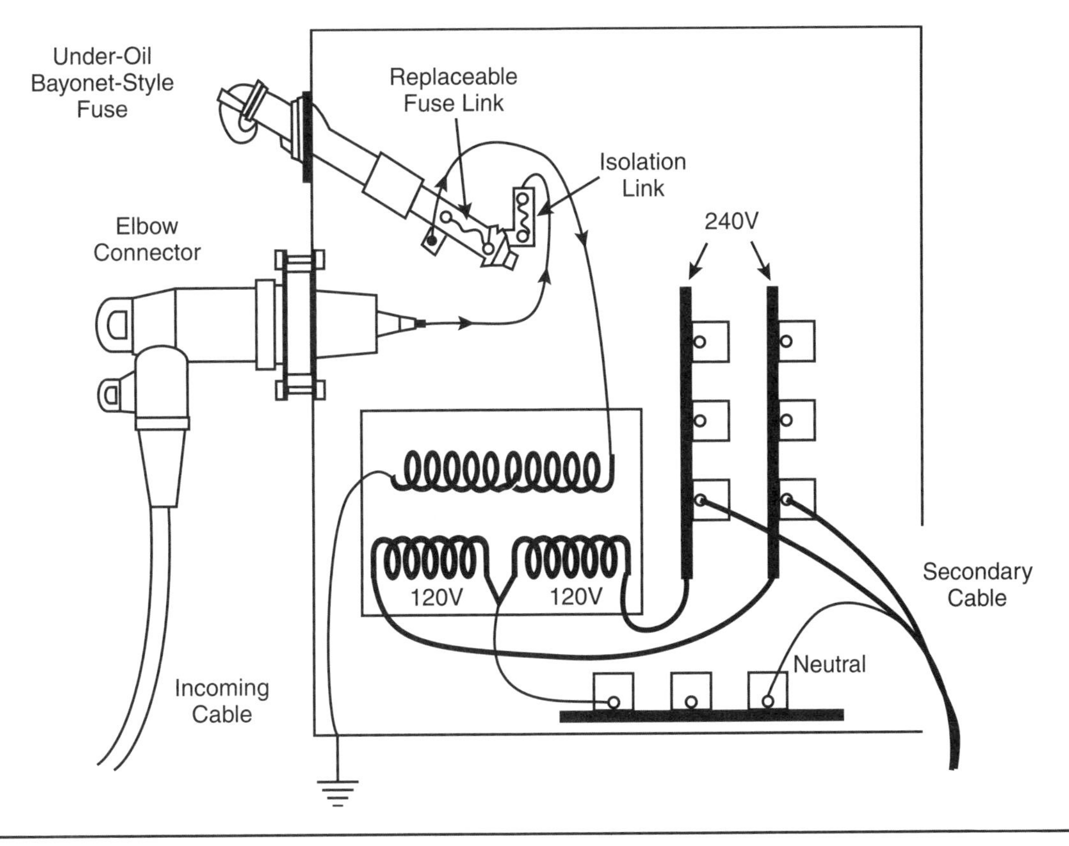

Figure 10–10 Pad-mount transformer fusing.

The surge arrester is installed ahead of the primary transformer terminal; however, utilities differ on whether to install the arrester ahead of or after a fuse cutout. An arrester protects best when it is installed as close as possible to the equipment it is protecting, but there is a concern about nuisance fuse blowing when the arrester is installed after the fuse cutout. There is usually a high follow-through current associated with a voltage surge, and a surge arrester causes the current to bypass the cutout when it is installed before the fuse.

The rating of an arrester is specified at a slightly higher voltage than the circuit. If the rating is too low, continuous exposure to small surges will cause the arrester to deteriorate prematurely. If the voltage rating of the arrester is too high, a damaging voltage and current may not be bypassed. The ground wire leading away from the arrester must be as short as possible and at least the same size as the primary lead.

Neutral Connections and Ground Connections

10.5.6 The neutral and ground connections on a transformer may appear to be interchangeable because the two are interconnected. However, the neutral connections have one purpose and the ground connections have another. All of the connections of each system must be completed.

A neutral is a *grounded* conductor. Neutral connections are part of the electrical circuit needed for normal operation to carry current back to the source.

A ground wire is a *grounding* conductor. Ground connections provide a path for current under abnormal conditions, such as during a lightning storm when an arrester or an insulator sparks over. The ground connections also keep the transformer tank and related equipment at the same potential, which eliminates possible shock for workers.

10.6 Single-Phase Transformer Connections

Single-Phase Transformers

10.6.1 Single-phase transformers have one high-voltage primary coil with two high-voltage terminals. Only one high-voltage terminal needs to be insulated when the coil is connected between a phase and a neutral. Both high-voltage terminals need to be insulated when the coil is to be connected phase to phase.

In North America, 90 percent of all transformers are installed as single-phase transformers connected to supply a standard 120/240-volt service. A customer receives a three-wire service consisting of two 120-volt hot wires and a neutral. One 120-volt leg is the opposite polarity of the other. In other words, the current on one leg is moving in one direction while the current on the other leg is moving in the other direction. The voltage between the two legs is 240 volts. Because the two legs are not always equally loaded, the neutral carries the current difference of the two legs.

This type of service is referred to as a single-phase service by electrical-utility personnel because of the primary connection. The secondary, consisting of two 120-volt hot wires, is really two-phase power with each phase being 180 degrees out of phase with the other.

The two low-voltage coils can be wired up in parallel or in series. Figure 10–11 shows a standard 120/240-volt service; the positive end of one coil is connected to the negative end of the other coil, which means the two secondary coils are connected in series. The coils are connected in parallel when only 120 volts are needed.

In some other places in the world, where a standard 240 (or 220) volts are used, the service is fed from a bus supplied by a 240/415-volt three-phase transformer or from a single-phase 240/480-volt transformer. A residential customer would have a three-wire service with one active (hot) wire, one neutral wire, and one safety ground wire. Higher voltage would be available with a four-wire service.

Typical Nameplate for a Single-Phase Transformer

10.6.2 The nameplate shown in Figure 10–12 is for a standard single-phase transformer. A nameplate would also show whether the transformer had dual primary voltage or had voltage taps. Figure 10–13 shows a typical single-phase overhead transformer, and Figure 10–14 shows a typical single-phase pad-mount transformer.

Connections to Delta or Wye Primary Systems

10.6.3 The primary of a single-phase transformer can be connected in two ways to create a high-voltage potential between the H_1 and H_2 terminals.

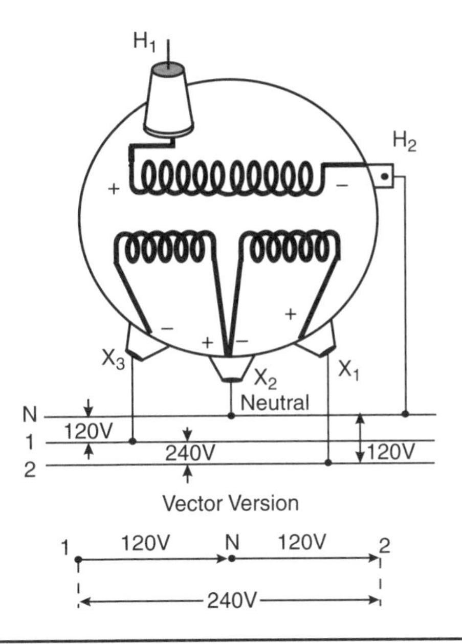

Figure 10–11 Single-phase transformer connections.

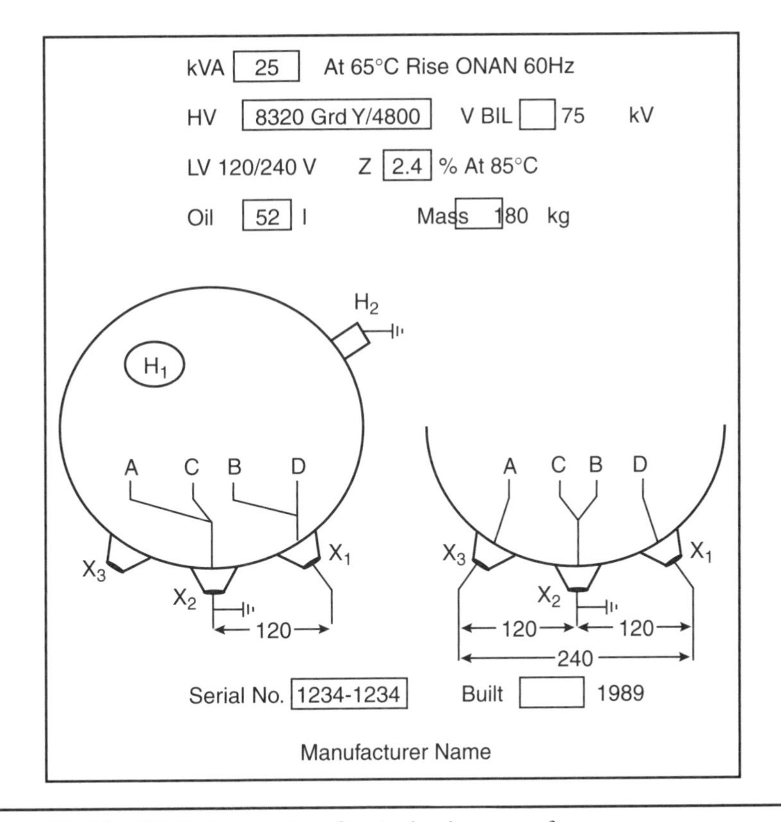

Figure 10–12 Typical nameplate for single-phase transformer.

Figure 10–13 Typical single-phase overhead transformer.

Phase-to-Phase (Delta) Systems

For a single-phase connection to a delta circuit, connect one high-voltage terminal to one phase and the other high-voltage terminal to another phase. The transformer must have two insulated high-voltage terminals for a delta connection. To protect the transformer, a fused cutout and a surge arrester are installed on both of the high-voltage primary leads.

Phase-to-Neutral (Wye) Systems

For a single-phase connection to a wye system, one high-voltage terminal is connected to the phase and the other is connected to the system neutral. Transformers intended for connection to wye systems can be constructed with only one insulated

Figure 10–14 Typical single-phase pad-mount transformer.

high-voltage terminal. The neutral connection to the high-voltage terminal of a *cover bushing transformer* is not insulated.

It is possible to connect a transformer with two high-voltage terminals phase to phase on a wye system, but the voltage rating of the transformer would have to be suitable. On an 8.3/4.8-kilovolt wye system, a transformer connected phase to phase (delta) must be rated as an 8.3-kilovolt system, and a transformer connected phase to neutral (wye) must be rated as a 4.8-kilovolt system.

Transformers Connected in Parallel

10.6.4 Two smaller transformers are sometimes connected in parallel to give the equivalent capacity of one large single-phase transformer. In Figure 10–15, the two secondary coils in each transformer are connected in parallel. One transformer feeds one leg at a positive polarity while the other transformer feeds the other leg at a negative polarity.

It is also possible to interconnect two transformers with the secondaries left connected in series. The positive terminals are connected to one leg of the bus, and the negative terminals are connected to the other.

The impedance of the individual transformers must be very close to each other because the transformer with the lowest impedance will draw the most current and can become overloaded. The difference in impedance should be less than 0.2 percent of each other. In other words, if one transformer has an impedance of 2 percent, then the impedance of the other transformer should be between 1.8 percent and 2.2 percent.

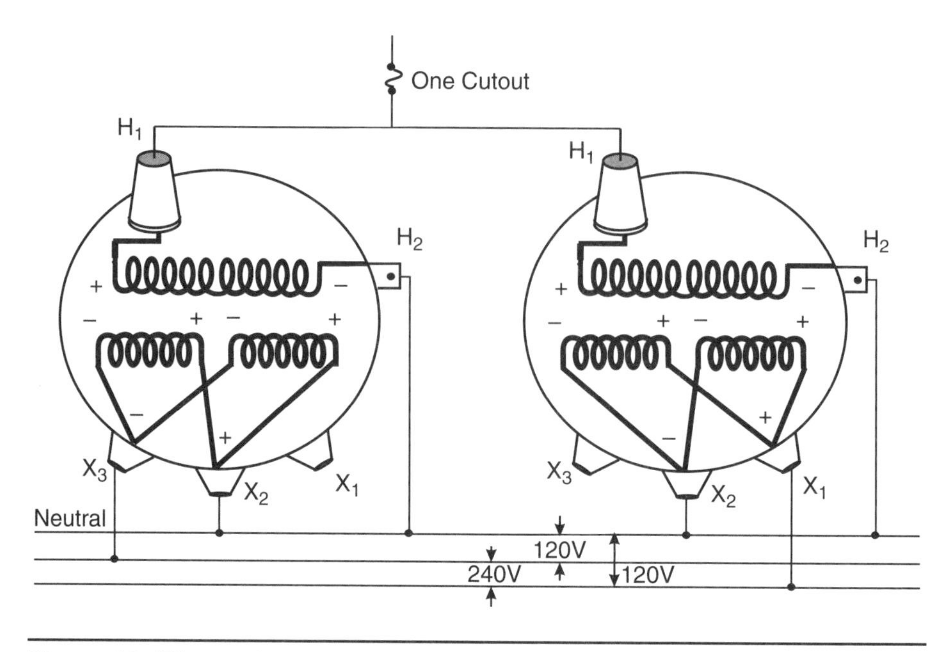

Figure 10–15 Single-phase transformers connected in parallel.

Secondary Network (Banked Secondaries)

10.6.5 Many utilities feed their secondary bus radially from one transformer. The secondary bus will have bus breaks installed to prevent any interconnection with other transformers.

A secondary bus can also be in a network system fed from many transformers connected in parallel to the same secondary bus (Figure 10–16). The most likely application would be an underground system in a city.

- The load is divided among all the transformers connected to the bus.
- Individual customers with peak loads are supplied by the greater available reserve capacity of the multiple transformers.
- Fuses on the secondary bus (network protectors) between transformers isolate faulted transformers and interrupt only the customers near the defective transformer.
- All of the transformers on a common bus are connected to the same phase and connected with the same polarity.
- Ideally, the secondary bus should form a complete loop.
- *When the primary of the transformer is opened, the primary terminal remains alive unless the network protectors are open.*

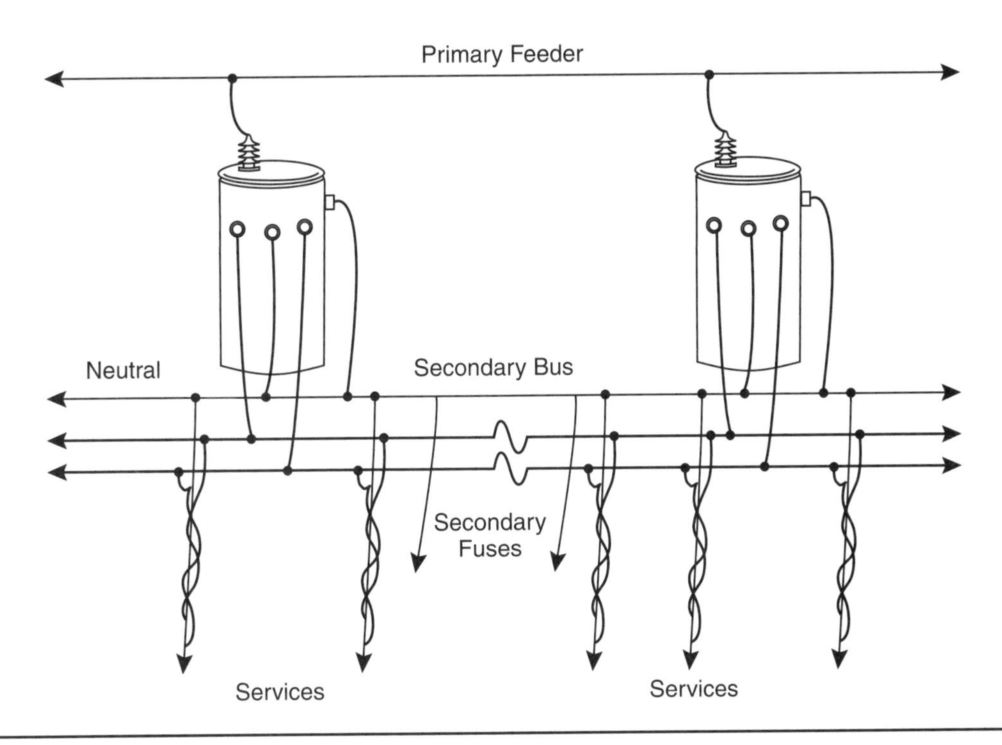

Figure 10–16 Secondary network.

10.7 Three-Phase Transformer Connections

Three-Phase Transformers

10.7.1 A three-phase service can be supplied by one three-phase transformer unit or by interconnecting three single-phase transformer units. One three-phase unit is smaller (Figure 10–17) than an equivalent size bank consisting of three single-phase units. One three-phase transformer tends to be used in underground vaults or as a pad-mount transformer (Figure 10–18).

An illustration of a three-phase transformer is shown in Figure 10–19. One three-phase unit is easier to install because the polarity and interconnections between the phases are fixed.

The use of three single-phase units is common in overhead distribution. When single-phase transformers are banked together, they can be interconnected to supply more than one type of service. For example, three transformers with 120/240-volt secondaries can supply a 120/208, 240/416, or a 240-volt three-wire service. Fewer specialized emergency spare transformers are needed when single-phase transformers are used.

Typical Nameplate for Three-Phase Unit

10.7.2 A transformer nameplate (Figure 10–20) should be checked to determine which three-phase configuration and voltage the transformer is able to supply.

Figure 10–17 Single-unit three-phase transformer.

Figure 10–18 Single-unit three-phase pad-mount transformer.

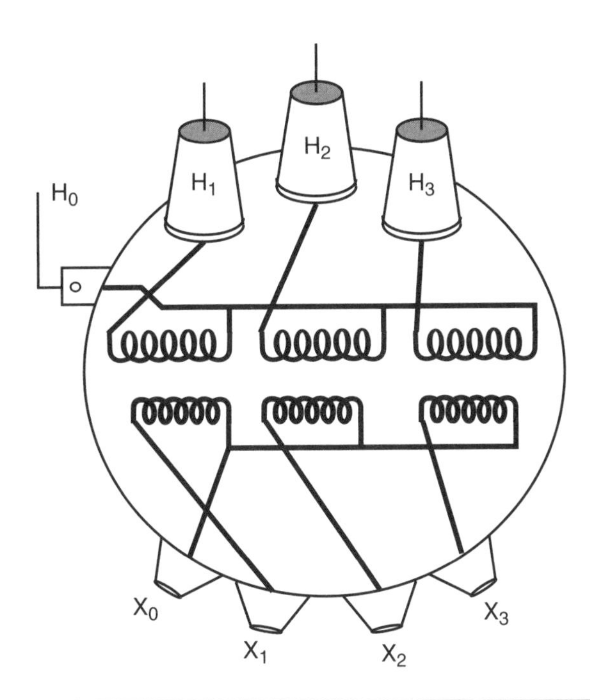

Figure 10–19 A three-phase wye-wye transformer.

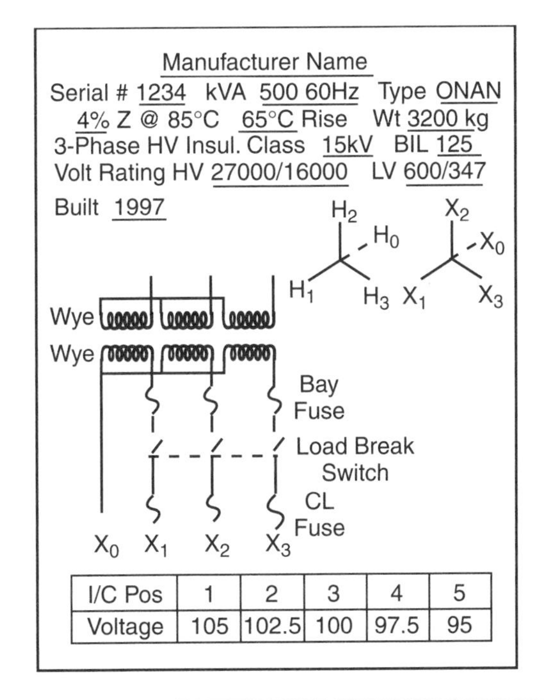

I/C Pos	1	2	3	4	5
Voltage	105	102.5	100	97.5	95

Figure 10–20 Nameplate for three-phase unit.

Figure 10–21 Three-phase transformer bank using single-phase transformers.

Banking Three Single-Phase Units

10.7.3 When interconnecting three single-phase transformers into a three-phase transformer bank (Figure 10–21), there are four specifications to look for when choosing a transformer:

1. The voltage rating of the transformer primary coil must be compatible with the applicable circuit. The voltage impressed across the primary coil will depend on whether the coil is connected in a wye (phase-to-neutral) or a delta (phase-to-phase) configuration.
2. The transformer must be able to deliver the needed secondary voltage. The supplied secondary voltage will be dependent on:
   - The voltage rating of the secondary coil.
   - Whether the transformer secondaries are interconnected in a wye or delta configuration.
   - Whether the secondary coils inside the transformer are connected together in series or in parallel.
3. If equipped with tap changers, the transformers must be on the same voltage tap. Dual-voltage transformers must be set on the proper voltage.
4. The impedance of the transformers in the bank should be within 0.2 percent of each other to avoid having the transformer with the lowest

impedance taking a greater share of the load. In other words, if one transformer has an impedance of 2 percent, then the impedance of the other transformer should be between 1.8 percent and 2.2 percent.

Wye or Delta Connections

10.7.4 A transformer coil must have a potential difference across it in order to operate. To have a potential difference across a transformer coil, the polarity of a terminal at one end of a coil is positive and the polarity of the terminal at the other end of the coil is negative. There are two ways to get a voltage across a transformer coil:

1. One way is to connect a coil between a phase and another phase. When each of the three transformers has its coils connected between the phases AB, BC, and CA, the transformers are interconnected in a delta configuration.
2. The second way is to connect a coil between a phase and the neutral. When each of the three transformers has its coils connected between a phase and a common neutral, the transformers are interconnected in a wye configuration.

Primary Delta Transformer Connections

The three ways to connect a transformer primary into a delta (phase-to-phase) configuration are shown in Figure 10–22. Note how labeling the polarity of the transformers on a drawing reduces confusion when making connections.

Each transformer coil in a delta primary or delta secondary is connected phase to phase. When two or more transformers are interconnected in a delta configuration, the coils are connected in series with each other. To connect a coil in series, each positive terminal of one coil is connected to a negative terminal of another coil. If a transformer primary is to be delta connected on a wye circuit, the voltage rating of the primary coil must be equal to the phase-to-phase voltage of the circuit.

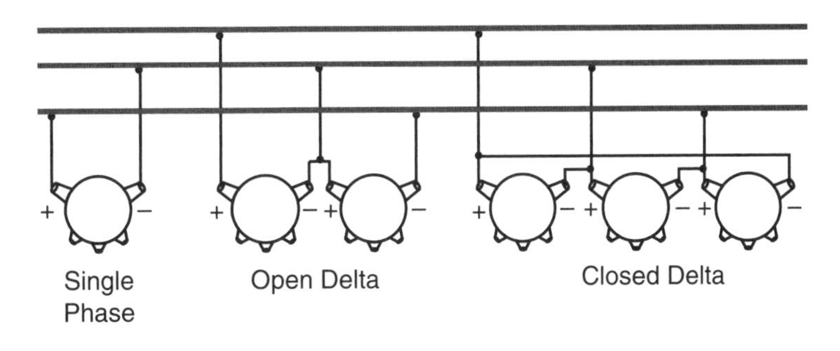

Figure 10–22 Three types of delta connections.

Primary Wye Transformer Connections

The three ways to connect a transformer primary into a wye (phase-to-neutral) configuration are shown in Figure 10–23. Note how labeling the polarity of the transformers on a drawing reduces confusion when making connections.

Each transformer coil in a wye primary or wye secondary is connected phase to neutral. When two or more transformers are interconnected in a wye configuration, the coils are connected in parallel with each other. To connect a coil in parallel, each positive terminal is connected to a phase and each negative terminal is connected to a neutral.

Wye-Wye Transformer Banks

Figure 10–24 shows the connections for a typical three-phase wye-wye transformer bank. A wye-primary-wye-secondary transformer bank can supply 120/208-volt, 240/416-volt, 277/480-volt, or 347/600-volt services. The phase-to-phase voltage is $\sqrt{3}$ or 1.73 times the phase-to-neutral voltage. The voltage across each transformer coil is equivalent to the phase-to-neutral voltage.

The primary neutral must be connected to the secondary neutral in a wye-wye transformer bank. This neutral connection provides a path for any fault current or current from an unbalanced load to get back to the source. There is a potentially lethal voltage between the primary and secondary neutrals if they are not connected together.

Delta-Delta Transformer Banks

Figure 10–25 shows the connections for a typical delta-delta transformer bank. A delta-primary-delta-secondary transformer bank supplies three-phase power at 120 volts, 240 volts, 480 volts, or 600 volts. The voltage across each transformer secondary coil is equivalent to the voltage supplied to the customer, which is the phase-to-phase voltage.

The load on a delta-delta transformer bank needs to be well balanced. Any unbalance will result in circulating currents within the service as the unbalanced

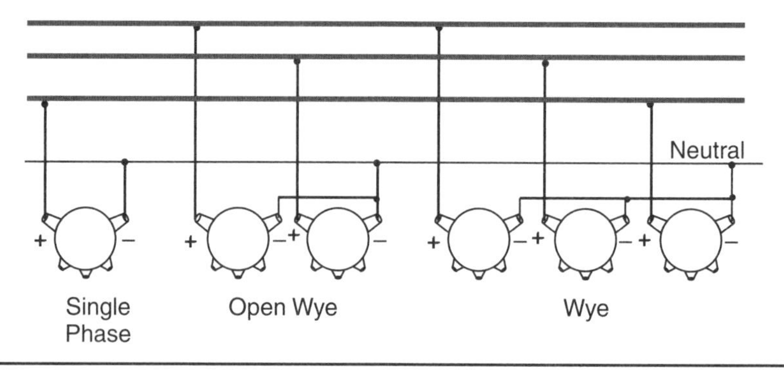

Figure 10–23 Three types of wye connections.

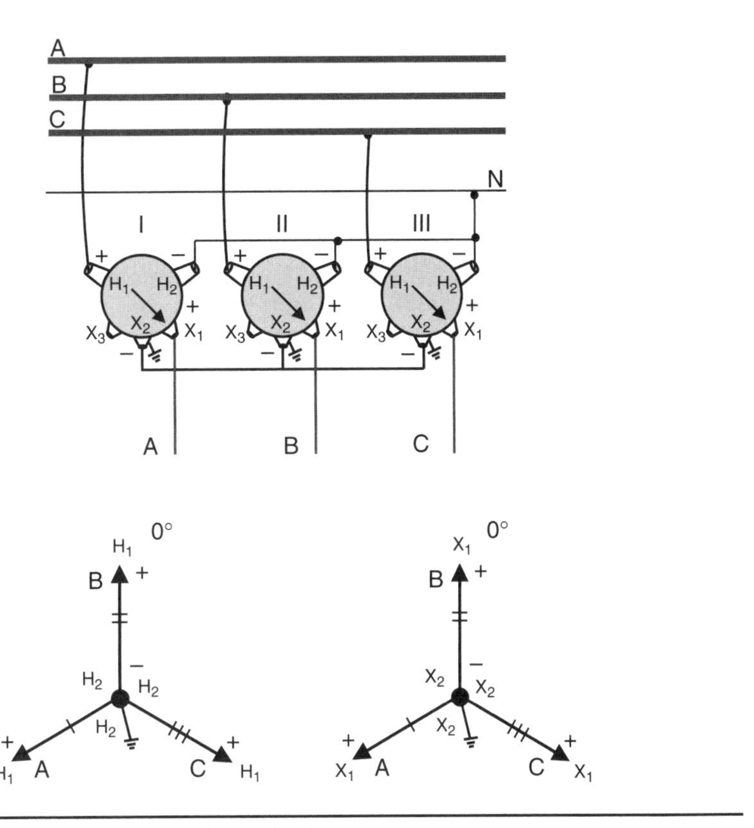

Figure 10–24 A wye-wye transformer bank.

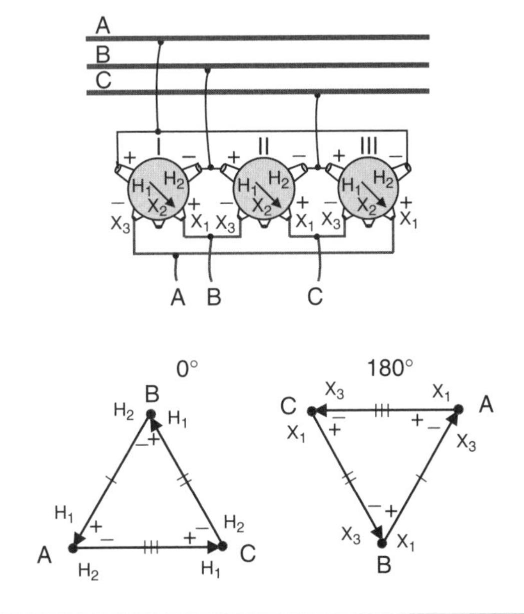

Figure 10–25 A delta-delta transformer bank.

current tries to find its way back to the source. To ensure that the utility is supplying a balanced voltage to the customer, the three transformers must have similar impedance and be on the same voltage tap.

Wye-Delta Transformer Banks

Figure 10–26 shows the connections for a typical wye-delta transformer bank. A wye-delta transformer bank supplies three-phase delta services at 120 volts, 240 volts, 480 volts, or 600 volts.

A wye-delta bank must have transformers with insulated H_2 bushings. The H_2 bushings are interconnected to each other, but they are not connected to system neutral or grounded. The neutral connection is left floating (ungrounded) and, therefore, can have a high potential on it. It must not be treated like a grounded neutral by anyone working on the transformer bank. If the H_2 bushings were connected to the system neutral, the transformer bank would carry extra current not related to the current needed to supply the normal service load.

If the primary wye circuit is unbalanced, extra current flows through the delta secondary as it tries to balance itself through the secondary of the transformer bank. If one phase on the primary circuit is faulted to ground, the high unbalanced current flows through the delta secondary.

When the H_2 is connected to the system neutral, the transformer bank automatically becomes a wye-open-delta transformer bank if one of the primary phases is opened. The two energized transformers continue to provide three-phase power but are subject to burnout because of overload. Two transformers now carry the load normally supplied by three transformers. This arrangement has a capacity of 57.7 percent of the capacity of three transformers.

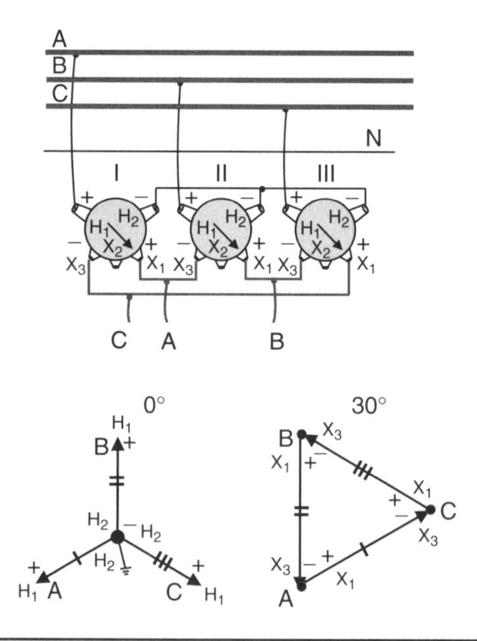

Figure 10–26 A wye-delta transformer bank.

Delta-Wye Transformer Banks

Figure 10–27 shows the connections for a typical delta-wye transformer bank. A delta-primary-wye-secondary transformer bank can supply standard wye services. The secondary neutral should be well grounded because of the unavailability of a primary system neutral.

Open-Delta Transformer Banks

A three-phase delta service can be supplied with two single-phase transformers. This type of service is called an open delta because the delta configuration is missing one side, preventing it from being a closed loop. Figure 10–28 shows the open-delta loop with the availability of three phases.

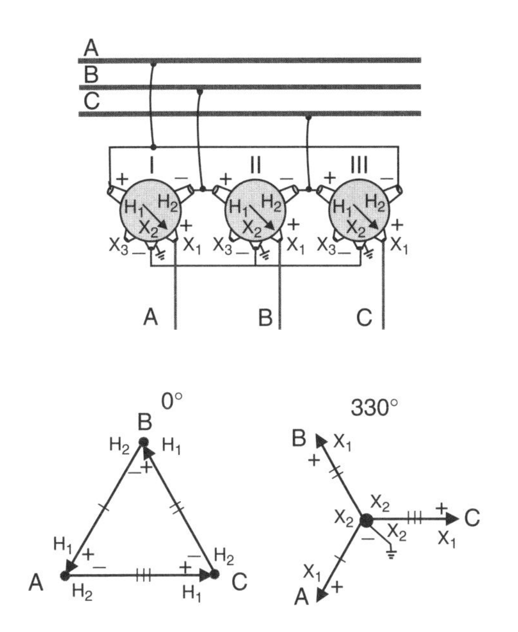

Figure 10–27 A delta-wye transformer bank.

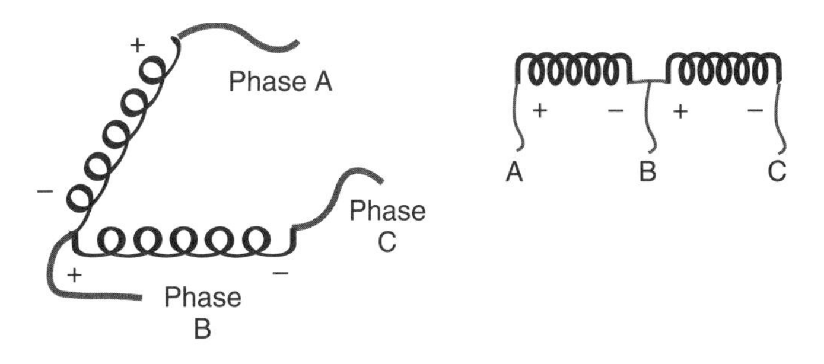

Figure 10–28 Open-delta configuration.

To feed an open-delta, three-phase secondary service, three primary wires are needed. A wye primary would need two phases and a neutral, and a delta primary would need three phases. This hookup is sometimes used as an economical way to feed a small three-phase delta load. One of the two transformers is often called the lighting transformer and is sized larger in order to feed the single-phase portion of the load. The smaller transformer is often called the power transformer and is there to help provide the three-phase load, typically a motor. A three-phase wye secondary service cannot be fed from two transformers.

An open-delta secondary provides three-phase power. The capacity of the two transformers is reduced to 86.6 percent of the nameplate rating. For example, two 100kVA transformers are 100 percent loaded when they supply:

$$0.866 \times 100 \times 2 = 173kVA$$

When one transformer of a normal three-phase delta-delta or wye-delta transformer bank is found to be defective, the connections can be changed to the configurations shown in Figure 10–29, which restores the service as an open delta. The customer should be told to reduce demand on service until the transformer is replaced, because the two good transformers will now only have the capacity to supply 57.7 percent of the capacity of three transformers.

Scott Connections

10.7.5 Scott-connected transformers provide two-phase power from a three-phase system. They can also be used to supply three phases from a two-phase system. Two special single-phase transformers are used. Each single-phase transformer has three primary bushings and special taps on the primary coil where the three-phase primary connections are made. The transformers can also feed a four-wire two-phase service.

Connecting Three-Phase Transformer Banks in Parallel

10.7.6 Three-phase transformer banks are sometimes networked together to a common secondary to add extra capacity and security to the service. Each transformer bank on the common secondary network must be similar:

- Each bank should have a similar impedance.

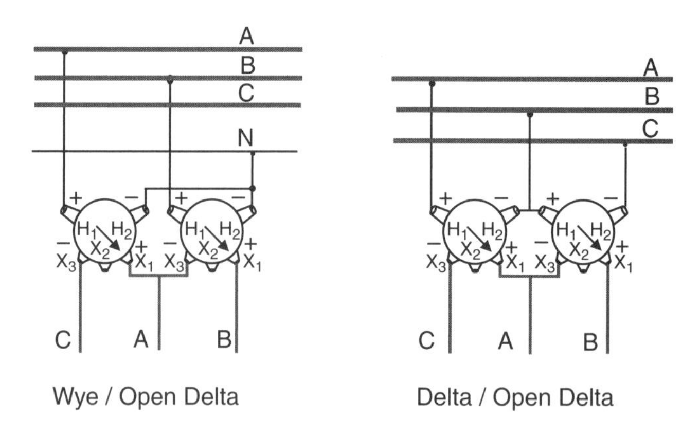

Figure 10–29 Open-delta transformer connections.

- Each bank must be on the same voltage tap setting.
- Each bank must have the same angular displacement or phase shift.

Angular Displacement of Wye-Delta and Delta-Wye Transformer Banks

10.7.7 Occasionally, it is necessary to know if there is an angular displacement or phase shift between the primary and secondary of a transformer. An example of an angular displacement is the difference between the X_1 and X_3 of a single-phase transformer. Even though each leg will read 120 volts from phase to neutral, the two legs would produce a dead short if they contacted each other. The X_1 is 180 degrees out of phase with the X_3.

Similarly, depending on the type of three-phase transformer bank and the way the secondary connections are made, there is an angular displacement or phase shift between the primary and the secondary. There is always an angular displacement or phase shift with a wye-delta or a delta-wye transformer bank. The secondary will be 30 degrees out of phase with the primary. This means that a wye secondary of a wye-wye bank, which has a 0-degree angular displacement, cannot be connected in parallel with a wye secondary of a delta-wye bank, which has a 330-degree angular displacement (Figure 10–30). There is no way a lines crew can switch secondary connections to allow these two transformer banks to be connected in parallel.

Similarly, a delta secondary of a delta-delta bank, which has a 0-degree angular displacement, cannot be connected in parallel with a delta secondary of a wye-delta

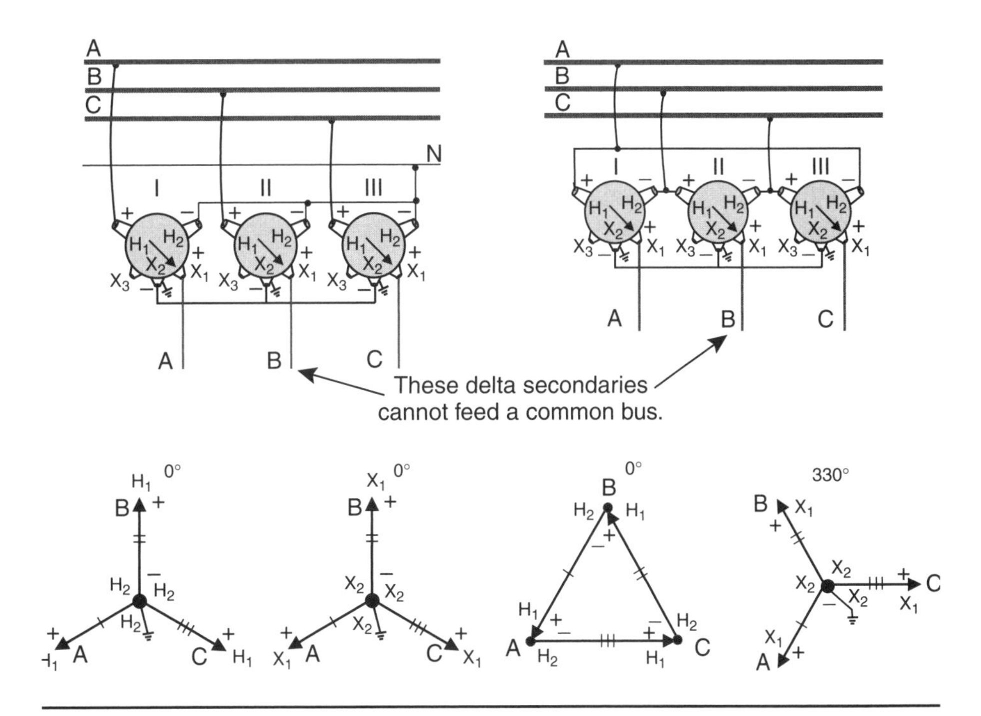

Figure 10–30 Angular displacement of wye secondary.

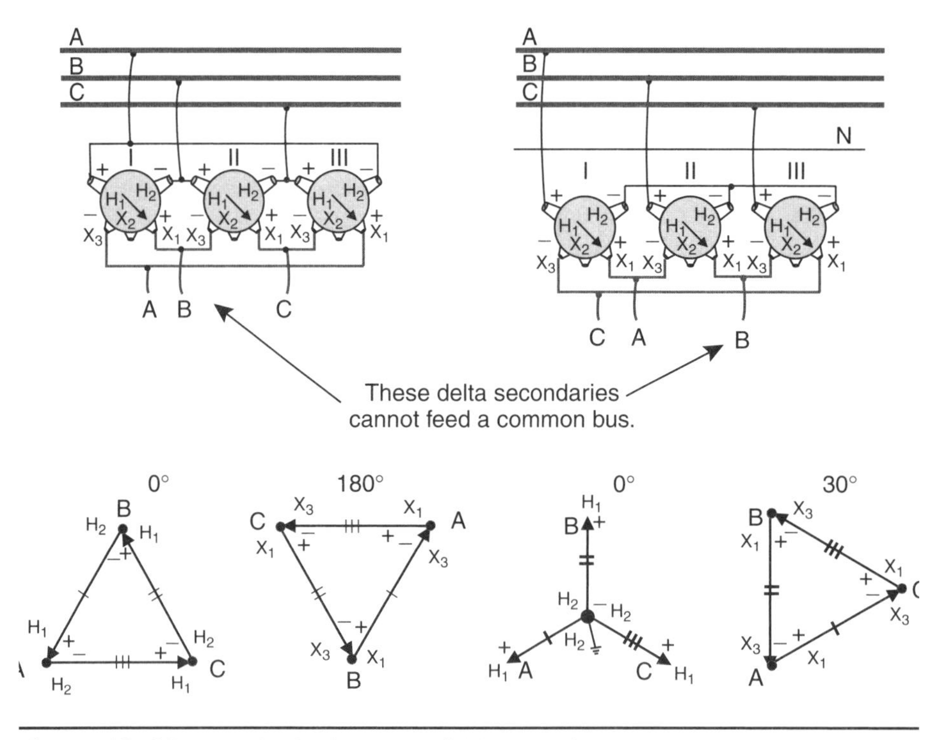

Figure 10–31 Angular displacement of delta secondary.

bank, which has a 30-degree angular displacement (Figure 10–31). There is no way a lines crew can switch secondary connections to allow these two transformer banks to be connected in parallel.

Angular Displacement of Substation Transformers

10.7.8 Control-room operators have to be aware of the existence of a phase shift between different feeders. A line fed from a substation, where the substation transformer is delta-wye, cannot be connected in parallel with a line fed from a substation where the substation transformer is wye-wye. Similarly, lines from a delta-delta substation transformer cannot be connected in parallel with a line from a wye-delta substation transformer.

To further add to the confusion, a wye-delta or delta-wye transformer bank can also be connected so that there is a 180-degree phase shift in addition to the 30-degree phase shift. A lines crew, therefore, should always check with operating control before closing a tie switch between two feeders fed from different stations. Phasing sticks can be used for a field test to determine if there is a voltage difference across an open tie switch.

Voltage Unbalance on a Three-Phase Service

10.7.9 A three-phase service supplied to a customer should not have a voltage unbalance exceeding one percent.

$$\%\ Voltage\ unbalance = \frac{max\ V \div min\ V - average\ V}{average\ V} \times 100$$

where the *average V* is the average of the three voltages and the *max* or *min V* are the voltages that have the greatest difference from the average.

An unbalanced secondary voltage can be caused by:

- An unbalanced customer load.
- An unbalanced primary voltage.
- Banked single-phase units with different kVA ratings.
- Banked single-phase units with different voltage tap settings.
- Banked single-phase units with different impedances.

10.8 Three-Phase Secondary-Voltage Arrangements

Various Voltages Available from a Transformer Bank

10.8.1 The secondary voltage from a three-phase transformer bank depends on more than just the transformer ratio. Three transformers with a given ratio can be interconnected to provide up to four different types of services. The secondary voltage is based on whether the transformer secondary is interconnected as wye or delta. The secondary phase-to-phase voltage is:

- Equal to the actual voltage across the transformer coil with a delta connection.
- Equal to 1.73 times the voltage across the transformer coil with a wye connection.

The secondary voltage is also dependent on whether a transformer with a center-tapped secondary coil has the two parts of the coil inside the tank arranged in parallel or in series. The output voltage of series-connected coils is double the output of two parallel-connected coils.

Secondary Coil Arrangements Inside the Tank

10.8.2 The turns ratio of a selected transformer is based on the desired output voltage. Transformers with a center-tapped secondary coil have a certain voltage induced across the full length of the coil and half of that voltage on each side of the center tap. When the two secondary coils are interconnected in series inside the tank, the secondary voltage is double the voltage of the two coils connected in parallel.

For example, on a 120/240-volt transformer, shown in Figure 10–32, the secondary provides 240 volts when the two coils are connected in series. Placing the two secondary coils in parallel allows the complete coil to be used to provide 120 volts.

Services Available

10.8.3 Table 10–1 shows three-phase voltages available from common distribution transformer secondaries in North America. In much of the world, the three-phase voltage available to the customer is equal to the standard voltage in that country × $\sqrt{3}$ or 1.732.

For example, common three-phase voltages are:

North America	*120 × 1.732 = 208V*
Europe	*220 × 1.732 = 380V*
Other countries	*240 × 1.732 = 416V*

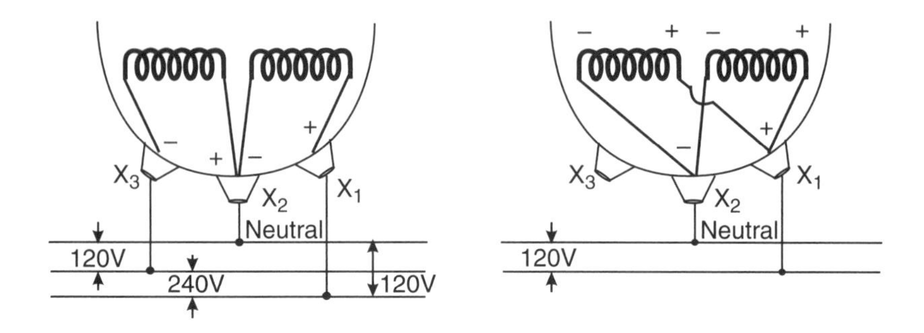

Figure 10–32 Series and parallel secondary connections.

TABLE 10–1 Standard North American Three-Phase Voltages

Transformer Secondary	Type of Three-Phase Service	Coil Arrangements Inside the Tank	External Configuration
120/240	120/208	Parallel	Wye
	240/416	Series	Wye
	120	Parallel	Delta
	240	Series	Delta
240/480	240/416	Parallel	Wye
	277/480	Series	Wye
	240	Parallel	Delta
	480	Series	Delta
277	277/480	NA	Wye
347	347/600	NA	Wye
600	600	NA	Delta

Three-Phase Voltages from a 120/240-Volt Secondary

Three single-phase transformers with a 120/240 secondary can supply a 120/208, a 240/416, a 240, or a 120-volt three-phase service.

A *three-phase 120/208-volt* transformer bank (Figure 10–33) has the secondary coils in each transformer internally connected in parallel to provide 120-volt output. The external secondary leads of the three transformers are interconnected as wye. Each phase-to-neutral voltage is 120 volts, and the phase-to-phase voltage is:

$$120 \times 1.732 = 208V$$

A standard single-phase 120/240-volt lighting service can be provided from a three-phase 120/208-volt service. In Figure 10–33, the internal secondary coils in the center transformer are left in series to provide a 120/240-volt supply. The kVA rating of the center transformer is usually increased to provide the capacity for the extra load.

A *three-phase 240/416-volt* service (Figure 10–34) has the internal secondary coils of each transformer connected in series to provide 240 volts. The external sec-

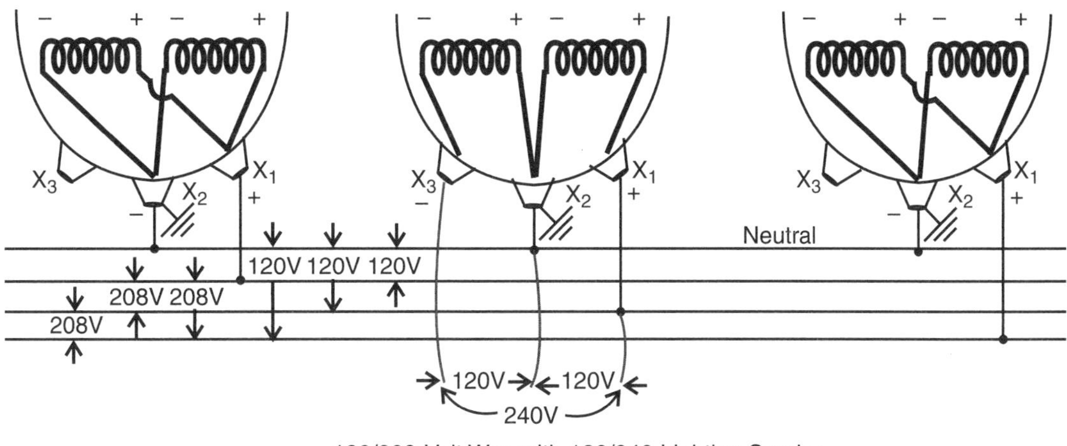

Figure 10–33 Three-phase 120/208V service.

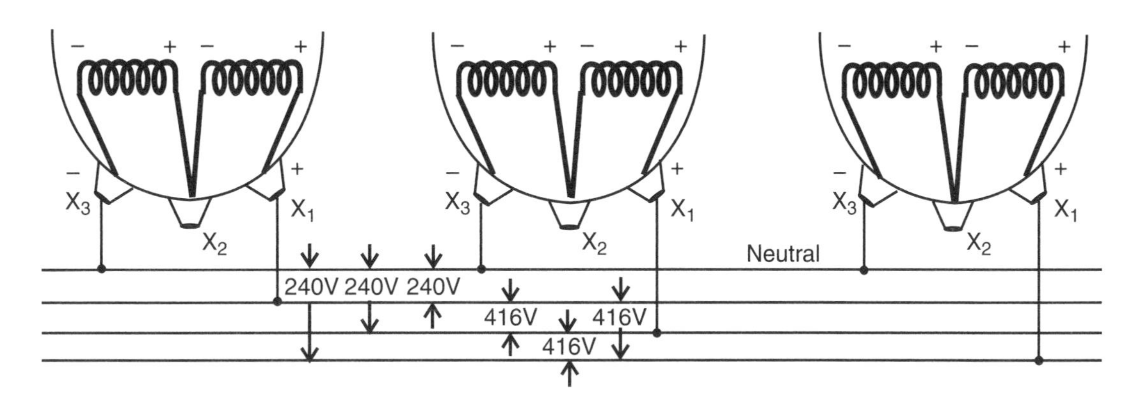

Figure 10–34 Three-phase 240/416V service.

ondary leads of the three transformers are connected wye. Each phase-to-neutral voltage is 240 volts, and the phase-to-phase voltage is:

$$240 \times 1.732 = 416V$$

The ground strap is removed from the X_2 terminal, and the terminal will be alive at 120 volts to ground.

A *three-phase 240-volt* service (Figure 10–35) has the internal secondary coils of each transformer connected in series to provide 240 volts. The center tap is left ungrounded. The external secondary leads of the three transformers are connected in delta, and 240 volts are available from phase to phase.

A phase-to-ground voltage reading is zero volts because there is no path or circuit back to the ungrounded delta. If there was a voltage from phase to ground, it

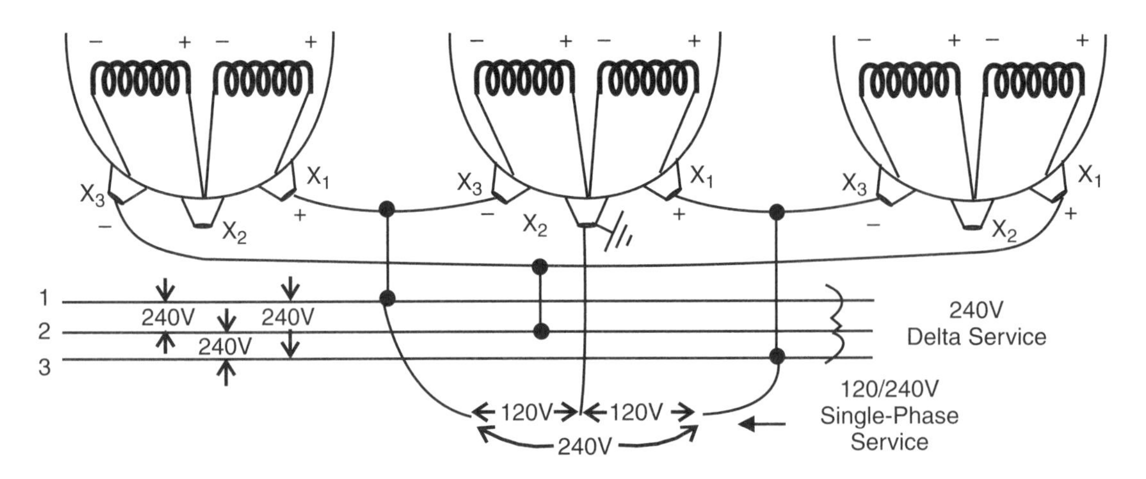

240-Volt Delta with a 120/240-Volt Lighting Service

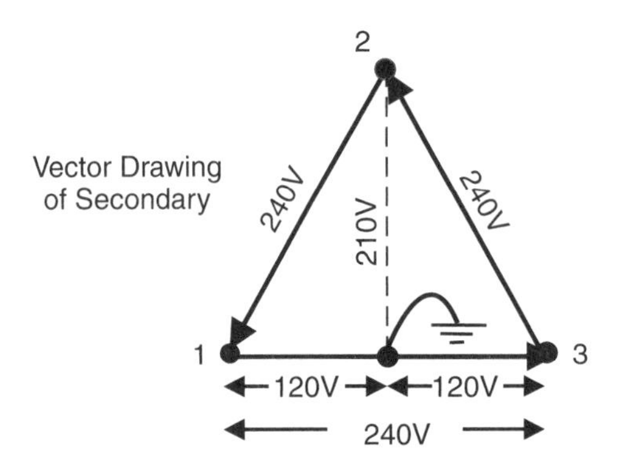

Figure 10–35 Three-phase 240V service.

would mean that there was a phase-to-ground fault somewhere or there was a lighting service from one of the transformers.

A phase-to-ground fault would mean that the earth was at 240 volts in relation to each of the two other phases. One phase-to-ground fault would not normally blow a fuse because there is no path for current to flow through the earth back to the source.

To get a 120/240-volt lighting load, the center tap of one secondary coil is grounded. There are 120 volts phase to neutral available from each of the two phases connected to the grounded transformer.

The remaining phase is sometimes called a *wild phase* and has a phase-to-neutral voltage of about 210 volts. This does not happen with a wye connection because the distances through the coil from each phase to the neutral center point are equal. A delta connection has one of the windings tapped in the middle, which leaves one phase of the transformer farther away from the neutral than the other two. This results in two phases with 120 volts to the neutral and one phase (the wild phase or *high* leg) at 87 percent of 240 volts.

The transformer with the grounded center tap will carry its full one-third share of the 240-volt three-phase load and two-thirds of the single-phase 120/240-volt load. A larger transformer is usually installed for the center-tapped transformer to handle the extra duty of the *power leg.* A lighting service from one of the transformers introduces a ground in the delta. One ground in a delta is not a short circuit; but if any other secondary phase becomes faulted to ground, the delta is shorted and a fuse should blow.

A *three-phase 120-volt* transformer bank has the secondary coils in each transformer internally connected in parallel to provide 120-volt output. The secondary leads are connected in a delta configuration similar to the 240-volt delta bank shown in Figure 10–35.

Three-Phase Voltages from 240/480-Volt Secondary

Three, interconnected single-phase transformers with 240/480-volt secondaries can supply a 240/416, a 277/480, a 480, or a 240-volt three-phase service.

A *three-phase 240/416-volt* service (Figure 10–36) can be supplied from three 240/480-volt transformers using the same secondary connections as the 120/208-volt transformer bank. The internal secondary coils are connected together in parallel, and the external leads are interconnected as wye.

A *three-phase 277/480-volt* service (Figure 10–37) can be supplied from three 480/240-volt transformers by having the internal secondary coils of each transformer connected in series to provide 480 volts. The external secondary leads of the three transformers are connected wye. The phase-to-phase voltage is 480 volts, and each phase-to-neutral voltage is:

$$480 \div 1.73 = 277V$$

The ground strap is removed from the X_2 terminal, and the terminal is alive at 240 volts to ground.

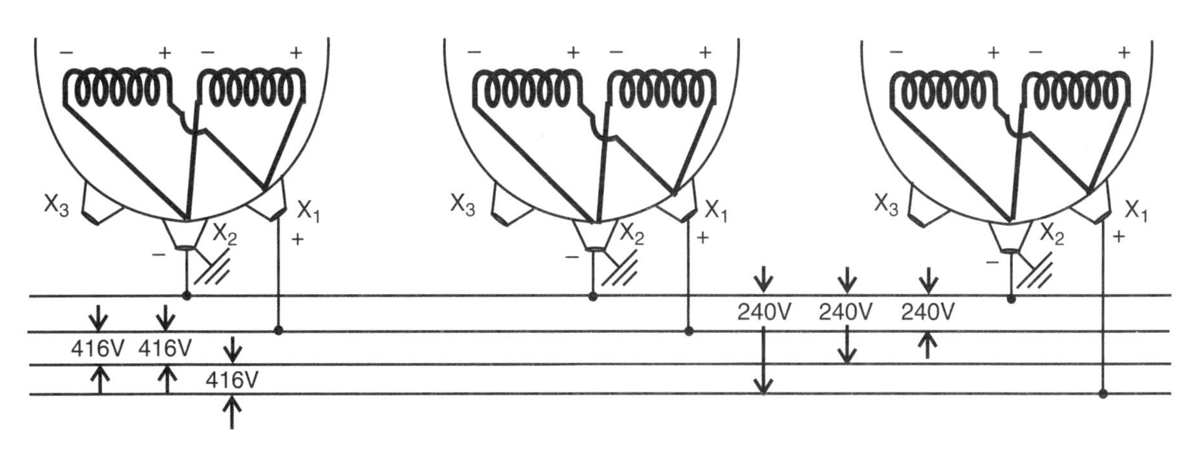

Figure 10–36 A three-phase 240/416V service.

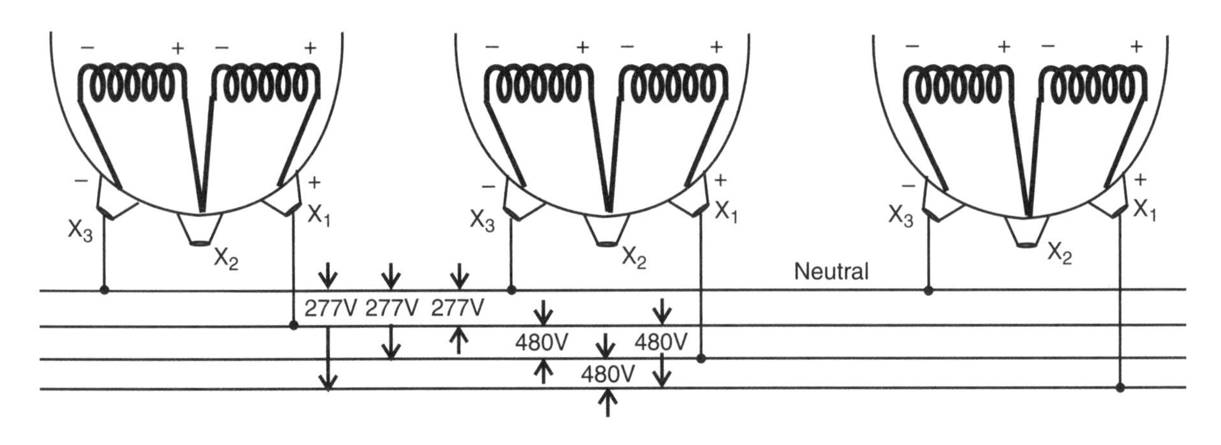

Figure 10–37 A 277/480V service.

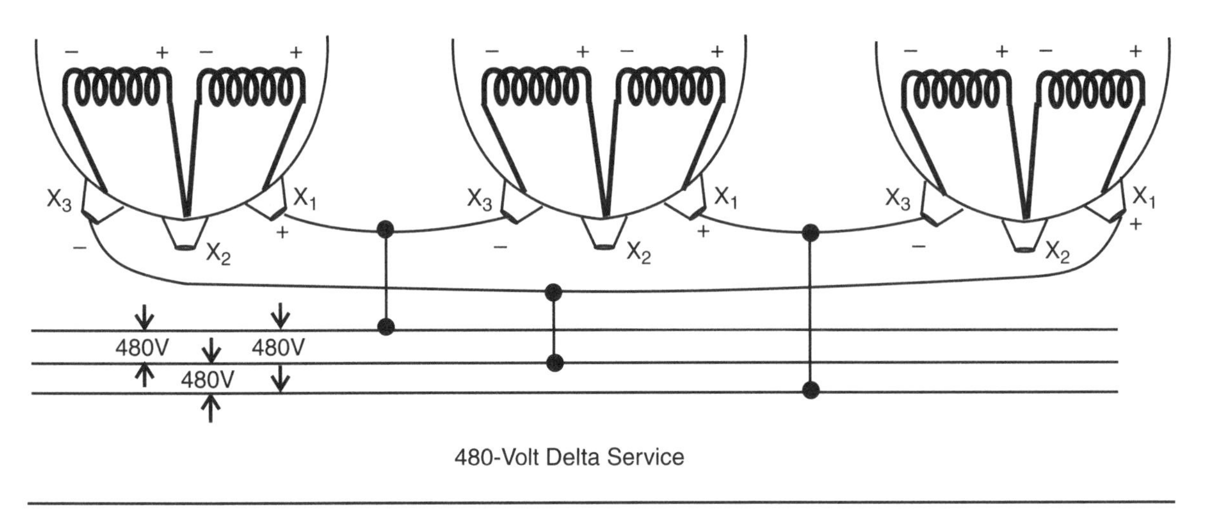

Figure 10–38 A three-phase 480V service.

A *three-phase 480-volt* service (Figure 10–38) is supplied from a 480-volt secondary. The secondary coils are connected in series inside the tank, and the external leads are connected in a delta configuration. The phase-to-phase voltage is 480 volts. A 240-volt supply could be made available by grounding the center point on one transformer.

A *three-phase 240-volt* service can be fed from a transformer bank with a 480/240-volt secondary. The secondary coils in each transformer are internally connected in parallel to provide a 240-volt output. The secondary leads are connected in a delta configuration similar to the 480-volt delta bank shown in Figure 10–38.

Three-Phase Voltages from a 347-Volt or 277-Volt Secondary

Three single-phase transformers with a 347-volt or 277-volt secondary coil can supply 347/600 volts and 277/480 volts, respectively. The connections shown in Figure 10–39 apply to both the 277/480-volt and 347/600-volt transformer banks.

A *three-phase 277/480-volt* transformer bank has the secondary connections in a wye configuration. Each phase-to-neutral voltage is 277 volts, and each phase-to-phase voltage is:

$$277 \times 1.73 = 480V$$

A *three-phase 347/600-volt* transformer bank has the secondary connections in a wye configuration. Each phase-to-neutral voltage is 347 volts, and the phase-to-phase voltage is:

$$347 \times 1.732 = 600V$$

Three-Phase Voltages from a 600-Volt Secondary

Three single-phase transformers with 600-volt secondary coils can supply a three-phase 600-volt delta service (Figure 10–40). The transformers are interconnected in a delta configuration. The phase-to-phase voltage is 600 volts.

10.9 Troubleshooting Transformers

Investigating the Secondary for Transformer Problems

10.9.1 Troubleshooting transformers (see Tables 10–2, 10–3, and 10–4) mainly involves checking for problems on secondary services. Many *no-power* or *partial-power* calls are due to internal customer problems. A utility responsibility normally ends at the service entrance. Opening the main switch and taking a voltage reading will determine whether the utility is the cause of the problem.

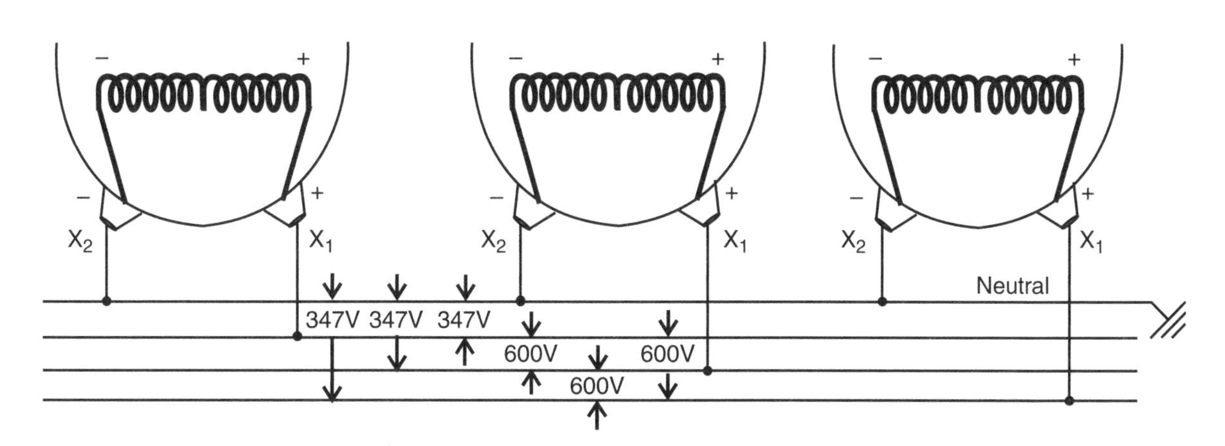

Figure 10–39 A three-phase 347/600V service.

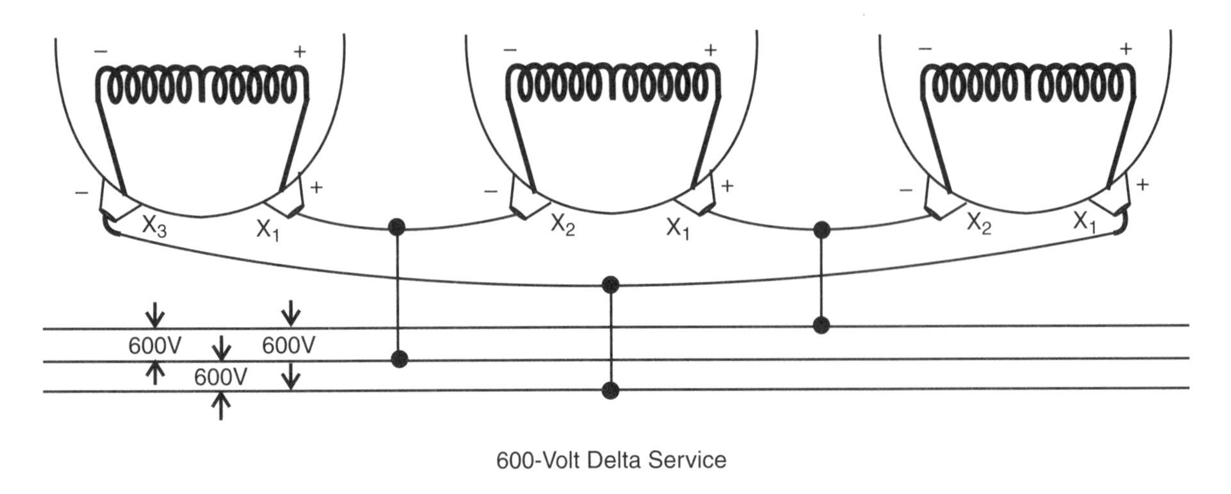

Figure 10–40 A three-phase 600V service.

TABLE 10–2 Troubleshooting a Single-Phase 120/240-Volt Service

If	Then
A customer has intermittent power and flickering lights.	Check for a voltage imbalance in the two 120-volt legs. If one 120-volt leg is two or more volts different from the other 120-volt leg, turn on a large 120-volt load. If the voltage increases on one leg and decreases on the other, then there is a poor neutral connection. A loose connection on either leg can increase resistance to current flow and result in arcing and intermittent power.
The lights in part of the customer's premises are very bright, and in another part the lights are very dim.	A poor or open neutral connection blocks the normal return path from 120-volt appliances. The current travels back to the source through the other 120-volt leg or through 240-volt equipment. The 240-volt equipment operates normally.
A customer is receiving half power. Some of the lights work and some do not. None of the 240-volt appliances work.	One of the main fuses at the service entrance is likely blown. A voltage reading of about 120 volts between the top and the bottom of the fuse indicates that there is a voltage difference and therefore the fuse is blown. If the power is off, a continuity tester or ohmmeter can also be used to check the fuse. If the fuses are good, check the connections back to the transformer.
A customer complains about erratic or low voltage.	A poor connection could cause low voltage at various times. Heavy-duty equipment used by a neighbor on the same bus can cause voltage problems for others on the bus.

TABLE 10–3 Troubleshooting a Wye Secondary (120/208, 240/416, or 347/600-Volt) Service

If	Then
A customer has an abnormal voltage.	Open the customer's switch and check the voltage. If the voltage on a phase to neutral reads zero, then check for a defective transformer or an open phase on the primary feeder. If the voltages on all three phases are balanced, close the customer switch and check the voltage. An unbalanced voltage indicates the customer's load is unbalanced.
Three-phase motors are overheating, and the thermal overload protection trips out the motor.	The usual cause of trouble on one phase is an unbalanced load. This type of service has single-phase loads. Unbalance beyond 3% can cause the low-voltage leg to draw more current and create heating of equipment. A faulty winding in a three-phase motor can cause a high current on a phase.
The load is balanced at the service entrance, but there is still a problem.	Open the customer's main switch and verify that the utility-supply voltage is correct. Close the customer's circuit breakers one circuit at a time and take voltage and current readings. A suspect circuit will have an unbalanced voltage or abnormal current readings for the equipment being fed.
All three-phase equipment and some single-phase equipment will not operate.	One phase is probably out. A phaseout can be due to a feeder problem, a transformer problem, or a blown fuse at the service entrance.
The single-phase equipment has intermittent power. The three-phase load is operating normally.	A poor or broken neutral connection will cause problems with single-phase loads. A customer-owned secondary (dry) transformer that steps down the higher 600 or 416-volt supply to 120/208 can be the cause of problems.

TABLE 10–4 Troubleshooting a Delta Secondary (600, 480, or 240-Volt) Service

If	Then
The customer has abnormal voltage. With the customer's switch open, the voltage on the supply side shows one or more of the phase-to-phase voltages at zero volts.	One or more phases feeding the customer is out of service. Check the transformers and the primary feeder. The customer's switch must be open for this voltage check because the backfeed in a delta service can show a voltage at the service entrance.
At the main panel, the phase-to-phase voltage readings are lower than normal and one phase-to-phase voltage reading is zero volts. For example, on a 240V service, the readings are 210V, 210V, and 0V.	When one phase of a primary feeder loses power, the voltage at a wye-delta transformer bank will have a reduced voltage on two of the phase-to-phase readings and zero volts on the third phase-to-phase reading. On a wye-delta transformer bank, the primary neutral is not connected to the system neutral or the secondary neutral, but it is left ungrounded or floating. If the transformer neutral was grounded and one primary phase is opened, the transformer bank would become an open-delta bank and continue to supply three-phase power at a 57% reduced capacity, exposing the bank to a burnout because of an overload.
At the main panel, all three phase-to-phase voltage readings are normal, but one phase to ground is reading very low or zero voltage.	A ground fault exists on a phase. When one phase of a delta circuit is faulted to ground, it does not blow a fuse because there is no path for a return current to the source. Both the phase and the ground are alive; therefore, a faulted phase-to-ground voltage reading will show a very low or zero voltage difference. Both the faulted phase and earth are alive in relation to the other two phases. A customer with a delta service often has a ground-fault-indicating light connected between each phase and ground. The light will go out when there is a ground fault and the ground becomes alive.

Review Questions

1. A transformer with a 240-volt secondary has a 60-to-1 turns ratio. What is the primary voltage feeding this transformer?
2. To conduct a turns-ratio field test on a transformer, why is it necessary to energize the high-voltage coil with a low voltage such as 120 volts?
3. What problem could be created when an off-load tap changer on a transformer is adjusted to improve the voltage to a customer?

4. Why is there a greater hazard working on live secondary close to a transformer than farther away?
5. Can a single-bushing transformer be installed on a delta circuit?
6. A transformer coil must have a potential difference across it in order to operate. What two types of connections are there to get a voltage across a transformer coil?
7. What would be the expected full-load current (100-percent loaded) on the 240-volt secondary of a 100-kilovolt-ampere, 7200/240-volt transformer?
8. When a secondary bus is fed from many transformers connected in parallel, what kind of secondary system is it?
9. When two or more transformers are interconnected with the positive terminal of one transformer connected to a negative terminal of another transformer, what kind of interconnection is it?
10. When two or more transformers are interconnected with the positive terminal of each transformer connected to a phase and all the negative terminals are connected together, what kind of interconnection is it?
11. Why should the primary neutral of a wye-delta transformer bank be left floating (ungrounded)?
12. Can two phases of a delta primary supply a three-phase open-delta service?
13. Can the wye secondary of a wye-wye bank be connected to the same bus as the wye secondary of a delta-wye bank of the same secondary voltage?
14. A center-tapped 120/240-volt secondary coil has the two parts of the coil inside the tank arranged in parallel. What would be the voltage between the X_1 and the X_3 terminals?
15. Name three types of services that three single-phase units with a 120/240-volt secondary can serve.
16. The lights in part of the customer's premises are very bright, and in another part the lights are very dim. What is the likely cause?
17. Three-phase motors on a 120/208-volt service are overheating, and the thermal-overload protection trips out the motor. What is the likely cause?
18. At the main panel, the phase-to-phase voltage readings of a delta 240-volt service are lower than normal and one phase-to-phase voltage reading is zero volts. For example, the readings are 210 volts, 210 volts, and 0 volts. What is the likely cause?

CHAPTER 11

Street-Lighting Systems

Topics to Be Covered	Section
Introduction	11.1
Luminaires	11.2
Streetlight Ballasts	11.3
Streetlight Circuits and Controls	11.4
Safe Maintenance	11.5

11.1 Introduction

Importance of Street Lighting

11.1.1 Street lighting was one of the earliest loads for central-station electric systems. The illumination of large areas has improved traffic safety, pedestrian safety, and security for people and property.

In some utilities, streetlight maintenance and troubleshooting form a large part of the work program.

11.2 Luminaires

Luminaires Defined

11.2.1 A luminaire (Figure 11–1) is a single, self-contained outdoor light. The luminaire consists of a ballast transformer, a lamp, a photocell, and (often) a capacitor.

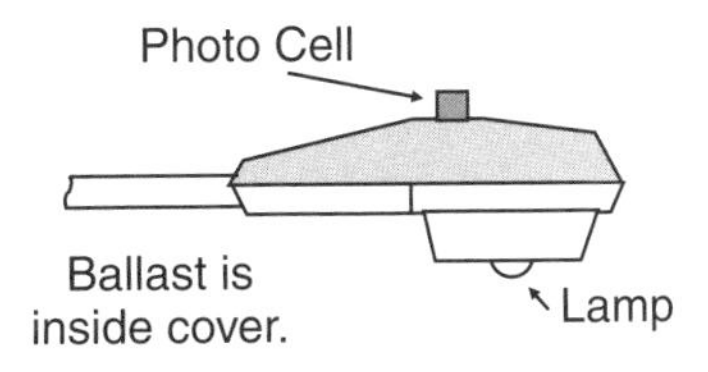

Figure 11–1 A luminaire.

Lamps and Lumens

11.2.2 Units used to measure light include the lumen, candela, foot-candle, and lux. A candela is a measure of luminous intensity or candlepower. A wax candle emits about one candela in all directions. One lumen is defined as the amount of light falling on a one-square-foot area surface one foot away from a one-candela source. In the metric system, one lumen per square meter is called a lux. One foot-candle equals 10.76 lux.

Powerline workers will probably continue to look for the watt rating when changing lamps. Lamps are designed to get as much visible light output with the least amount of energy input possible. The visible light output from a lamp is called luminor energy, and a lamp output is rated in lumens. The energy input of a lamp is rated in watts. In other words, the lumen is a measure of light output, not the wattage. The height and the intended pattern of a streetlight determine the size of the lamp. The type of lamp used determines the capital cost, the efficiency, and the lamp light.

Table 11–1 shows the efficiency (lumens per watt) and lamp life of the various types of lamps available.

TABLE 11–1 Efficiency and Lamp Life of Various Lamps

Type of Lamp	Watts	Lumens	Rated Life Hours
Incandescent	200	4,000	750
Fluorescent	40.7	3,250	12–20,000
Mercury vapor	400	23,000	16–24,000
Metal halide	400	34,000	7.5–15,000
High-pressure sodium	400	50,000	20–24,000
Low-pressure sodium	180	33,000	18,000

Incandescent Lamps

11.2.3 An incandescent lamp generates light when current flows through a piece of wire (filament). The filament is a resistor that heats up enough to glow white-hot or incandescent and generate light. The lamp is sealed in an oxygen-free bulb to prevent the filament from burning up. Depending on the design, an incandescent lamp can have a life of 2500 hours.

The filament in the most common incandescent lamps is made of tungsten, which is a metal that can withstand high temperatures. By changing the length or thickness of the filament, the temperature of the filament can be changed. A cooler filament would "boil" off more slowly, and the lamp would have a longer life, but it would produce more invisible infrared and yellow light and is less efficient for producing a visible white light. At a higher filament temperature, the lamp produces a whiter visible light. The tungsten "boils" off more quickly and the lamp has a shorter life.

The design of the lamp seeks to balance efficiency and lamp life. The efficiency of an incandescent lamp is about 14 lumens per watt for a 100-watt bulb and 18 lumens per watt for a 500-watt bulb.

Halogen Lamps

11.2.4 Halogen lamps are incandescent bulbs that operate at very high temperatures. The life of the tungsten filament is extended because iodine (a halogen) vapor is put in the bulb. When the tungsten boils off the filament, it combines with the iodine to form a tungsten-iodide gas. When this gas touches the filament, it decomposes; the tungsten is deposited on the filament again, and the iodine is released into the bulb. This "perpetual motion" could continue forever but the bulb eventually fails because the tungsten is not deposited perfectly even on the filament.

To withstand the high temperatures and allow the use of smaller bulbs, halogen bulbs are made of quartz instead of glass. Halogen lamps use about 40 percent less energy than the common incandescent lamps.

Gaseous Discharge Lamps

11.2.5 Gaseous-discharge lamps have become the lamp of choice for street lighting. They produce more lumens per watt than incandescent lamps. Light is produced when electric current passes through a gas. When a permanent arc is established in a gas-filled lamp, the light produced is referred to as an electric discharge or gaseous discharge.

Types of gaseous-discharge lamps include: fluorescent, compact fluorescent, low-pressure sodium, and high-intensity discharge. Mercury-vapor, metal-halide, and high-pressure sodium-vapor lamps are included in the high-intensity-discharge category.

Fluorescent Lamps

Fluorescence is a process whereby the light from the electric discharge in a gaseous-discharge lamp is made more visible and changed to a different color. For example, the color of an electric discharge through mercury-vapor gas is an invisible ultraviolet color. It is after the energy from the ultraviolet light contacts a phosphor powder that the ultraviolet light is converted to a visible white light. The phosphor powder coats the inside of the lamp.

Although there are many types of fluorescent lamps, the common fluorescent lighting found in most commercial buildings is a bulb or tube filled with argon gas and a small amount of mercury vapor. The tube has a tungsten filament electrode at each end. The inside of the tube is coated with a fluorescent material called phosphor. Electric current heats the filament at each end of the tube. A cloud of electrons of opposite polarity gathers around each electrode. The electrons then form an arc that travels from one electrode to the other, alternating back and forth with the AC.

The electrons in the arc bump the atoms of argon gas and mercury vapor, which in turn produces an invisible ultraviolet light. The energy from the ultraviolet light causes the phosphor coating in the tube to emit visible white light. The makeup of the phosphor coating determines the color of the light output.

Fluorescent lamps are more efficient than the common incandescent lamps and can convert more energy to a visible white light at cooler temperatures. Fluorescent lamps have a long lamp life of about 12,000 hours. However, because

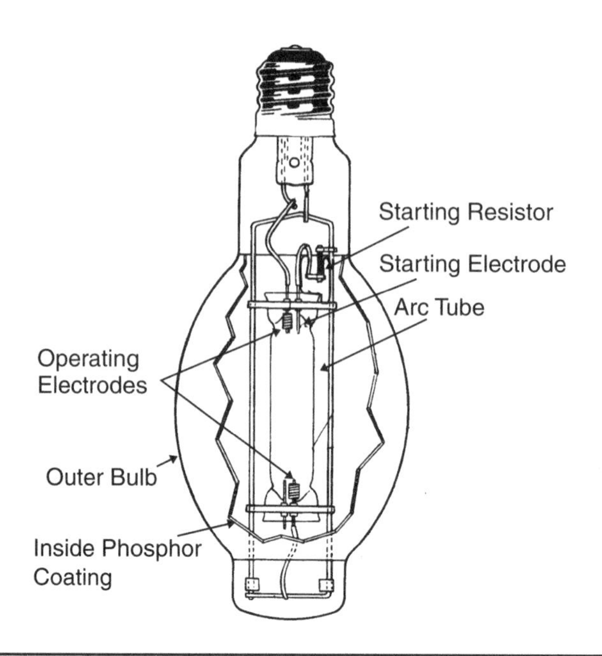

Figure 11–2 Mercury vapor lamp.

the electrodes decay, less energy is transferred through the mercury vapor and less and less light is emitted as the lamp gets older. Fluorescent lamps should have scheduled replacement rather than waiting for the lamp to quit. The efficiency of a fluorescent lamp is about 50 lumens per watt.

Mercury-Vapor Lamps

A mercury-vapor lamp is a high-intensity (HI) gaseous-discharge lamp. It has an inner arc tube made of quartz that contains argon gas and a small amount of liquid mercury. When an arc is established within this inner bulb, the gas inside the lamp produces an ultraviolet light. The inside of the outer bulb is coated with phosphor, which converts the ultraviolet light into visible light.

A mercury-vapor lamp does not start immediately. First, an arc must be established across a set of starting electrodes within the inner bulb; then, the arc is transferred to the main electrodes within the bulb (Figure 11–2.) To establish the arc in the bulb, between 200 and 350 volts are needed across the starting electrodes. This sets up an initial glow in the argon gas. The heat generated in the inner bulb causes the liquid mercury to vaporize. Then an arc travels between the two main electrodes, through the argon and vaporized mercury, to produce the ultraviolet radiation. An average 400-watt mercury-vapor lamp has an output of 50 lumens per watt and a lamp life of 24,000 hours.

High-Pressure Sodium-Vapor Lamps

A high-pressure sodium-vapor (HPS) lamp is a small cylinder arc tube. Because of its small size, there is no room for installing starting electrodes. Instead; the ballast supplies a high voltage of approximately 2500 volts to start the arc in the lamp. After the lamp is started, the voltage supply drops to a lower operating level.

High-pressure sodium-vapor lamps are commonly used for streetlights because they are very efficient. A 400-watt high-pressure sodium-vapor lamp has about a 100-lumens-per-watt rating and a 24,000-hour lamp life.

Low-pressure sodium-vapor lights are even more efficient than high-pressure sodium-vapor lights, but they were not well accepted as streetlights because of the very yellow color of the light emitted. High-pressure sodium-vapor has a more acceptable color, although it is still quite yellow.

Metal-Halide Lamps

A metal-halide lamp is similar to a mercury-vapor lamp but with other metallic elements, which results in a good-quality white light. A metal-halide lamp is about 35 percent more efficient than a mercury-vapor lamp, but it has a shorter lamp life. A 400-watt metal-halide lamp has a 75-lumens-per-watt rating with a lamp life of about 10,000 hours.

11.3 Streetlight Ballasts

The Function of a Ballast

11.3.1 Just as ballast is needed in a ship to stabilize it from the waves on the ocean, a ballast is needed to stabilize the voltage and current in a gas discharge lamp.

In an incandescent lamp, the filament is a fairly constant resistor and, therefore, the voltage and current are constant. In a gaseous discharge lamp, there is no fixed resistance and the lamp needs a variable voltage supply. It needs a very high voltage to strike an arc between the two electrodes within the vacuum tube. After the arc is struck, the resistance decreases and the normal 120-, 240-, 277-, or 480-volt rating will maintain the arc. Because an arc is like a short circuit, the ballast limits the current and prevents the current from continuing to increase and eventually burn out the lamp. The ballast automatically adjusts to supply the proper voltage over the lamp's life. As the lamp gets older, the ballast will supply the needed higher voltage to maintain an arc.

A ballast is a large inductor or reactor. It takes the place of a resistor and limits current flow while it lowers the voltage as the lamp tube begins to conduct. A ballast consists of a wire coil, an iron core, and sometimes a capacitor.

Reactor Ballasts

11.3.2 A common fluorescent light that can be started without needing a voltage boost still needs a ballast to control the current. A reactor ballast consisting of a coil of wire connected in series with the lamp acts as a choke coil and controls the current (Figure 11–3). As the current increases, more inductive reactance is generated. The inductive reactance of the coil acts as a self-limiting load that increases as the current increases.

A capacitor is not necessary, but when it is connected in parallel, it will improve the power factor of the load without canceling the effect of the inductive reactance of the coil connected in series.

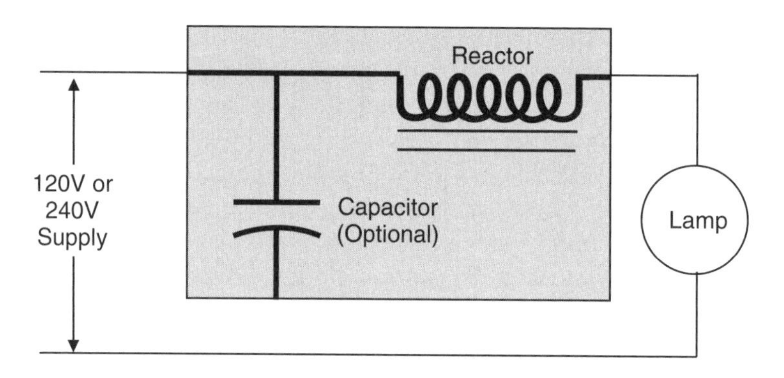

Figure 11–3 Reactor ballast.

Autotransformer Ballast

11.3.3 An autotransformer ballast (Figure 11–4) is used with high-intensity gaseous-discharge lamps to provide the voltage boost needed to get the initial arc to strike in the lamp. The secondary coil of the autotransformer is connected in series and therefore acts as a choke coil to control the current. A capacitor is not necessary but can be added to improve the power factor of the load.

Regulator Ballasts

11.3.4 The regulator ballast (Figure 11–5) is a transformer with the primary and secondary isolated from one another. The secondary magnetic circuit is operated in saturation, which keeps the secondary voltage at a constant level. A capacitor in series with the lamp sets up a capacitive reactance, which limits the current. The lamp, therefore, is almost at a constant wattage.

11.4 Streetlight Circuits and Controls

Energizing and Switching Streetlights

11.4.1 Each luminaire on a street needs to be supplied with the proper voltage and current. Streetlights can be fed from three different types of sources:

1. Use existing secondaries with the lights connected in parallel
2. Use a separate secondary with the lights connected in parallel
3. Use a high voltage with the lights connected in series

Street lights are turned on and off individually or in groups using a control switch and a control circuit. They can be:

1. Individual lamp controls
2. A control switch with a pilot wire
3. A cascading relay system

Controls that turn the lights on and off can be photoelectric, time-clocks, or manual switches. Photoelectric controls have taken over from manual switching or, in most cases, time-clock controls.

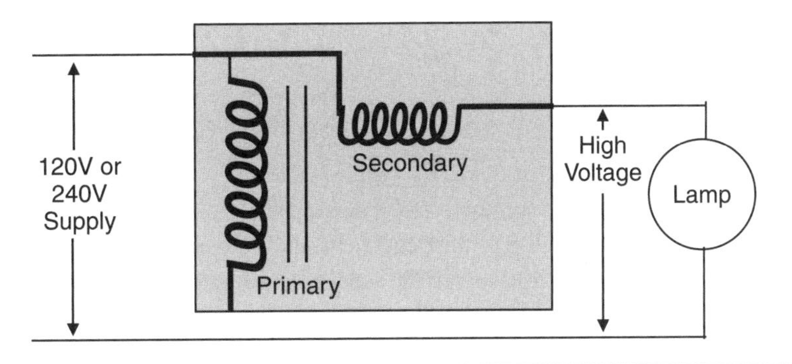

Figure 11–4 Autotransformer ballast.

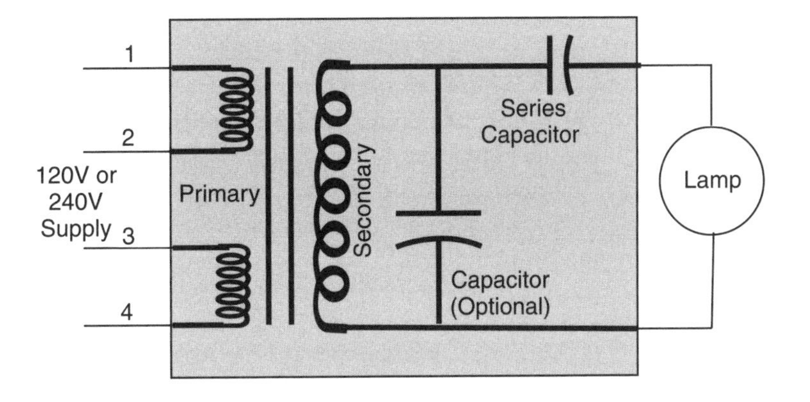

Figure 11–5 Regulator ballast.

When light strikes a photocell, a current is generated. The current from the photocell travels through a relay coil that keeps an armature from closing the relay contacts. When there is not enough light to generate current, the relay armature is released and the contacts close. A time-delay feature prevents the photocell from opening the contacts due to temporary lights, such as lightning or car headlights. Photoelectric controllers are made "fail safe" so that a component failure or dirt on the photocell will keep the contact closed and the light will stay on all day (day burner).

One control combined with relays can switch on entire streets at a time; however, it is also common to have a photoelectric control in each light.

Series Streetlight Systems

11.4.2 At one time almost all streetlights were installed in a series circuit, but, because they are most suited to the more inefficient incandescent lamps, they are generally being replaced. In a series circuit, many lights can be fed with one relatively small continuous wire that starts at the source transformer, forms a loop through the streets being illuminated, and goes back to the transformer.

The series circuit wire is fed at a primary voltage and is insulated and placed in a primary position on overhead distribution poles. A high voltage is needed because in a series circuit the sum of the voltage drop across each individual light is equal to the source voltage. The number of lights in the circuit determines the amount of source voltage needed to feed the lights. To build a transformer with the exact voltage output needed for each series lighting circuit would be impractical. A special constant current transformer, also called a regulating transformer, is used to raise and lower the voltage automatically depending on the number of streetlights in the circuit. The voltage will go as high as needed (up to several thousand volts) to supply a constant current through the circuit. As light units are added, the primary voltage will go up in order to continue to provide a constant current (typically 6.6, 7.5, or 20 amperes) through the special series circuit lamps.

A break in a series lighting circuit will stop the current flow and will cause an outage to all the lights in the circuit. The biggest hazard to powerline workers is that the voltage across any break in a series circuit will be equal to the output voltage of the transformer. Maintenance should be done with the circuit isolated and grounded to ensure that a circuit is not opened, cut, or joined unless the break is "jumpered" to avoid an open circuit condition.

Because there would be a primary voltage across any open point in the circuit, the lamp and socket are designed not to break the continuity when the light element ruptures or a lamp is removed from the socket. The lamp is designed to automatically bypass the current. Figure 11–6 shows the special series circuit lamp with the thin film of insulation between two points on the lamp that is punctured by the high voltage across the break restoring the lighting circuit. The socket in most series light fixtures is designed so that removing a bulb will not open the circuit.

A lamp used in a series circuit has a heavier filament because it has to carry a relatively high current compared to lamps connected in parallel. The source, a constant current transformer, makes an adjustment and lowers the current automatically when one lamp burns out, otherwise because the total resistance of the lighting circuit becomes smaller, and the current would tend to increase.

Individual Streetlight System

11.4.3 Many streetlights and all yard lights are independently fed from any available 120/240V source nearby and controlled by an individual in-head photocell. This works well in residential areas where there are transformers and secondary bus. Because there is no interconnection with any other lights on a street, these lights are the easiest to troubleshoot since, when this light is not working, the source of the trouble will be at that light or supply.

Pilot Wire Control System

11.4.4 A pilot wire streetlight system is a system that uses a pilot wire (a separate wire strung to all the streetlights) to control many lights from one spot so that they all go on and off at the same time. Each individual light does not have to be supplied from the same source. It can be fed from any available 120/240V source nearby. This works well in residential areas where there are transformers and secondary bus. The pilot wire is strictly a wire to turn the lights on and off. It is ener-

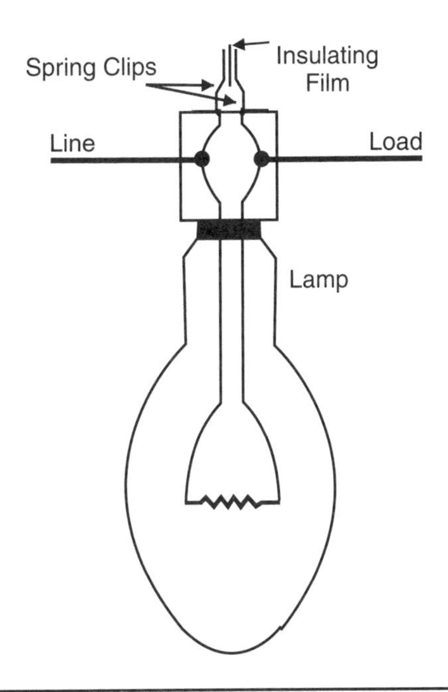

Figure 11–6 Special lamp for series circuits.

gized from a 120/240V supply through a photocell and will activate a relay when it gets dark.

One photoelectric control switch energizes the pilot wire when it becomes dark. When the pilot wire is energized, the normally open relay contact connected to the pilot wire closes. Each relay in the system energizes a number of lights. A relay is just a switch activated remotely by the pilot wire. It is also possible to have a photoelectric control deenergize the pilot wire when it becomes dark. This releases the contacts of the normally closed relays and the lights are energized. The normally closed relay system allows the lights to become day burners when the control switch or pilot wire becomes defective. Figure 11–7 shows normally open relays which are closed when the pilot wire is energized.

Normally open relay system:

1. One control switch, photoelectric or timed, energizes the pilot wire when it becomes dark.
2. The normally open relay contacts close when the pilot wire energizes the relay coil.
3. Each relay energizes a number of lights.

Normally closed relays:

1. One control switch, photoelectric or timed, deenergizes the pilot wire when it becomes dark.

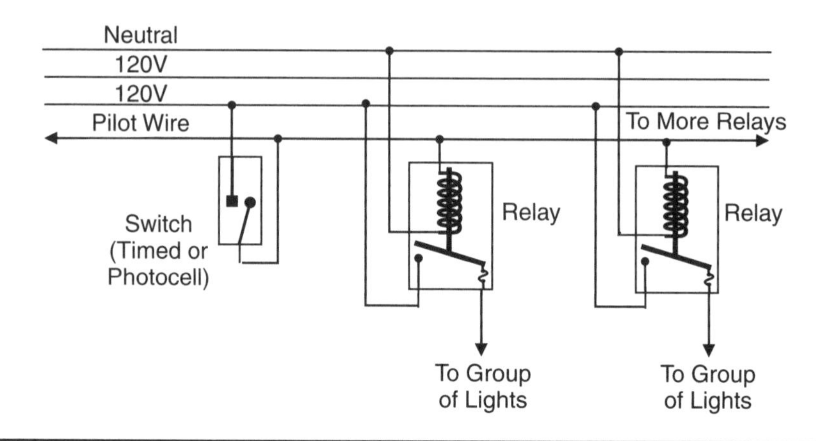

Figure 11–7 Pilot wire control system.

2. When the pilot wire deenergizes the coil in the normally closed relay, the armature is released and the contacts close.
3. Each relay energizes a number of lights.

The normally closed relay system allows the lights to come on when the control switch or pilot wire becomes defective.

Cascading Streetlight Control System

11.4.5 A cascading relay system energizes and controls all the streetlights from one transformer and one control. This works well in areas where there are few or no existing transformers or secondary bus. A streetlight wire is strung with relays installed along its length. To prevent having a very large relay (switch) to energize all the lights, each section of the lighting circuit activates the relay that energizes the next lighting circuit. Figure 11–8 shows that when a relay is energized it will energize a pilot wire to energize the next relay.

11.5 Safe Maintenance

Streetlight Hazards

11.5.1 Working with street lighting has its own special hazards. In addition to the usual environmental hazard of working in the vicinity of live circuits, there are also chemical, ultraviolet-radiation, and high-voltage hazards.

Ultraviolet-Radiation Hazard

The inner tubes of mercury-vapor and high-pressure sodium-vapor lamps put out an intense ultraviolet radiation. The outer phosphor-coated bulb protects people from exposure to the ultraviolet radiation. If exposed, the eyes will develop the same "sand-in-the-eye" symptoms as those from exposure to an electric arc or welding arc.

Even though the outer bulb is broken, it is possible for the inner bulb to continue to emit an unseen ultraviolet radiation without the outer bulb acting as a filter. The outer bulb can break while replacing a lamp and expose the worker to

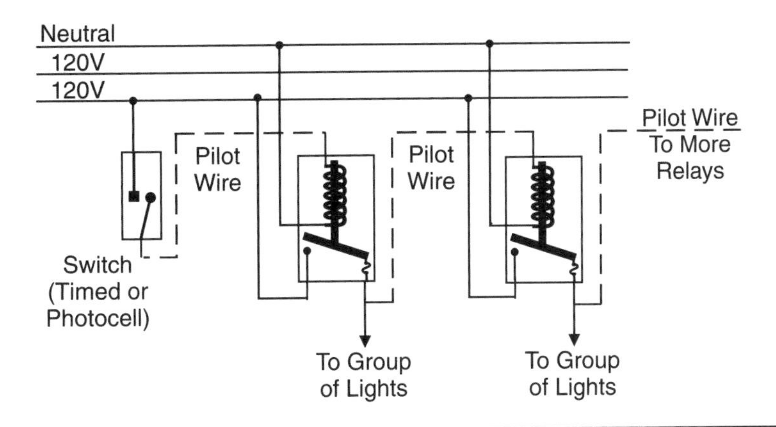

Figure 11–8 Cascading control system.

Table 11–2 Typical Operating Voltage for Gaseous-Discharge Lamps

Lamp	125W	150W	250W	400W	700W	1000W
Metal-Halide			100V	120V		250V
Mercury-Vapor	125V		130V	135V	140V	145V
High-Pressure Sodium-Vapor		100V	100V	105V		110V

ultraviolet burns from the inner bulb. Disconnecting the power source will prevent a person from receiving ultraviolet radiation burns.

High-Voltage Hazard

Gaseous-discharge lamps need a high voltage to start the initial electric arc. A ballast will step up the 120-volt supply voltage to provide a 750-volt to 4000-volt output. For example, a high-pressure sodium-vapor lamp needs high starting voltage of about 4000 volts. After the lamp starts, the operating voltage of the lamp will vary depending on the type of lamp and its wattage rating. Wait until the lamp is at normal brilliance (or approximately 30 seconds) before doing a voltage check on a gaseous-discharge luminaire. The voltage should settle to its normal operating voltage when the light is at normal brilliance. Typical operating voltages for lamps are found in Table 11–2.

Toxic-Chemical Hazard

Mercury, sodium, and phosphor are toxic chemicals. These chemicals can be found inside the outer bulbs of gaseous-discharge lamps and are released when the outer bulbs are broken. Staying outdoors and upwind when working with these lights reduces the risk of exposure.

Mercury is hazardous to the human body. Mercury is usually in a gaseous state because it vaporizes at −10°C (14°F). In the gaseous state, it can easily enter the body through the lungs. *Sodium* is a highly active poisonous chemical that oxidizes very quickly in air. Sodium in contact with water produces a chemical reaction that creates sodium hydroxide (caustic soda), which can damage the lungs when breathed in. *Phosphors* are poisonous and must not be breathed in.

Environmental Hazard

The capacitors used in older luminaires (before 1979) are impregnated with polychlorinated biphenls (PCBs). PCBs are not biodegradable and will stay in the environment and end up in the food chain. PCB waste must be disposed of properly according to regulatory requirements.

The mercury and lead found in lamps are toxic substances. There are companies that will separate and recycle lamp components such as glass, metal, mercury, and phosphor powder.

Review Questions

1. Using watts and lumens as units, how is the efficiency of a lamp defined?
2. Which of these lamps are high-intensity-discharge lamps?
   ( ) Fluorescent
   ( ) High-Pressure Sodium
   ( ) Halogen
   ( ) Mercury-Vapor
   ( ) Incandescent
3. What is the purpose of a ballast in a gaseous-discharge lamp?
4. Why can a break in a series lighting circuit be a hazard to a powerline worker?
5. If a streetlight stays on in daylight, what could be the cause?
6. If the outer bulb of a mercury-vapor lamp breaks while replacing the lamp, to what hazard is the worker exposed?
7. Why should a worker wait until the lamp is at normal brilliance, or approximately 30 seconds, before doing a voltage check on a gaseous-discharge luminaire?
8. What can be done to reduce exposure to toxic chemicals such as mercury, sodium, and phosphor found in the bulbs of gaseous-discharge lamps?

CHAPTER 12

Revenue Metering

Topics to Be Covered	Section
Introduction	12.1
Determining Cost to the Customer	12.2
Types of Revenue Metering	12.3
The Workings of a Meter	12.4
Single-Phase Metering	12.5
Polyphase Metering	12.6
Transformer-Rated Metering	12.7

12.1 Introduction

Metering an Electrical System

12.1.1 Every part of an electrical system is metered. There are meters found in transmission and distribution substations to record voltage, current, power, reactive power, and other data needed to operate and monitor the system. Much of the metering is telemetering used to monitor remote generating and substations. This chapter, however, deals with the "cash registers" for an electrical utility, namely, revenue metering at the customer.

12.2 Determining Cost to the Customer

Three Main Types of Charges to Customers

12.2.1 There are generally three types of charges that a customer can be asked to pay:

1. An *energy charge,* which is the actual kilowatt-hours (kWh) used
2. A *capacity* (or *demand*) *charge,* which can be the kilowatt (kW) demand or the peak kilovolt-ampere (kVA) demand
3. A *customer charge* that is not a function of either energy used or peak demand, but is a charge to cover the cost of the facilities to supply power

Utilities can have dozens of different rates that are variations on these basic charges.

Variables That Affect Cost

12.2.2 Customers supplied by an electrical utility can be residential, industrial, commercial, or another utility. The rates charged to these customers depend on the type of service, type of load, the quantity of load, and when the load is used. Some of these variables are measured through metering, and some costs are fixed when the service contract is negotiated.

Fixed charges:

- Minimum charge
- Supply voltage
- Location
- Interruptible power

Metered charges:

- Energy consumption in kWh (kilowatt-hours)
- Peak demand
- Power factor
- Load factor

Minimum Charge for Service

There is a fixed cost to a utility when it supplies power to a customer. A minimum charge relates to the wire and transformer needed to supply the energy a customer *may* require. The minimum charge is to cover the fixed cost of providing *voltage* regardless of the amount of current used. For a residential customer, the minimum charge saves a utility the extra expense of installing demand meters to track actual demand.

Energy Consumption

Energy consumption is the volts × amperes used by a customer. The voltage supplied is relatively constant; therefore, the amount of *current* used by the customer is the largest variable measured by a kWh revenue meter.

The largest portion of energy consumed by a customer is real or true power. A standard kilowatt-hour meter measures real power, which is equal to volts × amperes × hours × 1000. Some reactive power is also used by a customer, but it is not normally metered for a residential customer.

Peak Demand

The peak demand is the maximum rate of consumption used by a customer during a billing period. The demand for power can be very intermittent. Meanwhile, the utility has to generate, transmit, and transform the energy to supply the peak load when required.

Most utilities measure the peak demand for larger customers and not for residential customers. The reason most residential rates do not include demand charges is because the metering and tracking of residential demand add to a cost many utilities do not consider worthwhile. Residential consumers have very similar usage patterns so that a minimum charge can cover the fixed costs without much error.

Commercial and industrial customers have a more varied usage pattern. Their peak demand is measured so that a utility can recover the extra cost of supplying the capability to meet the demand. Measurement of peak demand encourages customers to spread out their need for energy.

1. *Demand* is a measurement of the instantaneous power used by a customer. For example, when ten 100-watt lights are all on at the same time, they are using one kW of demand at that moment.
2. *Peak demand* is the highest amount of instantaneous power used by the customer during a billing period. Peak demand for power is measured as peak kWs or peak kVAs. For example, if the ten 100-watt lights are the largest load drawn during the billing period, then the peak demand registered on a demand meter will be one kW. The load would have to be on for a given minimum time to register on the demand meter; for example, 15, 30, or 60 minutes are commonly specified demand intervals.
3. *Energy* is the amount of power used over time measured in kWhs. For example, the energy used by ten 100-watt lights after one hour will be one kWh.

Power Factor

Industrial or commercial customers frequently have large inductive loads that the utility needs to supply. A low power factor means that the utility must generate, transmit, transform, and distribute extra power to supply the inductive load.

A normal kilowatt-hour meter measures only the resistive load. For customers with a possible low power factor, a meter is installed to measure the kilovolt-ampere (kVA) peak as well as the kilowatt (kW) peak. The ratio of the kilowatt peak to the kilovolt-ampere peak is the power factor.

One method of billing is to charge for the highest of either 100 percent of the kilowatt peak or 90 percent of the kilovolt-ampere peak. Usually, a customer with a low power factor takes corrective action to keep the power factor above the level at which there would be extra billing.

Load Factor

An electrical utility needs an equal amount of equipment to supply a customer that consumes a relatively constant supply of power 24 hours a day and a customer that uses a similar amount of power for relatively short periods a day. The customer that uses the power 24 hours a day has a higher load factor and is billed at a lower rate.

The load factor is a percentage indicated by the ratio of the power consumed to the power that could have been consumed if the power had been used continuously.

$$\% \, Load\ factor = \frac{total\ consumption\ (kWh)}{peak\ demand\ (kW) \times hours\ a\ month}$$

Supply Voltage

Residential, small industrial, and commercial customers are supplied with their utilization voltage. Large customers are usually supplied by a line at subtransmission or transmission voltage feeding into a substation owned by the customer. There is less cost to the utility to only supply the circuit feeding the customer without supplying transformation. The customer is, therefore, billed at a lower rate.

Customer Location

It is reasonable for a utility to charge a higher rate to a remote cabin than to a home on a residential street. The cost to get power to the customer is reflected in rates. The rate structure takes into account the customer density in a location. Rural rates are higher than city rates.

Interruptible Power

Large industrial customers can get a better rate if their contract includes an agreement to reduce their demand for power during times when the utility is in short supply. During a utility's peak-load periods or in storm conditions, customers with an interruptible-power contract are asked to cut back their demand for power to the previously negotiated levels.

Time-of-Use (TOU)

Valley-hour rates or off-peak discounts can be attractive to some industries or residences. When load is shifted to off-peak hours, the electric utility benefits because the utility delays the need to build more generation and energy is being sold at times when generators would otherwise be underutilized. One variable to time-of-use metering is to have a lower-peak-demand penalty if the peak demand is during off-peak hours.

12.3 Types of Revenue Metering

Three Kinds of Power Are Measured

12.3.1 There are three kinds of electrical power:

1. Real power
2. Apparent power
3. Reactive power

TABLE 12–1 Units of Measure for Billing Purposes

Kinds of Power	Power	Energy
Real Power	Peak kW	kWh
Apparent Power	Peak kVA	kVAh
Reactive Power	Peak kVAR	kVARh

Revenue metering involves the measurement of one or more of the three kinds of power and/or energy used by a customer (see Table 12–1).

Power is a measurement of energy use at a given instant; for example, real power is the voltage × amperage used by a customer at a given instant. Energy is the amount of power used over a given period of time:

$$energy = power \times time$$

There are meters available to measure a variety of combinations involving the three types of power/energy.

Self-Contained Meters

12.3.2 A self-contained meter is a meter that can be installed without the use of current or potential transformers. The meter can carry the actual load current through its current coil and can use the actual service voltage in the meter potential coil.

Most customers take power at a secondary voltage. At this voltage, a self-contained meter can measure the actual voltage at the customer entrance. The voltage rating of a self-contained meter is normally 240 volts or less, but 600-volt meters are available. Self-contained meters are generally 200 amperes or less, although there are some higher-rated meters on the market.

Transformer-Rated Meters

12.3.3 When current or voltage is too high to be carried by a self-contained meter, instrument transformers are used to send a representative current and voltage to the meter. A *transformer-rated meter* is used to measure a representative current from the current transformer and a representative voltage from the potential transformer.

The outward appearance of a transformer-rated meter is similar to a self-contained meter. The transformer-rated meter, however, has a lower voltage and current rating than the self-contained meter. For example, a transformer-rated meter used for a primary service could be rated at 120 volts and 5 amperes. The values measured by a transformer-rated meter are multiplied by the ratio of the current and potential transformers.

Primary or Secondary Metering

12.3.4 Some customers buy power at a primary voltage and use their own transformers to step it down to a utilization voltage. A transformer-rated meter and its associated potential and current transformers are used to send a representative voltage and current from the primary to the meter.

Metering Demand

12.3.5 Demand meters measure the maximum rate at which electricity was used and can be measured as kilowatts (kW), kilovolt-amperes (kVA), or kilovolt-amperes reactive (kVAR). The meter is generally a combination energy and demand meter.

The electromechanical demand meter illustrated in Figure 12–1 shows that a pointer attached to a bimetal strip is heated by a heating element within the meter. A bimetal strip has two metals bonded together. When heated, the metals expand at different rates, which forces the strip to bend. The bimetal strip bends, and it moves a pointer across the meter dial. The pointer attached to the bimetal strip is called a pusher pointer and, in turn, pushes a maximum-demand pointer. The maximum-demand pointer stays in the farthest position it was pushed. The position of the pointer is the maximum demand for power used by the customer between meter readings.

A peak demand that only lasts a short time is not recorded on the demand meter. The heater element has a time-response feature so that the element takes time to reach its maximum heat. Two common elements are a 10-minute element and a 16-minute element.

Metering Power Factor

12.3.6 A customer's power factor can be measured as the power factor during peak demand or as the power factor of the total energy used.

1. One way to measure the power factor during the peak demand is through the use of a kW demand meter and a kVA demand meter. When the peak

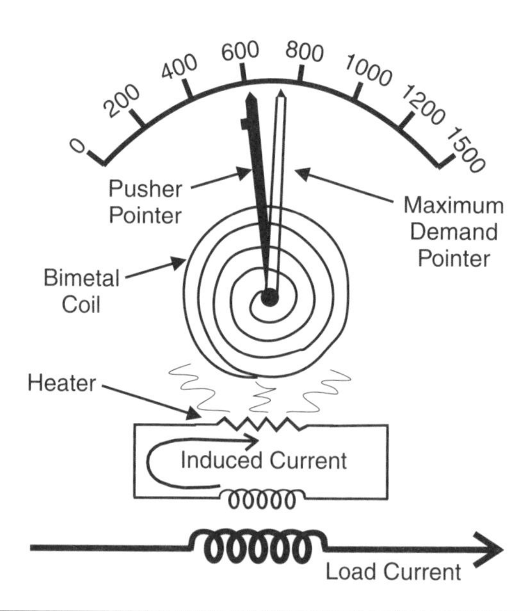

Figure 12–1 Demand meter.

demand for real power and the peak demand for apparent power are known, then the power factor during the peak can be calculated.

2. One way to ensure that a customer will pay for all the energy used is to meter the apparent energy. A VAR-hour meter is used when a customer is being billed for apparent energy. A VAR-hour meter is often identified by a large R on the meter face.

Time-of-Use (TOU) Metering

12.3.7 Time-of-use metering offers a customer a lower price for energy used during nonpeak hours. The meter has the ability to record the time when the power was used. Many residential customers have not been offered this service because of the expense of the needed metering change. Offering time-of-use metering to large industrial and commercial customers increases the demand on the utility valley hours. Electronic revenue meters are easily programmed for the various TOU options a utility may offer to customers.

Prepayment Metering

12.3.8 Prepayment metering is like using a vending machine. A token or a magnetic card is inserted in the slot of the meter to make payment before the energy is used. The meter switches off the power when the payment runs out.

Automatic Meter Reading (AMR)

12.3.9 Automatic meter reading is the remote collection of meter data by way of some kind of communication link from a central location. Each meter has an individual identification and is modified to receive and send signals. The communications link can be a telephone line, a powerline carrier, a radio frequency, a coax cable, or a satellite. Each system is competing for establishment in electrical utilities.

A powerline carrier system, for example, modulates the AC voltage wave to send a signal to the meter, and the AC wave is modulated to send the signal back to the central computer. In addition to reading the meter, an AMR system can be programmed to monitor customers for over- and under-voltage, momentary outages, meter tampering, peak demand, and time-of-use. Outages can be mapped, and system loads can be monitored and controlled. When a communications system is set up, it can be expanded to include the reading of other meters, monitoring burglar alarms, monitoring fire alarms, and operating equipment at remote sites.

12.4 The Workings of a Meter

Two Types of Meter Construction

12.4.1 There are two main types of meter construction, the electromechanical meter and the electronic meter. An *electromechanical* meter relies on a disk being turned in the same way that an electric motor turns an armature. The more voltage or current, the faster the disk rotates. The disk is connected by way of gears to a clock that registers the amount of power used.

An *electronic* meter measures the current and potential going through it without any moving parts. Features such as time-of-use and peak-demand measurements can be programmed more readily in an electronic meter. An electronic meter is very adaptable to automatic-meter-reading (AMR) systems.

The Current Coil in a Meter

12.4.2 The current to be measured flows through a current coil in the meter. The coil is made up of a small number of turns of large wire. In a self-contained meter, all the load current flows through the current coil. The few turns in the current coil indicate that there is very little inductive reactance created; therefore, the magnetic flux induced on the meter disk is in phase with the load current and the line voltage.

The nameplate on a meter shows the range of load for which the meter is designed. For example, a meter commonly referred to as a "100-amp" meter shows a range of 0.75 to100 amperes on the nameplate.

The Potential Coil in a Meter

12.4.3 To measure the voltage in a service, there is a potential coil inside the meter. The coil is made up of many turns of fine wire. The voltage in the coil induces a magnetic field onto the meter disk. The strength of the magnetic field is dependent on the magnitude of the voltage. The potential coil is an inductive load, and the magnetic field it induces on the meter disk lags the field produced by the current coil by 90 degrees.

The voltage-coil ratings match the various standard voltages to be measured. Typical voltage ratings are 120, 240, 480, and 600 volts. These voltages can come directly from the secondary service being measured or from a potential transformer that steps down the voltage of a subtransmission or a primary circuit.

Why the Disk Turns

12.4.4 The magnetic field from the potential coil induces an electromotive force (emf) on the metal meter disk. The induced emf on the disk has nowhere to go because the disk is not part of a circuit. The emf, therefore, causes current to flow in the form of circular (eddy) currents on the disk.

The north and south poles on the potential coil are 90 degrees out of phase with the north and south poles of the current coil. As seen in Figure 12–2, the interaction of the eddy currents being attracted and repelled between the poles of the potential and current coils causes the disk to turn.

The permanent magnet in the meter provides a magnetic brake on the aluminum disk. Without the damping effect of the "breaking" magnet, the interaction of the electrical fluxes would tend to cause a continuous acceleration of the disk. Although a magnet would not normally affect an aluminum disk, the magnetic field of a permanent magnet will affect the eddy currents that rotate the disk. The retarding torque or drag applied to the disk is proportional to the speed. Moving the magnet outward or inward on the disk, or installing a shunt that bypasses part of the flux of the magnet field, can make adjustments to the braking effect. Changing the braking effect is known as the "full-load" meter adjustment.

Elements of a Meter

12.4.5 One current coil and one potential coil working together to turn one disk define one element in an electromechanical meter (Figure 12–3). One element can

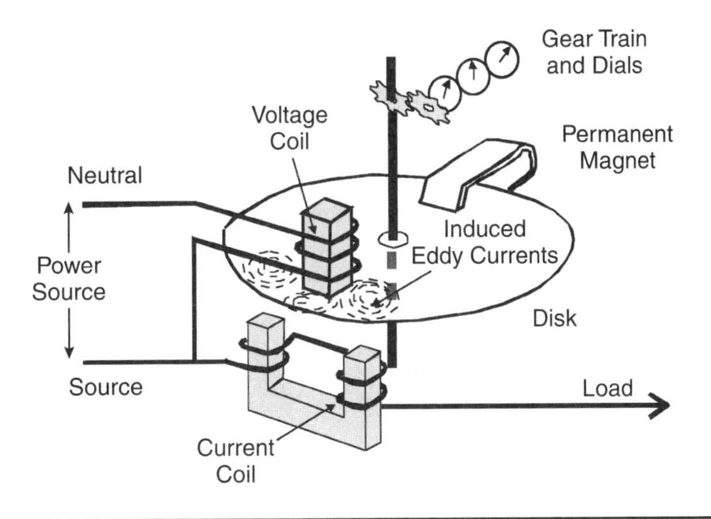

Figure 12–2 Turning the meter disk.

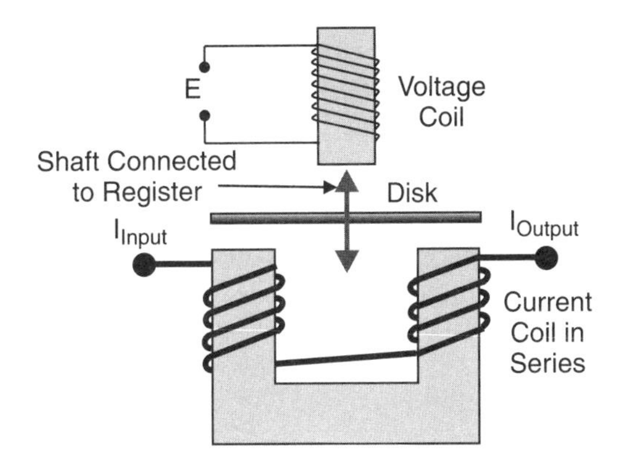

Figure 12–3 One element of a meter.

measure the current and voltage of a two-wire service. Three- and four-wire services require more meter elements. The general rule of thumb is that there is a need for one less element than the number of wires in the service to be measured.

In an electronic meter, the measuring element produces an output pulse. When the energy flowing through the meter increases, the number of pulses increases.

Kh Constant

12.4.6 A meter is designed to have a disk rotate within a certain optimum speed to save wear and tear on the meter. A meter is rated for a certain range of expected load, and the gearing between the disk and the register is set up to control the disk speed.

The *Kh constant,* also known as the test constant, is a constant that relates the number of disk rotations to the load being registered. The value of the Kh constant is the watt hours per turn of the disk. For example, when a meter nameplate shows a Kh of 7.2, it means that 7.2 watt hours have been used with one rotation of a disk, or 1000 revolutions of the disk = 7.2 kilowatt-hours. In an electronic meter the Kh constant is a given value in watt-hours of one pulse issued by the meter.

Kr Constant

12.4.7 While the Kh constant regulates the speed of the disk rotation, the *Kr constant* regulates the speed of the register. A meter is designed to have the register show the amount of energy used over a certain period of time. A meter is rated for a certain range of expected load, and the register should show the amount of energy used without repeating between-meter-reading intervals.

The Kr constant, also known as the meter multiplier, is a number that represents the ratio between the register gear train and the value registered on the meter dial. A Kr or multiplier of 10 would mean that the amount of energy consumed by the customer is ten times the amount shown on the register.

A-Base Meter

12.4.8 A revenue meter must be connected into a service so that the current coils are in series and the voltage coil is in parallel. A bottom-connected or a front-connected meter or an A-base meter must have the input and output leads wired into the meter (Figure 12–4). When connecting an A-base meter under live conditions, it is critical to connect the current in series. A parallel connection would put a voltage across the low-resistance current coil and result in an explosive dead short. An A-base meter is readily removable. Most A-base meters are used with instrument transformers.

S-Base Meter

12.4.9 A socket-connected meter—S-base—is plugged into a meter base that has been prewired to match the lugs of the meter to be installed (Figure 12–5). Most self-contained meters are S-base.

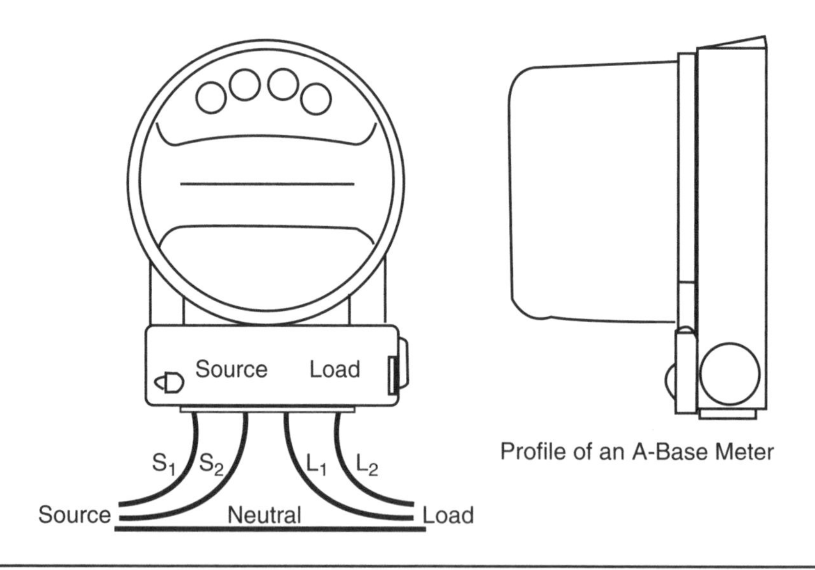

Figure 12–4 A-base meter.

The meter is often installed with the top incoming lugs alive. If the meter base is not wired up correctly or if there is a short circuit in the customer's wiring, the installation of the meter will energize the short circuit. The integrity of the customer wiring should be tested to avoid having the meter explode during the installation.

12.5 Single-Phase Metering

Single-Phase Meter

12.5.1 Most residential and smaller businesses are supplied with a single-phase service. Single-phase meters can be self-contained, transformer type, demand, A-base, or S-base. Figure 12–6 shows the typical workings of a single-phase two-wire A-base meter.

Two-Wire or Three-Wire Service

12.5.2 On a standard North American single-phase service, the service and the meter are either two-wire or three-wire. A two-wire service, as metered in Figure 12–7, consists of a 120-volt leg and a neutral. There is no 240-volt supply on a two-wire service. The voltage coil in the meter is energized at 120 volts. Only one current coil is needed.

A three-wire service is a 120/240-volt service. The voltage coil in the meter is connected across 240 volts. The load on the two legs of the service will not normally be equal, and, therefore, two current coils are needed, one for each leg of the service.

Checking Load on a Meter to Allow Safe Removal

12.5.3 An uncontrollable electrical flash can occur when a socket-based meter is removed under load. Some utilities allow a meter to be removed if the load is 10 kilowatts or less.

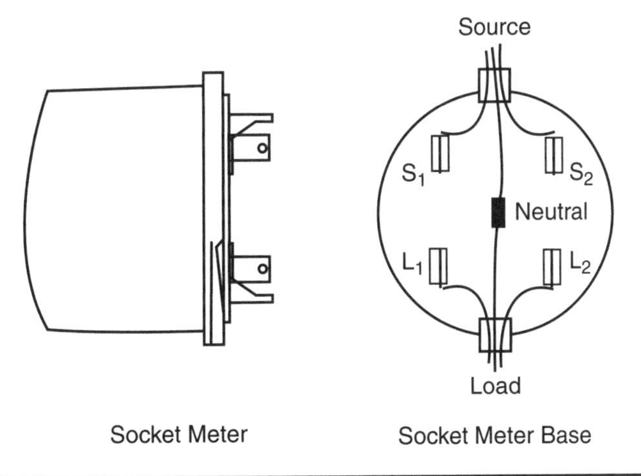

Figure 12–5 Socket-base meter.

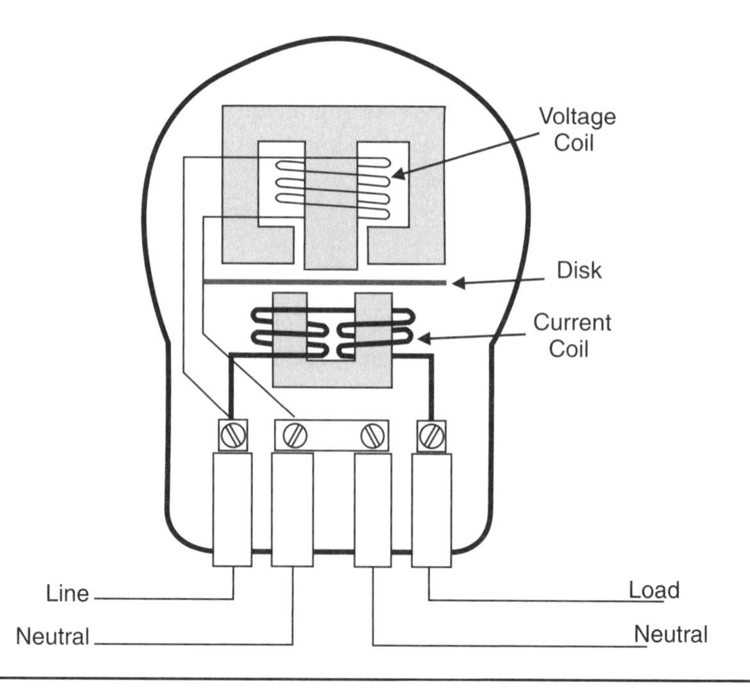

Figure 12–6 Single-phase two-wire A-base meter.

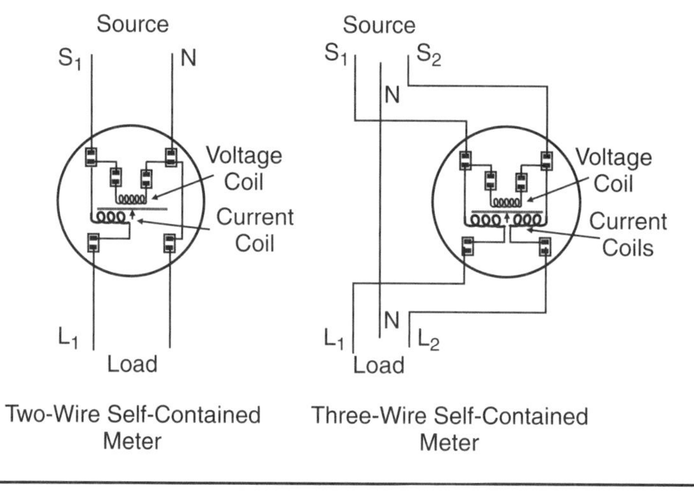

Figure 12–7 Two- and three-wire meters.

Table 12–2 can be used to determine whether a load is more or fewer than 10 kilowatts. Count the number of disk revolutions in 30 seconds, and, with reference to the meter Kh, a higher number of disk revolutions than shown on the chart means that the load is more than 10 kilowatts.

For example, the disk of a meter with a Kh of 7.2 will rotate eleven times in 30 seconds to register a load of 10 kilowatts. When the load is more than 10 kilowatts, the customer's main disconnect should be opened before removing the meter.

Table 12-2 Ten-Kilowatt Load

Kh	Time in Seconds	Disk Revolutions
.36	30	231
.6	30	138
.66	30	126
.72	30	115
2.0	30	41
3.0	30	27
3.33	30	25
3.6	30	23
6.0	30	14
7.2	30	11

Calculating Load Using the Meter Spin Test

12.5.4 There may be occasions when it would be beneficial to know the actual instantaneous power demand used by a customer. The instantaneous kW demand being registered by a self-contained meter can be determined by the formula:

$$kW = \frac{3.6 \times Kh \times rev}{sec}$$

where 3.6 = the number of seconds per hour ÷ 1000
Kh = the meter constant found on the meter nameplate
Rev = the number of disk revolutions; choose 10 for easier calculations
Sec = time in seconds for the total number of revolutions counted

The meter multiplier (Kr constant) on a self-contained meter does not affect the spinning of the disk and is therefore not applicable to this formula. To determine the load registered by a transformer-rated meter, multiply the answer in the preceding formula by the product of CT ratio, VT ratio, and meter multiplier.

Tests before Installing a Meter

12.5.5 Testing should be done to ensure there is no short circuit in the customer's wiring before energizing a service with a socket-base meter. A *500-volt insulation tester,* which is like an ohmmeter, can be used at the meter base to test for a short circuit. If the customer's main disconnect is open, there should be an infinity reading between L_1 and L_2, between L_1 and neutral, and between L_2 and neutral (Figure 12–8).

A *voltmeter* can also be used at the meter base to check for a short circuit. With the source side alive and the customer's disconnect open, there should be no voltage between the source terminals and the load terminals. To check for a short between the load wires and the neutral, install a temporary jumper between L_1 and the neutral. The voltage between S_1 and L_2 should read zero volts.

12.6 Polyphase Metering

Three-Phase Metering

12.6.1 Polyphase metering refers to energy metering on a service of more than one-phase. A powerline worker normally refers to it as three-phase metering.

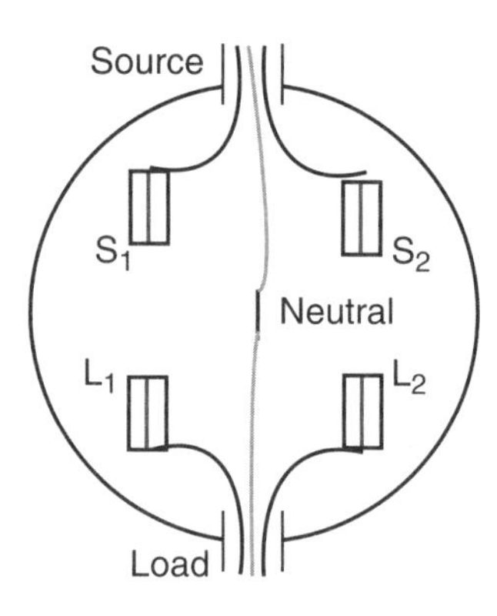

Figure 12–8 Socket-base meter.

Three-phase energy could be measured by metering each phase separately and calculating the resultant total amount of energy used. Three-phase power, however, is normally measured with one unit, which saves space, connections, and calculation of the total energy from three separate readings.

The principle of a three-phase meter is similar to that of a single-phase meter. The meter has three current coils, one for each phase, connected in series with the load. It has at least one voltage coil connected between two of the phases. Three-phase meters, like single-phase meters, can be self-contained, transformer-rated, demand, primary, secondary, A-base, or S-base.

Elements in a Three-Phase Meter

12.6.2 One current coil and one potential coil working together to turn one disk define one element in a meter. Two-, three- and four-wire services require more meter elements. The general rule of thumb is that energy can be metered with one less element than the number of wires in the service. A four-wire three-phase service can be measured with a three-element meter, which has three current coils, three potential coils, and three disks.

Another variation for metering a four-wire three-phase service is a 2½-element meter, as shown in Figure 12–9. This meter measures the current in all three phases, but the voltage is measured from two phases. The voltage would need to be reasonably balanced in all three phases for this metering to be accurate.

Element Arrangement

12.6.3 The elements in a meter can be vertically arranged as seen in Figure 12–9 or horizontally arranged as seen in Figure 12–10. The arrangement can also be set up to turn three, two, or one disks. Figure 12–10 shows a horizontal, three-element meter, which can register the energy of a three-phase four-wire service with a single disk.

Making Metering Connections

12.6.4 The connections to a meter follow certain principles. There are variations depending on the manufacturer and regulatory agencies. Drawings should be consulted when making connections to a three-phase meter because a mistake can lead to awkward circumstances for a utility.

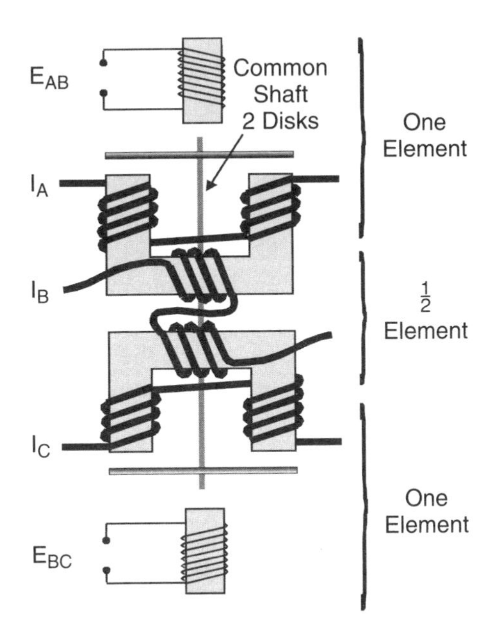

Figure 12–9 Four-wire, 2½-element meter.

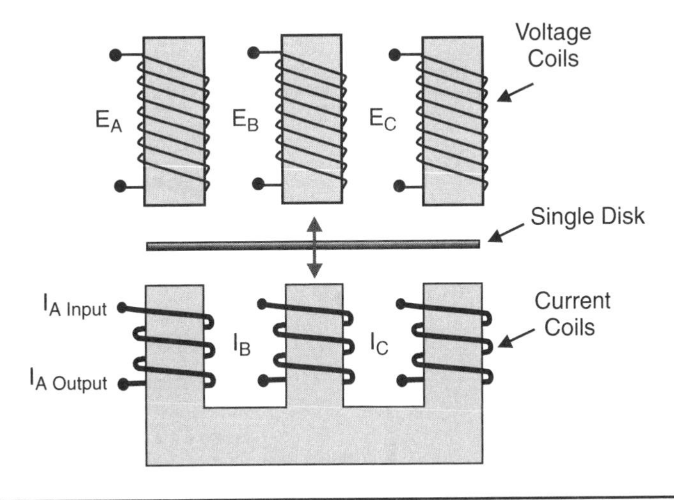

Figure 12–10 Single-disk three-element horizontal meter.

- For each phase, the source wires to the current coil and the voltage coil must be the same polarity.
- The load wires from the potential coil must be an opposite polarity to that of the current coil. In other words, depending on the type of service and meter, the load side of the potential coil must be connected to another phase or to the neutral.
- Each current coil must be connected in series with the load.

- Each potential coil must be connected in parallel, like a voltmeter.
- The color coding of the wiring must be followed.
- The three phases must be in a proper phase rotation.

Electronic Metering

12.6.5 The electronic (static or solid-state) revenue meter will eventually take over from the electromagnetic meter. A multitude of measurements can be programmed into an electronic meter. It is also more suitable for automatic-meter-reading systems.

In an electronic meter, the current and voltage act on solid-state (electronic) metering elements to produce an output pulse. A pulse initiator produces contact closures (pulses) proportional to the watt-hours being measured. The electronic meter illustrated in Figure 12–11 is a three-element four-wire transformer-rated meter. A typical electronic meter can be programmed to carry out many functions:

- It can be a multi-rate meter with more than one readout, each becoming operative at a specified time corresponding to a different tariff rate.
- It can be used to check the voltage and current in each phase, the power factor on each phase, and instantaneous kW and kVA for each phase.
- The meter can check itself for proper function much like a computer checks itself out. It can also check itself for proper installation such as phase rotation.

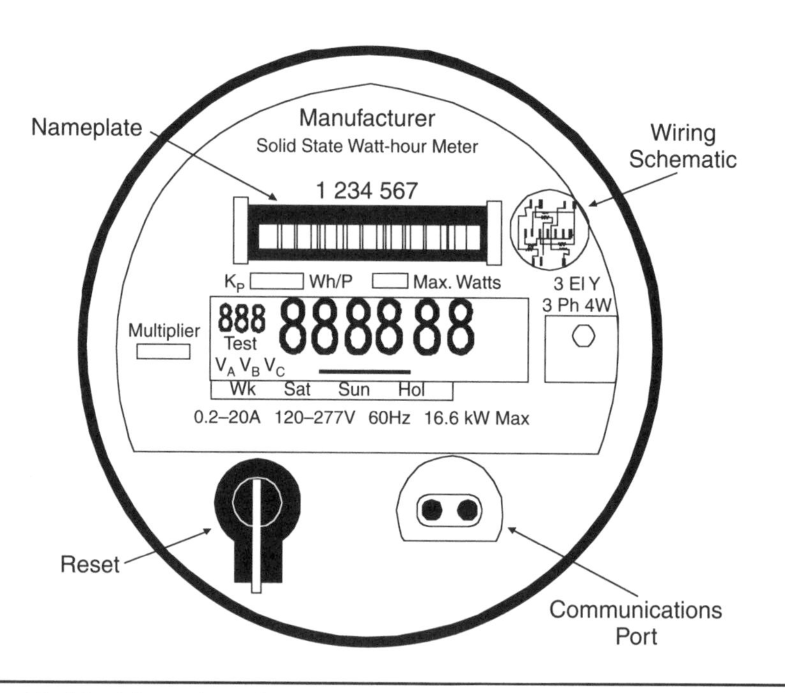

Figure 12–11 Three-phase electronic meter.

- It can register the time-of-use along with taking into account holidays and weekends.
- It can register kW demand and kVA demand.
- It is easily adapted to all the benefits associated with automatic meter reading by access through the communications port.

12.7 Transformer-Rated Metering

Transformer-Rated Meter Use

12.7.1 A transformer-rated revenue meter is used where the voltage and/or current is too high or impractical for a self-contained meter. A transformer-rated meter measures a current and voltage that have been stepped down by instrument transformers. The secondary of a voltage transformer (VT or PT) and a current transformer (CT) is an accurate representation of the primary voltage and primary current being measured.

The installation shown in Figure 12–12 has a current transformer in series with each phase. It has two voltage transformers that measure the phase-to-neutral voltage on two phases. There is a test block that provides access to do metering tests and to provide fuse protection. Figure 12–13 shows a three-phase 120/208-volt transformer-rated metering installation with current transformers.

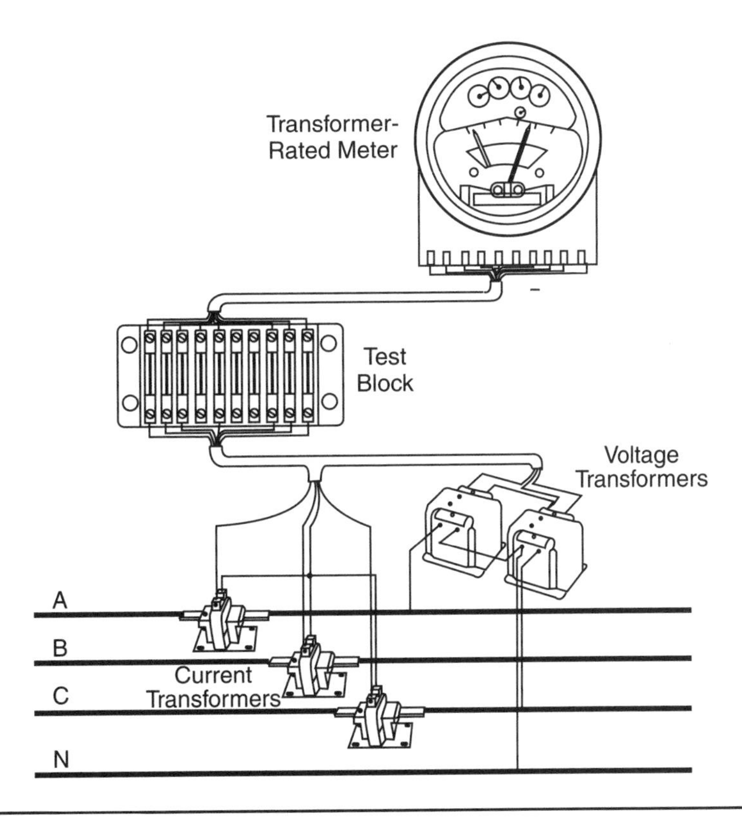

Figure 12–12 Transformer-rated meter installation.

Figure 12–13 Three-phase metering cabinet.

Voltage Transformer

12.7.2 A voltage transformer (VT) is used when the energy to be measured is supplied at a high voltage, such as subtransmission, primary, or high secondary voltage. A voltage-transformer output supplies a voltage at a manageable level to a revenue meter, typically rated at 240 or 120 volts.

A multiplier equal to the turns ratio of the transformer calculates the actual voltage of the circuit being measured. For example, a 4800/120-volt VT has a multiplier of 40:1.

Figure 12–14 shows how a voltage transformer and a current transformer are shown in technical drawings.

Current Transformer

12.7.3 A current transformer reduces the current to a transformer-rated meter. A transformer-rated meter is typically designed to operate at a maximum of 5 amperes. A low current allows the use of small wires with minimal losses in the metering circuit. Current transformers for metering are two-wire or three-wire, bar-type, donut-type, or bushing designs.

The two-wire CTs shown in Figure 12–15 have one primary and one secondary winding. The conductor running through the donut CT is considered the primary winding. This type of CT is installed in each leg of a 120/240-volt service or on each phase of a three-phase service.

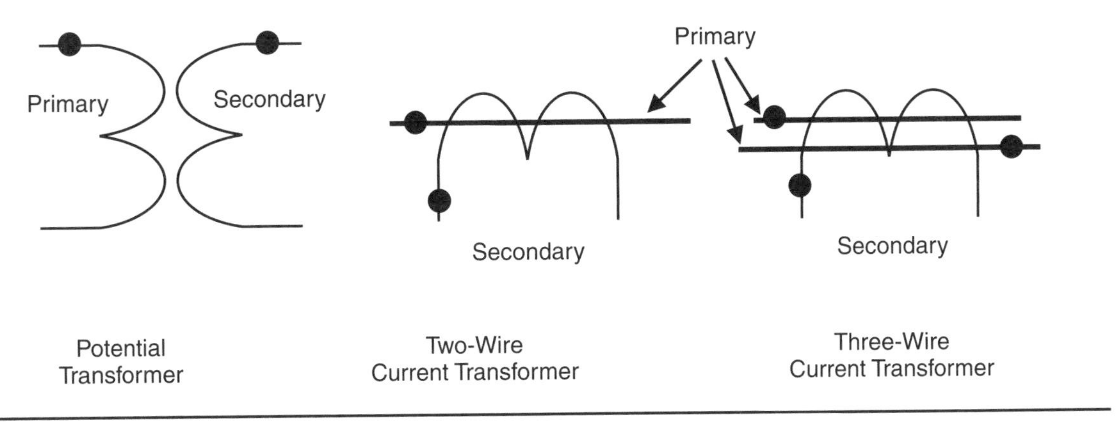

Figure 12–14 Drawings representing instrument transformers.

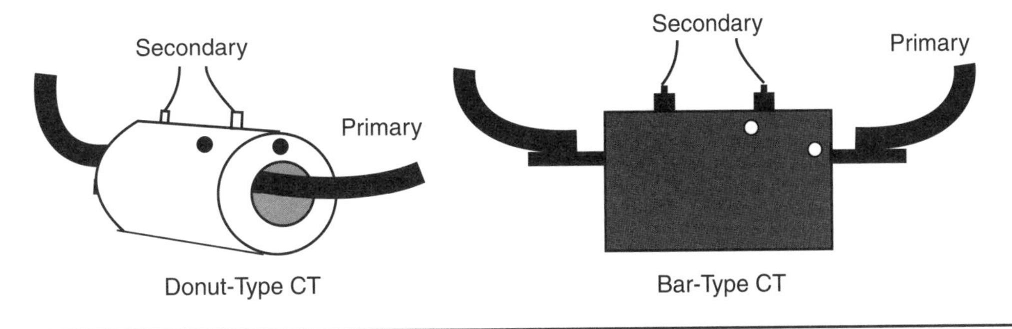

Figure 12–15 Two types of two-wire current transformers.

Three-Wire Current Transformer

One CT can be used with a three-wire 120/240-volt service. Both 120-volt conductors go through one CT. To have the *current* in the two opposite-polarity 120-volt legs go through the CT in the same direction, the two wires go through the CT from opposite directions. The current from the two legs is added together to double the secondary current. For example, a 200/5 CT becomes a 200/10 CT with this setup.

A three-wire CT used on a 120/240-volt service has two primary windings and one secondary winding. In Figure 12–16, the three-wire CT is actually a two-wire CT with two wires going through it acting as two primary windings. The secondary current from the CT represents the sum of the two 120-volt legs of the service.

The type of service shown in Figure 12–17 is a 120/240-volt service metered at a customer transformer pole. This central metering installation measures the energy at one location and then allows for services to go from this pole to a house, a barn, and other buildings owned by the customer. With this type of service, a customer can install a smaller service entrance panel in each building instead of installing one large-capacity service entrance to feed all the buildings.

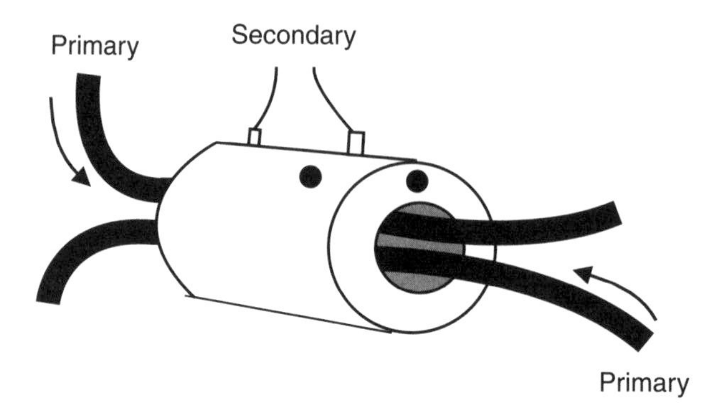

Figure 12–16 Three-wire current transformer.

Polarity Markings

As with all transformers, the polarity of an instrument transformer needs to be known and marked on the transformer. For example, when the current enters at the marked primary terminal, the marked secondary terminal is in phase with the primary. The polarity of the primary and secondary terminals of instrument transformers is normally marked with a large dot.

High-Voltage Hazard with Current Transformers

A current transformer is like other transformers:

$$\textit{the input } V \times I = \textit{the output } V \times I$$

When the current is stepped down, the voltage is stepped up at the same ratio. A voltage drop brings the voltage to a safe level when the secondary is shorted or connected to the meter current coil.

A high-voltage hazard exists when there is current flowing through the primary of a current transformer while the secondary is open circuited. This high voltage can puncture the insulation of the current transformer, and it is a dangerous shock hazard to any person in contact with the secondary circuit.

When there is current flowing through the primary and the meter is not installed, the secondary must be kept short circuited. This can be done by closing the shorting device at the secondary terminals of the current transformer. The circuit can also be kept closed automatically by a shorting device in the meter base. The meter base for a socket-mounted transformer-rated meter often has an automatic shorting device that will close the metering circuit when the meter is removed.

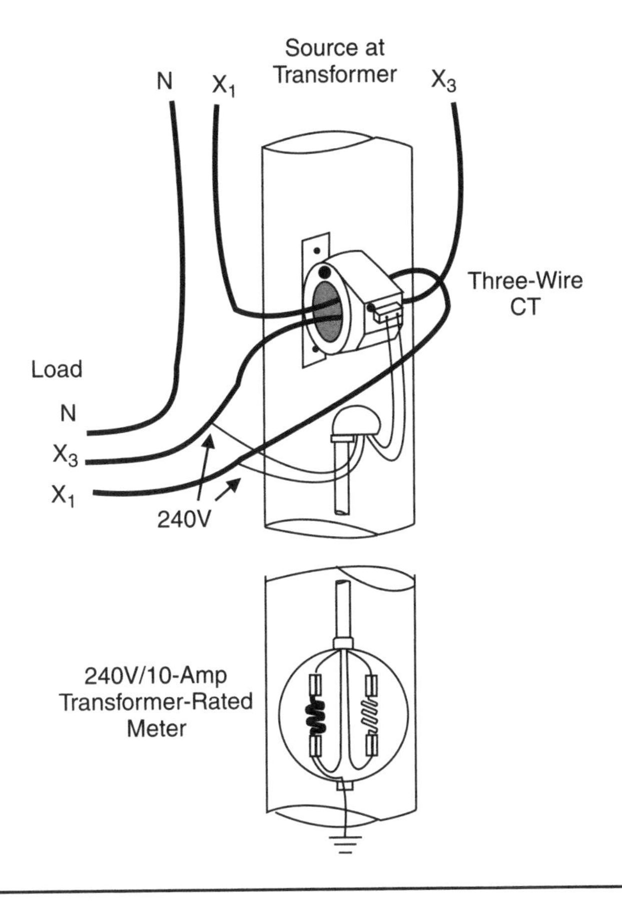

Figure 12–17 Three-wire CT on 120/240*V* service.

A Multiplier on a Current Transformer

When an instrument transformer is used, the meter register reading must be multiplied by a meter multiplier to get an actual status of the energy used. The CT nameplate shows the transformer ratio. When one wire goes straight through a CT, then the multiplier is the same as the transformer ratio. For example, a CT with a ratio of 400:5 (80:1) will have a multiplier of 80.

A CT multiplier can be reduced by looping the primary through the CT more than once. For example, when the primary is looped through the CT twice, as shown in Figure 12–18, the secondary current is doubled. A 400:5 CT becomes a 400:10 CT (200:5 or 40:1), which changes the multiplier to 40. The secondary of a 400:5 CT is rated to 5 amperes; therefore, a double-looped primary should only be used where the primary is 200 amperes or less to limit the secondary to 5 amperes.

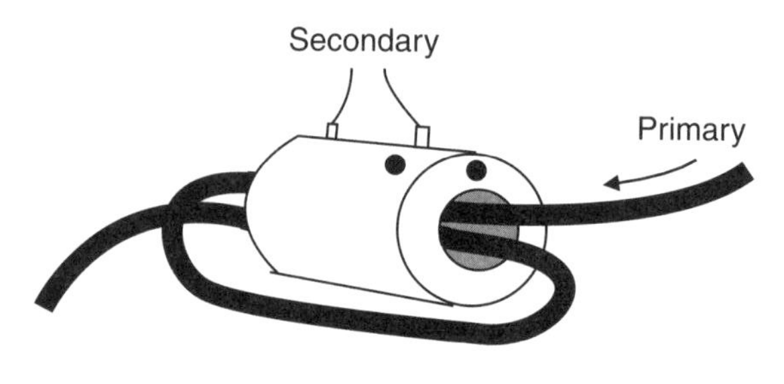

Figure 12–18 Double-looped primary.

Review Questions

1. Energy consumption is the volts × amperes used by a customer. What is the largest variable measured by a kWh revenue meter?
2. What is meant by peak demand?
3. Why would a utility charge extra for peak demand?
4. Revenue metering involves the measurement of one or more of the three kinds of power and/or energy used by a customer. What are they?
5. Why is a transformer-rated meter used for some services?
6. How many elements are in a meter that is normally used in a three-wire service?
7. The disk of a meter with a Kh constant of 3.6 rotates 100 times. How much energy was used?
8. How much power is being used by a customer when a meter with a Kh of 7.2 has 10 rotations of its disk in 40 seconds?
9. How can one CT be used to meter a 120/240-volt three-wire service?
10. Why is there a high-voltage hazard when there is current flowing through the primary of a current transformer while the secondary is open circuited?

INDEX

F

T

U

V

W